S0-BWW-843

PERSPECTIVES ON

LAKE ECOSYSTEM MODELING

PERSPECTIVES ON

LAKE ECOSYSTEM MODELING

edited by

DONALD SCAVIA　　ANDREW ROBERTSON
Research Scientist　　Head

Chemistry-Biology Group
Great Lakes Environmental Research Laboratory
National Oceanic and Atmospheric Administration
Ann Arbor, Michigan

ANN ARBOR SCIENCE
PUBLISHERS INC
P.O. BOX 1425 • ANN ARBOR, MICH. 48106

Copyright © 1979 by Ann Arbor Science Publishers, Inc.
230 Collingwood, P. O. Box 1425, Ann Arbor, Michigan 48106

Library of Congress Catalog Card No. 77-93388
ISBN 0-250-40248-3

Manufactured in the United States of America
All Rights Reserved

PREFACE

Since the pioneering work of Riley in the 1940s, mathematical modeling of aquatic ecosystems has expanded to where it is now considered an important aid to understanding and managing such systems.

There already exist a number of books that describe the techniques of aquatic model construction and testing, especially for resource management purposes, and the broad concepts for modeling diverse ecosystems; less has been done to document the utility of aquatic models in augmenting basic ecological research. We believe that modeling should be an integral part of both basic research and applied management programs. With this broader perspective in mind, we organized a special symposium on ecological modeling at the 20th Conference on Great Lakes Research at the University of Michigan in 1977.

Because we feel that modeling should be an integral part of aquatic ecosystem study rather than an independent discipline, we invited participants with expertise in experimental research as well as those involved in modeling. We asked all participants to consider model usage as well as possible model improvements and new directions for development. We specifically asked that model descriptions not be included unless necessary to the discussion.

Since the symposium was part of a conference on the Great Lakes, much of this book concerns those lakes. This, however, is not a shortcoming since ecosystem modeling on the Great Lakes is as advanced as anywhere in the world, and the material is generally applicable. We did, however, solicit and include several papers not presented at the symposium which do not focus on the Great Lakes.

The book is divided into four sections, each dealing with an individual area of aquatic ecosystem research—the adequacy and improvement of specific aspects, the synthesis role, the applied aspects, and possible developments—and the role of mathematical models in that area.

Section One, IMPROVED MODEL COMPONENTS, covers the adequacy of and possible improvement of specific aspects of aquatic models (plankton patchiness, phytoplankton-decomposer interactions and zooplankton vertical migration). Improvements in these specific model components are tied strongly to increased understanding of the dynamics modeled; thus, these chapters illustrate how coupled modeling-experimental programs provide insight into ecological processes not attained by experiment alone.

Section Two, IDENTIFICATION OF RESEARCH NEEDS THROUGH MODEL STUDIES, addresses the synthesis roles of modeling. In the past, many modeling programs were justified in part by their synthesis of information and subsequent identification of areas requiring further study, though few programs followed through to that end. The chapters in this section identify research needs from different perspectives and with different techniques.

Section Three, MODELS IN MANAGEMENT, addresses the applied aspects of ecosystem modeling. As models of lake systems become more prevalent and hold greater promise for aiding resource management, the potential for their misuse also increases. Specific model development for management purposes and methods for testing model adequacy, both qualitatively and statistically, are described and demonstrated with examples. Also, a new approach to large-lake surveillance combines a stochastic lake model and estimation theory.

The final section, NEW DIRECTIONS IN ECOSYSTEM ANALYSIS, presents three avenues for development of lake ecosystem mathematical modeling: coupling three dimensional hydrodynamic calculations to an ecological model to examine the influence of physical processes on biological and chemical properties; using output from an ecological model to examine elusive ecosystem properties; and using organism length as an independent variable in an innovative model that simulates concentrations of a toxic substance (PCB) in the aquatic food web.

We hope that this book illustrates and explains the important role models can play in both research and management programs, the importance of requiring stringent tests of models to be used for environmental management, and the directions in which lake ecosystem modeling can develop.

All papers in this book have been reviewed by at least two referees. We would like to thank, in addition to the contributors, the following researchers who served as referees: H. E. Allen, J. R. Bennett, V. J. Bierman, Jr., R. L. Chambers, S. C. Chapra, J. G. Ferrante, C. W. Gehrs, S. P. Larson, C. F. Powers, T. J. Simons, W. C. Sonzogni, S. J. Tarapchak and H. A. Vanderploeg. We also express special appreciation to Dr. Joseph Shapiro for writing the introduction, where he discusses the importance of modeling as a part of ecosystem study in a far better way than we could have.

Donald Scavia
Andrew Robertson
Ann Arbor, Michigan

Donald Scavia is a Research Scientist in the Chemistry-Biology Group at the Great Lakes Environmental Research Laboratory, National Oceanic and Atmospheric Administration in Ann Arbor, Michigan. Previously, he was a researcher with Rensselaer's Freshwater Institute in Troy, New York.

He received both a BS and MS in Environmental Engineering from Rensselaer Polytechnic Institute where his graduate research included the development and implementation of numerical models of lake eutrophication. His current research interests include the use of models in the analysis of aquatic ecosystems to explore interactions among biological, chemical and physical processes.

The author of over 20 scientific papers and articles on mathematical modeling and aquatic ecosystems, he is a member of Sigma Zi-RESA, Tau Beta Pi, The American Society of Limnology and Oceanography and the International Association for Great Lakes Research.

Andrew Robertson received a BS in Chemistry from the University of Toledo in 1958 and a PhD in Zoology from the University of Michigan in 1964. He conducted his doctoral research while on a Research Fellowship with the Scottish Marine Biological Association in Edinburgh, Scotland. Following his education, he continued at the University of Michigan for four years as a Research Associate in the Great Lakes Research Division. He then served three years as Associate Professor of Zoology at the University of Oklahoma. In 1971 he joined the National Oceanic and Atmospheric Administration. Initially he was a member of the management and coordination group overseeing the International Field Year for the Great Lakes, a large interagency binational research effort on Lake Ontario. At present he is Head of the Chemistry-Biology Group of NOAA'S Great Lakes Environmental Research Laboratory. He has authored over 50 scientific papers and articles in the areas of limnology and oceanography, primarily dealing with the ecology of the Great Lakes.

Dr. Robertson has served as president of the International Association for Great Lakes Research and is also a member of the American Society of Limnology and Oceanography, the Ecological Society of America, *Societas Internationalis Limnolgiae* and other scientific organzations.

CONTENTS

ix

LIST OF CONTRIBUTORS

James A. Bowers. Great Lakes and Marine Waters Center, The University of Michigan, Ann Arbor, Michigan 48109.

Raymond P. Canale. Department of Civil Engineering, University of Michigan, Ann Arbor, Michigan 48109.

Carl W. Chen. Tetra Tech, Inc., Lafayette, California 94549.

Leon M. DePalma. The Analytic Sciences Corporation, Reading, Massachusetts 01867.

Joseph V. DePinto. Civil and Environmental Engineering Department, Clarkson College of Technology, Potsdam, New York 13676.

Carol J. Desormeau. Center for Ecological Modeling, Rensselaer Polytechnic Institute, Troy, New York 12181.

Theresa W. Groden. Center for Ecological Modeling, Rensselaer Polytechnic Institute, Troy, New York 12181.

Efraim Halfon. Canada Centre for Inland Waters, Burlington, Ontario, Canada L7R 4A6.

Donald C. McNaught. Department of Biological Sciences, State University of New York, Albany, New York 12222.

Richard A. Park. Center for Ecological Modeling, Rensselaer Polytechnic Institute, Troy, New York 12181.

William F. Powers. Department of Aerospace Engineering, University of Michigan, Ann Arbor, Michigan 48109.

Kenneth Howland Reckhow. Department of Resource Development, Michigan State University, East Lansing, Michigan 48824.

Andrew Robertson. National Oceanic and Atmospheric Administration, Great Lakes Environmental Research Laboratory, Ann Arbor, Michigan 48104.

Donald Scavia. National Oceanic and Atmospheric Administration, Great Lakes Environmental Research Laboratory, Ann Arbor, Michigan 48104.

Donald J. Smith. Tetra Tech, Inc., Lafayette, California 94549.

William J. Snodgrass. Departments of Civil and Chemical Engineering, McMaster University, Hamilton, Ontario, Canada L8S 4L7.

Robert V. Thomann. Department of Environmental Engineering & Science, Manhattan College, Bronx, New York 10471.

ECOSYSTEM MODELING—
A PERSONAL PERSPECTIVE

We are all modelers. There is not one among us who has not in his mind, if not on paper, pursued the question of the effects of changes in one component of an ecosystem on another component. Most of us have gone far beyond that. We have said that, "If that is so, then this will happen, but this will lead to that, and that will do such and such"—and at that point most of us have flung up our hands and reflected on the difficulties of prediction—and have sought refuge in the ecologists' lament, "you cannot do one thing." We are then like the mariners wooed by the Sirens, we listen to voices telling us how impossible it is to continue in the direction of understanding and predictability in ecology and we flounder on the rocks of despair. But true modelers stuff their ears, if not with wax, like Odysseus, then with paper, and press on.

How many times have I sat with my qualitative colleagues listening to a speaker present his model and heard someone beside me utter deprecatory remarks about the validity of certain assumptions or the values of certain coefficients, or even the presumption of the whole attempt to model ecological events. And how many times have I agreed! I agreed because for many years modelers seemed to warrant that treatment. Many of their assumptions *were* wrong, many of their oversimplifications *were* unrealistic, and a generally cavalier air put them beyong the pale for many ecologists—not all modelers, of course—but enough, particularly the ecosystem modelers. But, as they say, there is nothing quite like a reformed sinner, and I am reformed, or at least reforming. I have less patience now with those muttering colleagues.

All of this came about quite innocently, of course. Sinners rarely know when they are being reformed. But it came about inevitably once I had to determined to think about lake restoration in ecological terms and once I had realized that such things as eutrophication are complex and involve more of the aquatic ecosystem than a "phosphorus causes algal growth" relationship would indicate.

In my scientific career, I have seen phosphorus come out of nowhere, so to speak, and reach preeminence as a factor to be reckoned with, and of course there is a good basis for this. There is, or should be, no question of the importance of phosphorus in initiating the symptoms of eutrophication. But eutrophication is more than algal growth. A great many phenomena occur, many of which themselves tend to perpetuate the syndrome. Thus increases in algal growth lead to increases in pH which result in predominance of blue-green algae. This in turn leads to fewer resources for the zooplankton. At the same time more decomposition leads to anaerobic hypolimnia which no longer serve as refuges from fish predation for larger herbivorous zoo-plankters. So the latter disappear or become rare. In consequence the standing crops of algae become even higher. The anaerobic hypolimnia, in addition to precluding cold-water predatory fish, also provide sediment sources of phosphorus, supplanted by phosphorus and nitrogen pumped into the system by the macrophytes and abundant bottom-feeding fish resulting from receptive organic sediments and winter fish kills, respectively. Super-imposed on these are such things as algal competition for phosphate, algal antibiosis, the effects of zooplankton in stimulating blue-green algae by passing them through their guts, the effects of the heightened pH on zoo-plankton grazing, the interplay of blue-green algae and cyanophage, possible sustenance of herbivorous zooplankters on macrophyte debris, the apparent ability of *Daphnia* to foster *Aphanizomenon*, the possible role of inorganic cations brought in with the phosphate-containing effluents in altering patterns of algal composition by affecting phosphate uptake by the algae, and so on.

Many of these phenomena are unlikely to be reversed by removing phos-phate from the influent. But juggling all these phenomena to determine their importance is difficult—it is far easier to juggle only one ball, phosphorus, and so many of us do just that. The result is that not only might we delude our-selves, we are in danger of deluding those who depend on us, the public, the so-called managers, the engineers. *We* have oversimplified to the point of being unrealistic and *we* have presented a version of eutrophication and of ecosystem behavior devoid of all assumptions but one, that phosphorus is important. Well, what is the answer? The answer is that we too must plug our ears and sail past the Sirens. We must learn to juggle all of the balls—if not at once, then a few at a time. The papers in this collection do just that—some more, some less—but the process has begun and that is what is important.

Dr. Joseph Shapiro
Limnological Research Center
University of Minnesota
Minneapolis, Minnesota

SECTION ONE

IMPROVED MODEL COMPONENTS

Models of ecosystems are composed of submodels designed to simulate specific compartments or processes within the ecosystem. It is at this submodel level that close interactions between experimentation and the modeling process become important. The verisimilitude of submodels can only be advanced as the state of understanding of the modeled process is increased. Also, one often can evaluate and advance the present state of understanding by attempting to construct a "complete" model of the process. The process of building and testing submodels as part of experimental programs often leads to insights not attainable by experimentation or modeling alone.

The three chapters in this section deal with the modeling process in the context of improving specific components of submodels in concert with increasing knowledge concerning specific processes. In the first chapter, D. C. McNaught discusses the significance of plankton patchiness for modeling large ecosystems. He illustrates the space- and time-scale relationships among important components of aquatic ecosystems and suggests how one should interpret data from patchy systems for model studies. J. V. DePinto, in Chapter 2, discusses the importance of a component of ecosystem models that too often receives only cursory treatment, i.e., phytoplankton-decomposer interactions. In that chapter, results from field and laboratory studies are used to illustrate the importance of decomposers in controlling phytoplankton biomass and dynamics and thus to highlight the danger in a superficial treatment of the decomposers. In the third chapter, J. A. Bowers examines the relation between vertical migration and grazing in zooplankton. Laboratory and field data are used to illustrate and describe the influence of migration and several previously published models of migration are described.

These three chapters are quite specific to their respective subjects and, of course, many other topics could have been discussed just as profitably.

1

These chapters, however, provide good examples of the interactions between experimentation and the modeling process, as well as providing more specific details concerning patchiness, phytoplankton-decomposer interactions, and zooplankton vertical migration.

CHAPTER 1

CONSIDERATIONS OF SCALE IN MODELING LARGE AQUATIC ECOSYSTEMS

Donald C. McNaught

Department of Biological Sciences
State University of New York
Albany, New York 12222

THE PROBLEM

Both spatial and temporal heterogeneity, and the resulting scales in distribution of plants and animals in lakes, must be clearly understood to simulate aquatic ecosystems mathematically. The variable spatial dimensionality in the distribution of the limnoplankton must be known to determine the rate at which important physiological and behavioral processes occur. Certainly rates of nutrient uptake by the phytoplankton and grazing by the zooplankton are obvious examples of processes dependent upon resource concentration. If limnologists continue to measure the mean concentrations of chemical and biological constituents of such systems, then process models based upon such parameter estimations will be difficult to validate and systems models will be impossible. Basically this is because the uptake of various substrates is nonlinear (whether the substrate is a biochemical substance such as an amino acid of interest to a bacterium or a whole organism such as a zooplankter of interest to an alewife).

Specific objectives of this synthesis include: (a) the description of the spatial-temporal scales of phytoplankton, zooplankton and fish distributions in large lakes, including extremes in the patchiness of both biomass and those rate functions such as grazing and excretion that lead directly to further patchiness of phytoplankton and nutrients; (b) the

discovery of spatial-temporal interactions that underlie increased efficiencies of ecosystem functioning; (c) the search for evidence of coincidence
of spatial-temporal scales of predator-prey species necessary to such
increased efficiency; and (d) the application of information on heterogeneity in distributions of planktonic biomass and associated fluxes to
model calibration.

Four basic processes characteristic of open-water, planktonic systems
are considered: nutrient cycling, primary production and secondary production (herbivory and carnivory). Variations in fluxes characteristic of
these processes, as well as the magnitude of important rapidly cycling
biomasses, are described. First, the four processes are considered with
regard to spatial and temporal variations. Spatial variations are limited to
those expected in large lakes the size of the Great Lakes. Temporal
variations include the limits of important fluxes and the size of rapidly
changing populations of organisms over the time of days (diel variation)
as opposed to seasons. Special consideration has been given to those
processes best documented in the literature, or for which I have a body
of unpublished data. These include the spatial and temporal distribution
of phytoplankton in the sea as well as in large lakes, diel variations in
uptake of nutrients by phytoplankton, scales of aggregations of zooplankton,
as well as typical grazing rates, and spatial and temporal variations in the
biomass of predatory fishes, including the temporal nature of their
feeding rates.

Modelers are often provided with mean estimates of biomass for a
particular season and, in unusual cases, with estimates of fluxes. Since
most uptake processes are characterized by high nonlinearity between
process rate and resource concentration, it is critical that modelers have
estimates of variability for such processes available, accompanied by the
results of behavioral-physiological studies that indicate whether organisms
such as zooplankton are able to locate aggregations of phytoplankton,
thus feeding at very high cell concentrations, or whether they feed at
resource concentrations approximating the environmental mean
concentration.

SCALES OF PLANKTON DISTRIBUTION

Observed Scales of Phytoplankton Distribution

Scales of phytoplankton patchiness have been determined both in
marine and freshwater environments. Various oceanographers have
observed patches 20 x 100 km, but the most common sizes were
elliptical streaks no larger than 5 x 20 km. Often it is difficult to

differentiate patches caused by upwelling from those caused by Langmuir circulation or internal waves. Patches of phytoplankton with dimensions of 10 km are thought to be created by both biological and physical forces. Thus early observations were important for envisioning scale and developing theory.

The use of spectral analysis has brought needed precision to studies of scale. Spectral analyses have been used to determine dominant relationships between clumps of organisms in instances where suitable (time or distance) data sets were available. Such analyses are appropriate, as nonlinear relationships show periodic behavior in time (*e.g.,* the vertical migrations of zooplankton) and space (*e.g.,* the regular distances separating clumps of zooplankton in Langmuir cells). In spectral analysis the variance of a series of numbers (as densities of plankton) is partitioned into contributions at frequencies that are harmonics of the length of the data set (Platt and Denman, 1975). For plankton samples of high density collected within and between Langmuir convergences, the results are intuitive, in contrast to the outcome at low frequencies where this analysis becomes a necessity.

Platt (1972) found that the distribution of the variance of phytoplankton abundance followed a consistent power relationship, at least over distances of 1,000 m. Platt and Filion (1973) measured both productivity (P) and biomass (B) of the phytoplankton in a small embayment. Differences in growth (P/B ratio) were evident over distances of hundreds of meters and were especially obvious during calm conditions, which accentuated small-scale fluctuations. These observations obviously implicate turbulence.

Denman and Platt (1975) collected pumped samples (chlorophyll) along transects (52-80 km) in the North Atlantic. Samples were integrated over a distance of 300 m. Distances between clumps (wavelengths of 2 and 8-9 km) were evident from coherence spectra. By reducing the distance over which chlorophyll concentration was averaged to 3.2 m, clumps on the order of 100 m were detected. At these scales, biomass estimates (chlorophyll) varied by a factor of five (Table I). The authors suggested that nutrient uptake, growth and grazing were factors accounting for such clumps. It is significant that scales below 100 m were not evident; certainly new techniques for determining small-scale patchiness of phytoplankton must be developed.

In smaller freshwater lakes, spectral analysis has been used to define scales of patchiness. Richerson *et al.* (1975) observed horizontal spatial heterogeneity at scales of 1.2-2.5 km, and smaller scales of 450 and 225 m (wavelength). When a 6.85 km transect, over which samples were collected at 68.5 m intervals, was subdivided to determine a variance

Table I. Compilation of Scales of Horizontal Aggregation for Various Organisms

Trophic Level	Scale	Author	Maximum/Mean Observation $x_{max}/\bar{x}$
1. Phytoplankton	5-10 km	Steele (1974)	--
	100 m, 2-8.5 km	Denman and Platt (1975)	5x
	225, 450 m; 1.2-2.5 km	Richerson et al. (1975)	--
	< 15 m	Nichols (1977)	--
2. Zooplankton	4.5, 8 m, 30 m; 160 km	McNaught (unpublished)	3.6 -- 6.3x
	20-40 km	Patalas (1969)	--
3. Fishes	10 m	McNaught and Hasler (1961)	--

spectrum, only the shorter wavelengths of 450 and 225 m were significant. On a smaller scale, I have observed but not measured streaks of phytoplankton separated by only centimeters. These streaks in Lake Mendota appeared to be related to surface micro-Langmuir circulation associated with a temporary thermocline formed during warm calm weather.

Possibly the smallest published scales of phytoplankton distribution have been observed by Nichols (1977) in a small pond (30 m x 30 m) at Cornell University. He found species associations on a horizontal scale using samples collected approximately 51 m apart. Five groups composed of two to eight species followed natural taxonomic lines and were not related by size alone. The genera *Chlamydomonas* and *Coccochloris* often occurred together, but not with other species. Thus phytoplankton aggregations may occur on a very small scale, where biological interactions may lead to the creation of clumps (due to grazing or inhibition by algal metabolites) largely independent of physical aggregating forces. These observations probably represent one end of a spectrum of aggregation and are probably limited to small ponds where turbulence is minimal.

Observed Scales of Zooplankton Distribution

The largest scale of zooplankton distribution might in theory include an entire lake. Using original data collected at 44 stations in southern Lake Huron, contour maps of the distribution of most crustaceans have been constructed. The map for copepod nauplii suggests that they filled one entire basin in June 1974, comprising one large clump (140 x 160 km). Calanoid copepodites fit this same description. Other species were limited in their distributions, even at their seasonal maxima. While the adults of *Diaptomus oregonensis* (Figure 1) were found everywhere in November 1974, they were predominantly clustered in two large aggregations, one in the mouth of Saginaw Bay (14 x 29 km) and the other in the central lake north of Port Huron (16 x 56 km). Adults of *Cyclops bicuspidatus* were likewise everywhere in June 1974, but a large clump (40 x 105 km) was evident in the main lake. In contrast, species that erupted during periods of rapidly increasing temperature often formed aggregations in inshore waters. We found *Eubosmina coregoni* in two small patches (5 x 29 km and 13 x 19 km) along the eastern Michigan coast near Harbor Beach in June.

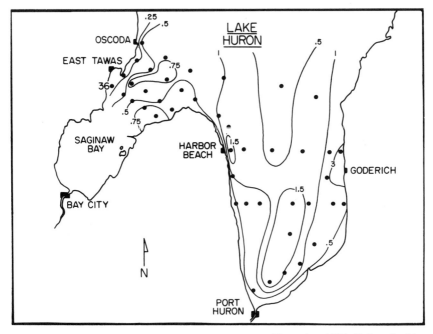

Figure 1. Distribution of *Diaptomus oregonensis* in November 1974 (0-5 m hauls); (•) station locations.

For Lake Ontario, Patalas (1969) observed the development of large aggregations over a period of months. Within these patterns, large clumps were evident (20 x 40 km for *Bosmina longirostris;* 30 x 45 km for *Ceriodaphnia* and 25 x 55 km for *Diaptomus*). One might guess that large-scale patches are due to the rapid growth of at least the cladocerans. Their stability is presently a matter of conjecture. In the sea, Cushing and Tungate (1963) observed large patches for weeks. Platt *et al.* (1977) suggested that patches of 10-m scale would persist for 10 min, while larger patches of 100 m would exist for 1.5 hr; patches at both scales would eventually be destroyed by diffusion. Clumps larger than 100 m should be stable against diffusion, with scales of 1 km stable for 10 hr and 10 km for 3 days.

Detailed records of small-scale patchiness of zooplankton are rare in the literature. With the development of acoustical techniques (McNaught, 1969), we have been able to detail small-scale clumps of particles (0.4-3 mm dia) over distances of 100 m. During August 1975 we observed uniform clumps of particles, which were most likely zooplankton, sep-arated by distances of 4.5 m at a station in Lake Huron (McNaught, unpublished). During October 1975 extremely uniform clumps were evident, with patches occurring regularly every 8 m. Those patches most certainly occurred in convergences or divergences of the Langmuir circula-tion. Whether they occurred within the former or latter depends upon the current velocities in the zones of upwelling or downwelling (Stavn, 1971) and upon whether the zooplankton are in an up- or downswimming mode.

Scales of Spatial Heterogeneity—Summary with Phase-Space Diagram

Basically both the phytoplankton and zooplankton are limited in their distributions by the size of the lake they occupy. While one clump may, in theory, fill a large lake and persist for days, biologically interesting interactions occur at smaller scales (tens of meters) and over shorter periods of time (minutes and hours), as depicted in the accompanying phase-space diagram (Figure 2). Those data used in constructing Figure 2 are from the following observations and sources. Phytoplankton popula-tions might persist for 40 days and occupy, at the longest scale detectable, most of a lake basin. The associated temporal scales are presently theo-retical and based on physical factors (Platt *et al.*, 1977). Within a basin, intermediate-sized patches may rapidly develop in areas where zooplankton have regenerated nutrients through grazing-down earlier patches. These intermediate patches (0.2-12 km) may, in turn, be further concentrated by purely physical forces, such as currents associated with internal waves. The converging currents of the smaller-scale Langmuir circulation create streaks, separated by distances of 5 to 10 m; however, to date phyto-plankton have not been associated with these. Even smaller micro-Langmuir

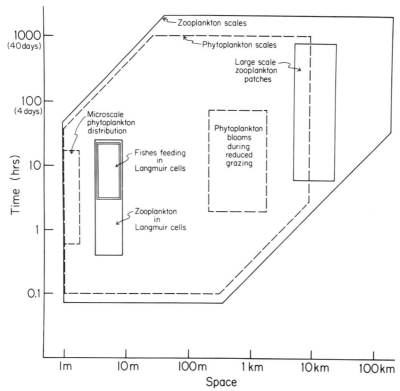

Figure 2. Phase-space diagram depicting interacting horizontal spatial-temporal scales for phytoplankton, zooplankton and fishes. The spatial axis represents the distance (wavelength) between clumps, while the temporal axis represents the period.

cells containing bluegreens (dimensions less than 1 m) have been observed by the author in highly eutrophic lakes. Their scale enables food to be sought out and grazed upon for a significant time by zooplankton.

Zooplankton are likewise limited in scale by the lake they occupy. Large clumps have been observed in the Great Lakes; in Huron by McNaught *et al.* (in press) and in Ontario by Patalas (1969) and McNaught *et al.* (1975). Recently smaller scales of zooplankton distribution have been identified, the most common being associated with the Langmuir distribution (Figure 2). Finally, schools of plankton-eating fishes have been observed foraging on such clumps, presumably greatly increasing their foraging efficiency (McNaught and Hasler, 1961). Feeding was restricted spatially to the convergences between the Langmuir cells and temporally to those times of day when vertical heterogeneity (through vertical migration of prey) was optimal for the predator.

The envelopes encompassing all observed phytoplankton and zooplankton spatial-temporal relationships depicted in the phase-space diagram (Figure 2) are similar. That is, aggregations of organisms at large scales do not have brief persistence times (lower right) and conversely, small-scale aggregations do not persist for long times (upper left). Further research will indicate whether these envelopes are, in fact, hexangular or ellipsoidal. In any case, they permit us to define the limits of spatial-temporal relationships in aquatic environments.

SPATIAL-TEMPORAL INTERACTIONS

Nutrients and Primary Producers

The observation that the rate of nutrient supply varies throughout the day in the sea (Dugdale, 1967) has led to a number of investigations detailing the underlying relationships that involve the uptake of nutrients by algae and remineralization by zooplankton. Spatial-temporal variations in excretion can be analyzed in these terms. Spatial excretion rates should be proportional to zooplankton biomass and vary by a factor of six times the mean. But temporally, nighttime excretion is three times the daytime rate (Hargrave and Geen, 1968) and is further accentuated by a tenfold increase in herbivores in surface waters through vertical migration (McNaught, 1966). Thus, fluxes of phosphorus and nitrogen in surface waters may vary by a large factor over 24 hr.

However, at least two feedbacks may operate. Animals will be less concentrated if Langmuir circulation is not prevalent at night. Furthermore, zooplankton excretion may be relatively reduced when food is abundant (Martin, 1968). So spatial-temporal factors that increase variability in the distribution of nutrients could, at maximum, create concentrations 180 times the mean (or 6 x 3 x 10), but we might expect to find somewhat less (Table II, nutrient cycling). In the same fashion, phytoplankton uptake should vary on a horizontal basis relative to large-scale aggregations of cells, but also temporally because of differential uptake (Eppley et al., 1971; Chisholm and Nobbs, 1976). These horizontal spatial variations may thus interact to lead to large (10x) spatial-temporal variations in the concentration of nutrients.

Phytoplankton standing crops can be interpreted in the same manner. Scales vary in space (Table I) but also temporally due in part to synchronized cell division (Doyle and Poore, 1974). Combined spatial-temporal

Table II. Magnitude of Variation of Important Standing Crops and Fluxes in Time and Space

Process	Spatial	Temporal	Spatial-Temporal Interactions
1. Nutrient Cycling			
a. Standing Crops	Unknown; large variations expected	Diel variations relative to excretion in surface waters at night (Hargrave and Geen, 1968)	Replenishment of nutrients in trophogenic zone
b. Fluxes	Spatial excretion rates by zooplankton, proportional to biomass (6x)	Diel excretion (Hargrave and Geen, 1968) of P, N by zooplankton (3x) accentuated by vertical migration (10x) or (30x)	Nighttime excretion (180x) greater than daytime in localized areas; 180x magnitude reduced to 30x in absence of Langmuir circulation and because of feedback relative to algal abundance (Martin 1968)
	Spatial uptake by phytoplankton proportional to biomass (5x)	Temporal uptake by phytoplankton of P, N (Eppley, 1971; Chisholm, 1976) (2x)	Diel uptake in surface waters (10x); possibly related to niche separation (Stross and Pemrick, 1974)
2. Primary Production			
a. Standing Crops	Spatial scales from large to small (Table I) (5x)	Diel division of cells with increase in numbers at specific time (Lorenzen,1970; Doyle and Poore, 1974) (2x)	Spatial temporal increases in cell numbers at surface at specific time (10x)
b. Fluxes	Spatial differences in carbon fixation related to product of biomass and flux expressed as areal flux (5x)	Fluxes of C vary temporally, but occur only in daylight; therefore vital whether physical factors causing spatial differences are temporal	Spatial variations in carbon fixation (5x)

Table II (Continued)

Process	Spatial	Temporal	Spatial-Temporal Interactions
3. Secondary Production (herbivores)			
a. Standing Crops	Spatial differences in zooplankton abundance (McNaught, unpublished) in Lake Huron related to physical factors, especially Langmuir circulation (6x)	Temporal differences in standing crops related to vertical migration of zooplankton (McNaught, 1966) (10x)	Nighttime concentrations of zooplankton could exceed 60x daytime; likely offset by lessened incidence of Langmuir at night (20x)
b. Fluxes	Spatial differences in grazing rates on areal basis (6x)	Temporal differences in ingestion rates (McNaught, unpublished), wherein nighttime rates exceed daytime by 2x	Spatial-temporal interactions lead to increased community grazing rates (12x)
4. Secondary Production (fishes)			
a. Standing Crops	Aggregations of fishes feeding on clumps of zooplankton (McNaught and Hasler, 1961), with heavy concentration of fish biomass (10x)	Temporal aggregations feeding at dawn and dusk only	Spatial-temporal variations in fish biomass (10x)
b. Fluxes	Grazing rates of fishes will vary spatially related to concentrations of zooplankton (McNaught and Hasler, 1961) (10x)	Grazing rates of fishes vary temporally and occur when zooplankton concentrated on diel basis (10x) or horizontal basis (McNaught and Hasler, 1961) (6x)	Grazing rates of fishes upon zooplankton vary spatially and temporally (60x)

factors may lead to clumps of concentrations 10 times the mean. However, small-scale clumps of phytoplankton apparently do not occur. This apparent fact may be related to the observation that herbivores should function more effectively in large clumps (MacArthur and Pianka, 1966). Fluxes of primary production apparently show even less small-scale variation, but these are not well documented (Table II).

Zooplankton and Fishes

Considerable evidence indicates that zooplankton patches are related to physical factors. However, temporal variations through active vertical migrations rhythmically enhance, and then reduce populations in surface waters. Thus, considerable variation (20x) may be found in space and time. This general result does not contradict the magnitude of variation observed in Lake Huron (Tables III-V). Since the zooplankton are heterogeneous in distribution, so is grazing. Temporal variations (2x) in ingestion rates of herbivores have been observed in Lake Huron (McNaught, unpublished). Together, spatial-temporal interactions suggest that grazing rates might vary greatly (12 times the mean).

Fishes can readily be observed feeding on clumps of zooplankton (McNaught and Hasler, 1961). They may forage during limited times of the day, when prey are concentrated vertically by migrations and horizontally by physical factors. This interaction leads to greatly increased predation rates (60x) relative to those calculated from mean values (Table II, Fishes).

Evolution of Coincident Scales

Scales of horizontal variations may reflect the influence of physical factors on smaller organisms as well as the development of feeding behavior in higher forms. Small-scale variations apparently have not been documented for phytoplankton, except in small ponds. Thus an intermediate scale of 200 m as suggested by Richerson *et al.* (1975) is depicted in Figure 3. In contrast, the zooplankton are characterized by small-scale (s.s.) variation on the order of 7-10 m, associated with the Langmuir circulation, as well as large-scale (l.s.) variations possibly created by internal seiches (McNaught, unpublished).

It is not unlikely that variations in phytoplankton density lead to a portion of the population existing at a concentration below the grazing threshold of the zooplankton and thus in a spatial refugium from grazing. The zooplankton are clumped (Figure 3, middle) and two or more scales are evident. But such clumps seem to bring no obvious benefit to the grazers; thus one might argue that these same physical forces also lead to

Table III. Mean Density ($\bar{x}$), Maximum Density (x_{max}), and $x_{max}/\bar{x}$ for Total Crustacean Zooplankton of Southern Lake Huron (from 28 April 1975 to 23 October 1976)

	Date											
	28 April	14 May	4 June	16 June	17 July	26 Aug	8 Oct	10 Nov	10 April	6 Aug	23 Oct	MEAN
$\bar{x}$	13740	27357	56814	183618	95060	68657	81522	64883	28983	64646	36010	
x_{max}	31364	85092	339364	901273	362530	182126	191733	137654	86556	225972	111640	
$x_{max}/\bar{x}$	2.3	3.1	6.0	4.9	3.8	2.7	2.4	2.1	3.0	3.5	3.1	3.3

Table IV. Mean Density ($\bar{x}$), Maximum Density (x_{max}), and $x_{max}/\bar{x}$ for Most Common Cladoceran, Cyclopoid Copepod and Calanoid Copepod in Southern Lake Huron[a]

Species	28 April	14 May	4 June	16 June	17 July	26 Aug	8 Oct	10 Nov	10 April	6 Aug	23 Oct	MEAN
Cladoceran												
$\bar{x}$	E.c. 112	E.c. 1219	B.l. 4281	B.l. 66170	E.c. 43766	E.c. 4364	E.c. 10061	E.c. 3963	E.c. 15	E.c. 7247	E.c. 5498	
x_{max}	257	18847	58259	498291	152025	42582	43279	35470	247	70390	34563	
$x_{max}/\bar{x}$	2.3	15.5	13.6	7.5	3.5	9.8	4.3	9.0	16.5	9.7	6.3	8.8
Cyclopoid												
$\bar{x}$	C.b. 666	Cop 2118	Cop 7718	Cop 17510	Cop 6273	Cop 5491	Cop 15801	Cop 17793	Cop 895	Cop 5970	Cop 11272	
x_{max}	3365	20699	21210	62287	19062	24329	36557	43868	3106	38805	27279	
$x_{max}/\bar{x}$	5.1	9.8	2.8	3.6	3.0	4.4	2.3	2.5	3.5	6.5	2.4	4.0
Calanoid												
$\bar{x}$	D.s. 587	D.s. 1446	D.s. 5241	D.s. 11858	D.s. 8662	D.s. 16996	D.s. 18296	D.s. 11350	D.s. 1501	D.s. 7626	D.s. 5360	
x_{max}	3732	4752	21210	49836	26942	38573	64793	41296	4145	18288	32806	
$x_{max}/\bar{x}$	6.4	3.3	4.1	4.2	3.1	2.3	3.5	3.6	2.8	2.4	6.1	3.6

[a]E.c. = *Eubosmina coregoni*, B.l. = *Bosmina longirostris*, C.b. = *Cyclops bicuspidatus*, Cop = *cyclopoid copepodites*, D.s. = *Diaptomus sicilis*.

Table V.   Mean Density (x̄), Maximum Density (x_max) for *Rarest* Cladoceran, Cyclopoid Copepod and Calanoid Copepod in Southern Lake Huron (from 28 April 1975 to 23 October 1976)

	28 April	14 May	4 June	16 June	17 July	26 Aug	8 Oct	10 Nov	10 April	6 Aug	23 Oct	MEAN
Cladoceran (*Leptodora*)												
$\bar{x}$	0	0	1	5	17	78	48	4	0	81	15	
x_{max}	–	–	–	121	247	454	403	108	–	908	151	
$x_{max}/\bar{x}$	–	–	–	24.2	14.5	5.8	8.4	27	–	11.2	10.1	14.5
Cyclopoid (*Mesocyclops*)												
$\bar{x}$	3	7	14	61	17	246	517	160	3	326	211	
x_{max}	13	74	120	1027	200	1983	2269	832	838	3328	1261	
$x_{max}/\bar{x}$	4.3	10.6	8.6	16.8	11.8	8.1	4.4	5.2	279	10.2	6.0	33.2
Calanoid (*Limnocalanus*)												
$\bar{x}$	52	109	47	46	1	8	57	35	25	4	0	
x_{max}	176	694	322	362	34	136	1224	378	92	32	–	
$x_{max}/\bar{x}$	3.4	6.4	6.9	7.9	34	17	21.5	10.8	3.7	8	–	10.9

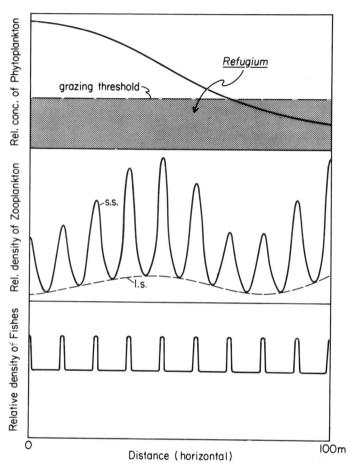

Figure 3. Coincidence of heterogeneity in horizontal distribution of phytoplankton (top), zooplankton (center) and fish (bottom). (s.s. = small scale, l.s. = large scale.)

small-scale (10 cm-1 m) aggregations of phytoplankton. However, MacArthur and Pianka (1966) have argued that herbivores should be more efficient in large patches, and my conclusions drawn from the literature may be correct.

Higher organisms, chiefly fishes, feed in clumps of zooplankton, effectively increasing their grazing efficiency. Schooling fishes like the white bass (McNaught and Hasler, 1961) actively feed precisely within the convergences of Langmuir cells. This highly developed foraging behavior is so dominant that it must constitute a sizeable fraction of all of their activities. Thus, an evolutionary pattern of foraging may be evident; vertebrates actively forage in patches. Crustacean zooplankters may be carried passively by the currents *or* actively seek areas rich in food resources. Phytoplankters are purely drifters, but utilize temporal distributions in resource availability in stratifying their niches.

Vertical coincidence of predator and prey also occurs frequently. In the Great Lakes herbivores migrate vertically into patches of phytoplankton during darkness (Bowers, Chapter 3). Thus, zooplankton and their phytoplankton foods are coincident vertically, but possibly not horizontally. A number of herbivores in Lake Michigan have apparently adapted to reduce fish predation not only by migrating downward during lighted periods and then migrating upward into food-rich layers after dark but also by dispersing when in lighted layers where they are visible to predators (McNaught, 1966). In fact, the rate of vertical movement is closely tied to grazing rate (Haney and Hall, 1975). But only horizontal spatial-temporal relationships have been emphasized in the scaling diagram (Figure 2). Also, in resolving the actual coincidence of predator and prey (Figure 3), only horizontal overlap was depicted. It is obvious that both concepts would involve greater ecological complexity if they included a third axis or vertical scale. Then temporal scales could be related to both spatial scales. To avoid excessive pictorial complexity, I ask the reader to do this conceptually.

Spatial-temporal coincidence is one key to modeling predator-prey interactions in aquatic ecosystems. Hutchinson (1965) long ago demonstrated that temporal niche differentiation may involve mechanisms like vertical migration. To such highly evolved behavioral responses, I add passive spatial heterogeneity in a horizontal plane. Organisms are spatially separated in both fashions, as reflected in their total niche structure. Thus, the newer concept of contemporaneous disequilibrium (Richerson *et al.*, 1970) must be enlarged to involve both space and time. Development of a general theory of spatial heterogeneity as related to stability (Steele, 1974) should be further encouraged.

MODEL CALIBRATION AND SPATIAL-TEMPORAL HETEROGENEITY

Spatial Heterogeneity of Biomass and Model Calibration

Plankton sampling with nets or water bottles produces samples characterized by high variability, chiefly because of the relationship between the size (volume) of the samples and the many spatial scales involved. For model calibration the tendency has been to take the mean density of a species. Let's examine a number of real examples characterizing aggregations of scales common to the Great Lakes. These samples were collected over eleven monthly cruises on Lake Huron during 1975 and 1976 (Figure 1). Hauls from 5 m to the surface were taken with a large (0.8 m diam) net of 64-μm mesh apertures. The relationship between mean density and maximum density for each cruise will be examined.

Total densities (number per m^3) are shown for the crustacean zooplankton (Table III). Characteristically the maximum value (x_{max}) varies between 2.1 and 5.9 times the mean value, with a mean ratio ($x_{max}/\bar{x}$) of 3.3. If I were calibrating a model, I would use the *mean* in making estimates of processes carried out by the zooplankton, such as grazing, respiration and excretion, as well as in estimating standing crops. However, where these same animals are prey for fishes, results of behavioral studies (McNaught and Hasler, 1961) suggest the *maximum* value would better characterize densities at which predators feed. It is interesting to note that the ratio $x_{max}/\bar{x}$ was largest for June and July. What processes are common in that period that might account for such horizontal heterogeneity? Possibly both increased physical aggregation and biological factors, including resource availability and predation, are involved.

Further useful information can be obtained by examining this same ratio ($x_{max}/\bar{x}$) for individual species and life stages. The most common species (Table IV) should be sampled most precisely. The value $x_{max}/\bar{x}$ was largest for the Cladocera (8.8, range 2.3-15.5). These slow swimming organisms are easily trapped in convergences; on a larger scale they are found in inshore waters. Both observations may account for their high spatial heterogeneity. The cyclopoida, which are typically rapidly swimming predators, and the calanoida, fast swimming omnivores, had lower $x_{max}/\bar{x}$ values. Again I would use mean values for estimates of zooplankton processes and maximum densities in estimating fish predation on the zooplankton.

The predaceous zooplankton are relatively rare (Table V). *Leptodora*, a predator on cladocerans, as well as *Mesocyclops* and *Limnocalanus* (omnivores) have high $x_{max}/\bar{x}$ ratios (14.5, 33.2 and 10.9, respectively). These high values are, in part, related to the probability of finding these rare forms, as well as their aggregated nature. Their populations will always be difficult to characterize in a model.

Thus, we have a broad generalization: Mean density best estimates predator effectiveness, both for zooplankton grazing on algae and for fish capturing zooplankton, but maximum density is likely the best estimate of an organism's availability as prey.

Spatial-Temporal Estimates of Nutrient Uptake and Grazing

Both nutrient uptake by phytoplankton and grazing by zooplankton are highly nonlinear with regard to the concentration of substrate. Such rates are being determined more frequently *in situ*. It is of utmost importance to know how the food resource is distributed. We have estimated the ingestion rates of most crustacean zooplankton in the Great Lakes (McNaught *et al.*, in press). As an example, *Bosmina longirostris,* often the dominant cladoceran in Lake Huron, ingests algal foods above a minimum concentration of 25 mgC m^{-3} (Figure 4). Uptake is linear over the range A in this figure, and mean food concentrations above this range are not encountered in Lake Huron. *Bosmina* is most likely characterized by a rectilinear uptake curve (Mullin *et al.*, 1975), which for Great Lakes forms saturates at 10^5 cells ml^{-1} or about 466 mgC m^{-3}. At the saturation level, feeding becomes independent of resource concentration.

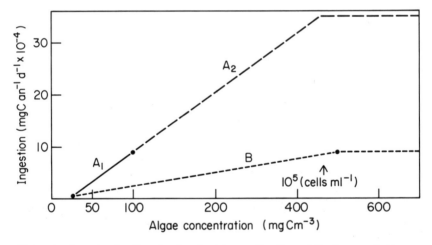

Figure 4. Ingestion by *Bosmina longirostris* as a function of food concentration. (A$_1$ = observed; A$_2$ = extension of rectilinear model; B = alternative curve, see text.)

Since phytoplankton densities in many large lakes never reach saturation levels, a consideration of the linear response curve is important. These determinations were made in the field. Algae were collected in a Niskin bottle, and therefore small-scale (< 10 cm) aggregations were probably averaged in sampling. If the zooplankton were feeding in a patch concentrated 5 times the mean density, the calculated rate of uptake (Figure 4, B) would be significantly different over a range of food concentrations.

Similar arguments could be made in calibrating nutrient uptake. Possibly the simplest solution to the problem is to assume an algal concentration factor of about 5 times, as observed by Denman and Platt (1975) in the Atlantic, when calculating uptake rates (Table I). Parameterization is especially critical when models are calibrated with uptake relationships determined in the laboratory and then used with average resource concentrations estimated in the field. Maximum concentrations should be used when warranted.

SUMMARY

Scales of phytoplankton and zooplankton distributions characteristic of large lakes have been described and the most common ones related to causal factors. Phytoplankton aggregations were characterized by scales of hundreds of meters and tens of kilometers. Examples from small lakes tentatively suggested the existence of smaller scales (meters and centimeters). Zooplankton distributions in the Laurentian Great Lakes were characterized by scales of meters, while longer scales have not been observed, probably because sampling programs that would detect them have not been carried out. Scales for fishes and zooplankton were similar, thus at least the physical mechanisms controlling their distributions are likely the same. Certainly the Langmuir circulation is dominant in creating patches of zooplankton to which fishes respond so dramatically. Evidence was lacking to indicate that phytoplankton were also aggregated within Langmuir convergences.

Spatial-temporal interactions may serve to magnify small-scale differences in plant or animal densities and associated fluxes. Physical factors such as currents lead to predominantly horizontal spatial heterogeneity in aquatic ecosystems, although growth and grazing are also involved. In contrast, the behavior of higher aquatic organisms creates vertical temporal heterogeneity, chiefly related to well-known patterns of vertical migration. In turn, both the physically caused horizontal and the behaviorally caused vertical heterogeneities are interrelated. Examples provided include hypothetical small-scale heterogeneity in inorganic nutrients created through physically driven differences in horizontal zooplankton distributions and patterns of vertical migration of herbivorous zooplankton, which lead periodically to high densities in surface waters. The animals in these nocturnal aggregations are, in turn, characterized by high grazing and excretion rates. Thus, they help provide a favorable nutrient environment for photosynthetic organisms the following day. Examples of such spatial-temporal interactions serve to expand the concept of temporal niche stratification, and underlie our current concepts of ecosystem stability. They must be considered in any attempt to model aquatic systems at the process level.

Coincidence in spatial-temporal scales is necessary for heterogeneity to lead to functional interrelationships between phytoplankton, zooplankton and planktonphagous fishes. For coincidence to occur, evidence of similar scale is not enough; however, actual coincidence has been demonstrated in the field with regard to zooplankton and fishes. Highly evolved foraging behavior of fishes may be the basis of such coincidence, although behavioral patterns of predator avoidance (migrations) by zooplankton may ameliorate it. This fact alone should be a great challenge to modelers, for aquatic ecosystems have likely evolved within the framework of physical and behavioral aggregation of phytoplankton and zooplankton, and the resultant aggregations of nutrients and predatory fishes.

Information on the scales of spatial-temporal heterogeneity for important trophic components is vital to model calibration. The mean density of zooplankton should be used in calibration of rate functions like grazing and excretion. However, when zooplankton densities are considered in determining food availability to fishes, the maximum observed density should be used to determine fish consumption rates. Due to operational characteristics of common sampling devices, even these maximal prey densities underestimate prey levels in clumps.

ACKNOWLEDGMENTS

Studies of zooplankton grazing, acoustical investigations of zooplankton spatial scales, and intensive net sampling of zooplankton in southern Lake Huron, all necessary to develop the ideas expressed herein, were supported by the U.S. Environmental Protection Agency (Grant 803178). Logistic support was provided by the crew of the research vessel *R. V. Simons.* The author would like to thank the Project Officer at E.P.A., Nelson Thomas, as well as David Griesmer, Michele Kennedy and Marlene Buzzard for their support in this project. An unidentified referee was especially helpful in clarifying certain concepts.

REFERENCES

Chisholm, S. W., and P. A. Nobbs. "Simulation of Algal Growth and Competition in a Phosphate Limited Cyclostat," in *Modeling Biochemical Processes in Aquatic Ecosystems,* R. P. Canale, Ed. (Ann Arbor, MI: Ann Arbor Science Publishers, Inc., 1976), pp. 337-356.

Cushing, D. H., and D. S. Tungate. "Studies on a Calanus Patch. I. The Identification of a Calanus Patch," *J. Mar. Biol. Assoc. U.K.* 43:327-337 (1963).

Denman, K. L., and T. Platt. "Coherences in the Horizóntal Distributions of Phytoplankton and Temperature in the Upper Ocean," *Mem. Soc. Roy. Sciences de Liège* 7 (6e Série):19-30 (1975).

Doyle, R. W., and R. V. Poore. "Nutrient Competition and Division Synchrony in Phytoplankton," *J. Exp. Mar. Biol. Ecol.* 14:201-210 (1974).

Dugdale, R. "Nutrient Limitation in the Sea: Dynamics, Identification and Significance," *Limnol. Oceanog.* 12:685-695 (1967).

Eppley, R. W., J. N. Rogers, J. J. McCarthy and A. Sournia. "Light/Dark Periodicity in Nitrogen Assimilation of the Marine Phytoplankters *Skeletonema costatum* and *Coccolithus huxleyi* in N-limited Chemostat Culture," *J. Phycol.* 7:150-154 (1971).

Haney, J. F., and D. J. Hall. "Diel Vertical Migration and Filter-Feeding Activities of *Daphnia,*" *Arch. Hydrobiol.* 75:413-441 (1975).

Hargrave, B. T., and G. H. Geen. "Phosphorus Excretion by Zooplankton," *Limnol. Oceanog.* 13(2):332-342 (1968).

Hutchinson, G. E. *The Ecological Theater and the Evolutionary Play* (New Haven: Yale University Press, 1965), 139 pp.

Lorenzen, H. "Synchronous Culture," in *Photobiology of Microorganisms,* Per Halldal, Ed. (London: Wiley Interscience, 1970), pp. 187-212.

MacArthur, R. H., and E. R. Pianka. "On Optimal Use of a Patchy Environment," *Amer. Nat.* 100(916):603-609 (1966).

Martin, J. H. "Phytoplankton-Zooplankton Relationships in Narragansett Bay. III. Seasonal Changes in Zooplankton Excretion Rates in Relation to Phytoplankton Abundance," *Limnol. Oceanog.* 13:63-71 (1968).

McNaught, D. C. "Depth Control by Planktonic Cladocerans in Lake Michigan," in *Proceedings of 9th Conference on Great Lakes Research* (Great Lakes Research Division, University of Michigan, 1966), Publ. 15, 98-108.

McNaught, D. C. "Developments in Acoustic Plankton Sampling," in *Proceedings of the 12th Conference on Great Lakes Research* (Int. Assoc. Great Lakes Research, 1969), pp. 61-68.

McNaught, D. C., M. Buzzard and S. Levine. "Zooplankton Production in Lake Ontario as Influenced by Environmental Perturbation," U.S. Environmental Protection Agency Ecological Res. Ser., EPA-660/3-75-021 (Washington, DC: U.S. Government Printing Office, 1975).

McNaught, D. C., M. Buzzard, D. Griesmer and M. Kennedy. "Water Quality and Zooplankton Population and Food Dynamics in Southern Lake Huron," U.S. Environmental Protection Agency Res. Ser. (in press).

McNaught, D. C., and A. D. Hasler. "Surface Schooling and Feeding Behavior in the White Bass, *Roccus chrysops* (Rafinesque), in Lake Mendota," *Limnol. Oceanog.* 6(1):53-60 (1961).

Mullin, M. M., E. F. Stewart and F. J. Fuglister. "Ingestion by Planktonic Grazers as a Function of Concentration of Food," *Limnol. Oceanog.* 20(2):259-264 (1975).

Nichols, S. "Phytoplankton Species Relationships at Short Time Scales," manuscript.

Patalas, K. "Composition and Horizontal Distribution of Crustacean Plankton in Lake Ontario," *J. Fish. Res. Bd. Can.* 26(8):2135-2164 (1969).

Platt, T. "Local Phytoplankton Abundance and Turbulence," *Deep-Sea Res.* 19:183-187 (1972).

Platt, T., and C. Filion. "Spatial Variability of the Productivity: Biomass Ratio for Phytoplankton in a Small Marine Basin," *Limnol. Oceanog.* 18(5):743-749 (1973).

Platt, T., and K. L. Denman. "Spectral Analysis in Ecology," in *Annual Revue of Ecology and Systematics,* vol. 6 (Palo Alto: Annual Revue, Inc., 1975), pp. 189-210.

Platt, T., K. L. Denman and A. D. Jassby. "Modeling the Productivity of Phytoplankton," in *The Sea,* vol. 6, *Marine Modeling,* E. D. Goldberg, I. N. McCave, J. J. O'Brien and J. H. Steele, Eds. (New York: John Wiley and Sons, Inc., 1977), pp. 807-856.

Richerson, P. J., R. Armstrong and C. R. Goldman. "Contemporaneous Disequilibrium, a New Hypothesis to Explain the Paradox of the Plankton," *Proc. Nat. Acad. Sci.* 67:1710-1714 (1970).

Richerson, P. J., B. J. Dozier and B. T. Maeda. "The Structure of Phytoplankton Associations in Lake Tahoe (Cal-Nev)," *Verh. Int. Ver. Limnol.* 19:843-849 (1975).

Stavn, R. H. "The Horizontal-Vertical Distribution Hypothesis: Langmuir Circulations and *Daphnia* Distributions," *Limnol. Oceanog.* 16(2):453-466 (1971).

Steele, J. H. "Spatial Heterogeneity and Population Stability," *Nature* 248:83 (1974).

Stross, R. G., and S. M. Pemrick. "Nutrient Uptake Kinetics in Phytoplankton: A Basis for Niche Separation," *J. Phycol.* 10:164-169 (1974).

CHAPTER 2

WATER COLUMN DEATH AND DECOMPOSITION OF PHYTOPLANKTON: AN EXPERIMENTAL AND MODELING REVIEW

Joseph V. DePinto

     Civil and Environmental Engineering Department
     Clarkson College of Technology
     Potsdam, New York 13676

INTRODUCTION

One of the great contributions to our understanding of biological and chemical processes in aquatic ecosystems has been the development and refinement of deterministic ecological models. These models are a vehicle by which knowledge of natural waters can be synthesized and evaluated. In this context ecological models provide an excellent research tool to test experimental findings on interactions in aquatic systems as well as to suggest new avenues of experimental investigation. Another important objective of developing these models, however, is to predict the outcome of a series of possible management alternatives for a given aquatic system. This goal has been accomplished with some success for certain aquatic systems, but it has been missed badly in others. It is this author's opinion that the advancement of predictive modeling has been slowed, to a certain extent, by the inadequacy of experimental methods in describing process mechanisms.

Deterministic phytoplankton models are basically an ordered framework of mechanistic or semiempirical submodels. Each submodel describes a particular process, whether it be the circulation patterns in a lake, the nutrient-limited growth rate of phytoplankton, or some loss rate of the phytoplankton population (such as sinking, zooplankton grazing or bacterial decomposition). The success or failure of a given model depends not only on the inclusion of all important submodels but on the way in which a given process is mathematically described.

The purpose of this chapter is to discuss the bacteria-mediated* decomposition of phytoplankton and phytoplankton-related organic matter and the subsequent remineralization of algal growth-regulating nutrients such as phosphorus and nitrogen in terms of our inadequate understanding of those processes and the resultant failure by modelers to describe them adequately. This chapter will review the experimental literature and the latest attempts by modelers to respond to this evidence in building submodels for death and decomposition of algae in the water column. In addition, suggestions will be forwarded for needed experimental study of the process. It is hoped that this review will spawn new experimental research aimed at quantifying these processes.

QUALITATIVE PROCESS DESCRIPTION

The following section contains a qualitative description of the biological and chemical processes which lead to the death and decay of phytoplankton biomass in the water column of lakes. Also, the role of these processes in nutrient regeneration within the water column, as distinguished from sediment nutrient release, will be discussed.

There are three main processes for the conversion of organic materials to inorganic nutrients in the water column of lakes: (1) zooplankton metabolism and excretion, (2) respiration and cellular release by viable algal cells, and (3) active decomposition by microbial decomposers. In addition to hydraulic washout and sinking, these processes represent the major mechanisms for loss of phytoplankton biomass from the epilimnion and, as such, should be included in a mass balance of phytoplankton-nutrient systems. The first processes, zooplankton grazing and regeneration of phytoplankton phosphorus, have been demonstrated by a number of authors (Pomeroy et al., 1963; Peters and Lean, 1973; Peters, 1975). While there certainly are interactions among all three processes (for example, zooplankton feeding on bacteria), the algal respiration and active decomposition modes are more difficult to separate.

Phytoplankton respiration represents the autooxidation of phytoplankton organic carbon to carbon dioxide. This process contributes to the decay rate of phytoplankton population biomass as well as the possible

---

*In this context and throughout this manuscript "bacteria" may refer to all possible microbial decomposers which may be important in the detritus food chain.

remineralization of algal nutrients. Algal respiration is theoretically independent of the microbial decomposer population. It is not, however, only species-specific, but is also probably dependent on physiological parameters such as the nutrient status of the algae or the cell age. The rate of biomass loss due to respiration is also a function of physical parameters such as temperature.

In addition to reducing the phytoplankton biomass, algal respiration and subsequent excretion also contributes to the pool of soluble and particulate organic matter available for decomposition by microdecomposers. The excretion of photosynthetic intermediates (such as glycollate) and end products (such as polysaccharides) by algae has been observed by a number of authors (Fogg, 1952, 1971; Hellebust, 1965; Watt, 1969; among many others), and such excretion is apparently not limited to the stationary or declining phase of growth (Watt and Fogg, 1966; Nalewajko and Lean, 1972). On the other hand, Sharp (1977) maintains that the only significant contribution by phytoplankton to organic matter in seawater is from dead and senescent algae.

In any case, the explanation for a positive bacterial response to phytoplankton blooms may come from the knowledge that additional metabolites become available when primary productivity is or has been high. In fact, in field studies Tanaka et al. (1974) found a close correlation between the depth distribution of algae and that of glycollate-using bacteria. One of the major extracellular metabolies found in natural waters is glycollic acid (Fogg et al., 1969), and it has been shown that aquatic bacteria in nonaxenic cultures of freshwater algae metabolize the glycollate excreted by the algae (Nalewajko and Lean, 1972). Wright (1975) determined that respiration (mineralization of ^{14}C-labeled glycollic acid by freshwater heterotrophs) accounted for 69% of the total glycollate uptake thus preventing its accumulation and, at the same time, returning inorganic carbon to the available carbon pool. It is certainly possible that the same process may lead to the mineralization of extracellular phosphorus-containing and nitrogenous organic compounds.

While there is no doubt that natural algal "death" rates must be included in phytoplankton models, there is strong evidence that heterotrophic microdecomposers play an active role that involves more than merely processing the remains of dead algae. There may be an actual parasitism-like interaction which leads to an increased rate of algal death. This rate is a function of the composition and state not only of the algal community but also of the bacterial, and perhaps fungal, communities as well. A simple process description might be as follows. As the algal crop increases, more carbonaceous material, both cellular and extracellular, becomes available for microbial decomposition. The heterotrophic microorganism community increases in response to this extra available energy but with some lag behind the algal development. However, because of the intrinsically more rapid growth rate of the heterotrophs, their ability to degrade the algal autolytic soluble and particulate material eventually surpasses the rate of production of these materials. At this point, those organisms capable of attacking live algae, albeit algae in a weakened physiological state, will do so at a rate which masks or even exceeds the algal growth rate. As a result a rapid decrease in the algal biomass ensues and is followed by a decline in heterotrophic activity. Since it is well known and has been demonstrated (DePinto and Verhoff, 1977; Golterman, 1973b) that only a small fraction (less than 20%) of the decomposed algal-related organic matter is converted to bacterial biomass (a considerable amount being needed for energy), most of the organic nitrogen and phosphorus is converted to inorganic forms. DePinto and Verhoff (1977) found that between 31 and 95% (mean, 74%) of algal phosphorus was converted to orthophosphate in bacteria-inoculated, dark batch-cultures, incubated for between 29 and 68 days.

The confirmation of the above scenario depends on our ability to prove that heterotrophic microorganisms can respond positively to primary production and can, in fact, parasitize live algal cells. It has long been established that a relatively good correlation exists between high bacterial activity and highly productive (eutrophic) lakes. Bere (1933), in a study of more than 50 American lakes, concluded that bacterial density was proportional to the concentrations of organic and inorganic compounds in one-half of the examined lakes and depended on the organic content alone in one-third of them. Henrici (1940) found that microscopic counts of bacteria in water from eutrophic lakes were at least an order of magnitude higher than similar counts from oligotrophic lakes. Possible discrepancies between bacterial activity and organic matter concentrations in lakes may be explained by the work of Kuznetsov (1949), which showed that bacterial counts depend not on the total quantity of dissolved organic matter but only on the assimilable part of it. More recent studies on the Great Lakes (Rao and Jurkovic, 1977) have shown a good correlation between the ratio of aerobic heterotrophic bacteria to total bacteria and the trophic state of the lakes, with Lake Erie having the highest ratio and Lake Superior the lowest.

While there is probably little controversy over the fact that, in general, bacterial activity is directly proportional to lake productivity, it has not been definitely established whether bacterial activity fluctuates in response to seasonal variations in phytoplankton density. Some authors regard algae as a source of toxic substances that retard growth of aquatic microorganisms. By inoculating meat-peptone agar dishes with live phytoplankton from several lakes, Razumov (1962) showed that many growing algal cells have sterile surfaces. He also found that incubating lake water without the phytoplankton increased bacterial density and shortened generation times in comparison to bacterial growth in unfiltered lake water. Drabkova (1965) supported these results by observing that algal blooms in a lake near Leningrad adversely affected the bacterial density.

Based on somewhat contradictory information to that given above, other authors believe that bacteria feed on organic products derived from or generated by algal metabolism and decomposition. Waksman *et al.* (1937) found that saprophytic bacteria respond positively to dead or dying phytoplankton in laboratory cultures; however, they also observed a negative response to the introduction of a live algal culture of *Nitzschia closterium*. Microscopic examination of dying plankton often reveals intense development of bacteria (Belyaev, 1967). Jones (1971) established positive correlations between chlorophyll *a* estimates and bacterial counts in a study of Esthwaite Water and Lake Winderemere. He found that the bacterial association varied with phytoplankton species (colonial green and blue-green algae having the closest association with bacteria), with an increased degree of attachment of bacteria to dying algal cells. Menon *et al.* (1972), on the other hand, observed in Lake Erie both positive and negative correlations between bacterial densities and chlorophyll *a*. Some of the mechanisms he proposed to explain these diverse relationships include: silica cell walls of diatoms being more resistant to bacterial attack; antibacterial substances produced by algae during blooms; positive bacterial response to algal excretion of degradation products; temperature, dissolved oxygen and pH effects; response to allochthonous organic material; zooplankton predation; and bacterial decomposition of dead algal cells.

Recent work in Lake Tahoe indicated a horizontal spatial relationship between primary productivity and heterotrophic activity measured by substrate uptake kinetics (Pearl, 1973; Goldman, 1974). These same studies showed a high correlation of heterotrophic activity with detritus (silts and decaying remnants of algae and zooplankton), which not only provides a source of direct nutrition but also a site for attachment of the bacteria. Overbeck (1975) confirmed the coupling between autotrophic and heterotrophic processes; however, his uptake data revealed that production and decomposition could be spatially separated in the vertical profile, with high heterotrophic activity occurring in the metalimnion and high primary productivity occurring in the epilimnion. Likewise, Niewolak (1974) found a thermocline maximum in density of heterotrophic bacteria during summer stratification to be a characteristic feature of Kortowskie Lake, Poland. He attributed this phenomenon to a response to the accumulation of greater quantities of decaying algae in the thermocline due to the gradient of specific gravity. Similar bacterial stratification was observed by Collins (1970) in British lakes and by Kuznetsov (1970) in Glubokoe Lake, USSR.

Obviously under certain conditions the surface of many phytoplankters are free of bacteria (Razumov, 1962), but some recent studies provide direct evidence of the parasitic attack of phytoplankton by certain micro-organisms. For example, Daft and Stewart (1973) have isolated and observed an aerobic bacterium (CP-1) from Scottish waters which causes lysis of a variety of blue-green algae. When cultures were observed, the bacteria were seen to attach themselves to algal cells and cause lysis of vegetative cells within 30 minutes. In fully lysed cells, they observed a loss of polyphosphate bodies and other cellular constituents. Other investigators, including Shilo (1970), Granhall and Berg (1972), Gromov *et al.* (1972) and Reim *et al.* (1974), also have found bacteria with the capability of lysing blue-green algae. Shilo (1971) also found viruses which attacked cyanophytes. Ho and Alexander (1974) reported the use of blue-green algae as food sources by several species of amoebae. Gunnison and Alexander (1975) provided field evidence that certain algal species were more susceptible to microbial decomposition than others, with blue-green algae being more likely to be susceptible. They also found the susceptibility of live algae to decomposition to be correlated with the digestability of algal cell walls and not necessarily to the production of toxins by the algae.

In summary, there is little doubt that certain aquatic bacteria can thrive on the extracellular organic secretions of phytoplankton and on the soluble and particulate remains of dead phytoplankton. The questions that remain are whether aquatic microorganisms can be causative agents in the termination of an algal bloom and should this process be dealt with explicitly in phytoplankton models. In other words, do bacteria, fungi, protozoans or viruses actually attack live algae in epidemic proportions, thus causing dramatic algal population crashes? In view of the observations discussed above and since we know that environmental factors can determine the outcome of any ecological competition, there is little doubt that conditions can exist in a lake which lead to microbial attack being a dominant factor in phytoplankton death rates. More often than not the time and space averaged nutrient, light and temperature levels in lakes (in contrast to laboratory cultures) establish phytoplankton growth conditions that are suboptimum. Natural phytoplankton populations are, therefore, often in a weakened physiological state and subject to microbial antagonism. As will be discussed in a later section, many modelers appreciate the importance of this process but lack the laboratory and field data necessary to quantify its contribution to phytoplankton dynamics.

QUANTITATIVE INFORMATION ON DECOMPOSITION
AND NUTRIENT REGENERATION IN THE
WATER COLUMN

Although more research is needed, the modeling community has bene-
fited from certain quantitative laboratory and field evidence on decomposition of algae and nutrient regeneration in the water column. This section
contains a review of research which points out the magnitude of these
processes and their significance in attempting to quantify phytoplankton
dynamics.

Field Studies

Field investigations, which attempt to quantify phytoplankton loss rates
or water column organic decay, often indicate that most phytoplankton
decay takes place in the water column as opposed to in the sediments.
On two different occasions, Kuznetsov (1939) found an obvious correlation
in the metalimnion of Lake Glubokoe between the decrease of a phytoplankton bloom and a rapid increase in bacteria, accompanied by a decline
in dissolved oxygen. Based on this and a number of other field studies,
Kuznetsov maintains that the breakdown of a phytoplankter in the water
mass begins only after its death and that 90% of the dead phytoplankters
undergo decomposition before leaving the water column (Kuznetsov, 1968).
Kajak *et al.* (1970) estimated that 63% of total primary production in
several Polish lakes was decomposed in the epilimnion. Of the remaining
37% that reached the hypolimnion, about half was decomposed before
reaching the sediments. Jassby and Goldman (1974), using primary productivity and phytoplankton biomass measurements in Castle Lake, estimated that specific loss rates of phytoplankton varied between 0.2 day^{-1}
and 0.8 day^{-1}. These loss rates could not be attributed to water transport,
sinking or grazing and were, therefore, considered the result of cell mortality and decomposition.

In their study of two polar lakes, Kalff *et al.* (1975) noted that the
differences between phytoplankton growth rates predicted by primary
production measurements and those computed from biomass changes
averaged 8%, indicating a loss of 8%/day. They further estimated that
80% of this loss rate was the result of decomposition in the water
column.

The development of oxygen deficits in lakes also seems to be related
to high rates of aerobic organic decay in the metalimnion and upper
hypolimnion. For example, Lasenby (1975) found that in 13 of 14
southern Ontario lakes the hypolimnetic areal oxygen deficits decreased
linearly as the upper limit of the hypolimnion of each lake was chosen

at successively deeper levels. This suggests, in agreement with the observations of Lund *et al.* (1963), that the upper layers of the hypolimnion
consumed more oxygen than the lower zones. In a detailed study of one
lake (mean depth, 19.7 m), Lasenby (1975) determined that only 27%
of the total oxygen uptake in the hypolimnion could be attributed to the
sediment. He also pointed out that a linear relationship (r = -0.85)
existed between the logarithm of the areal oxygen deficit and the logarithm
of the average summer Secchi depth for 26 lakes.

This same phenomenon, higher oxygen consumption in the upper strata
of the hypolimnion, was found in a moderately productive lake located in
the extreme northwest part of Adirondack Park, New York. Lake Ozonia
(mean depth 7.3 m, surface area 171 ha) is primarily phosphorus-limited
and on the borderline between oligotrophy and mesotrophy (Cangialosi,
1976; Lepak, 1976). Upon calculating an areal hypolimnetic oxygen
deficit, it was also discovered that a very good inverse linear relationship
existed between areal deficit and the depth chosen for the top of the
hypolimnion. The total oxygen deficit as a function of hypolimnion
depth is presented in Figure 1. The mean Secchi depth for the summer
was 3.7 m, so there is probably very little photosynthesis occurring below
6 m. Note that below a hypolimnion depth of 7 m there is an excellent
linear decrease in deficit. For this particular lake geometry, if one
assumed a constant total oxygen consumption in each strata throughout
the hypolimnion, the line in Figure 1 would not be straight but would
decline at an increasing rate with water depth. The actual data indicate
that either the mud surface or the open water in the upper hypolimnion
is producing a larger fraction of the total oxygen deficit than the lower
strata. According to Lund *et al.* (1963), Hargrave (1972) and Lasenby
(1975), however, there is not likely to be a large difference in sediment
oxygen demand (SOD) with depth as long as temperature does not vary
significantly. The temperature profiles for Lake Ozonia seem to negate
the possibility of differences in the SOD causing the effect. In fact if
one assumes a constant sediment demand of 0.02 mg-O_2 cm^{-2} day^{-1}
(Lund *et al.,* 1963), the open water accounts for 43% of the total areal
hypolimnetic deficit and virtually all of the observed decrease in deficit
with depth of the hypolimnion boundary (Figure 1). A possible explanation for the failure of the 6-7 m stratum to follow the pattern is that
for most of the summer a positive heterograde oxygen profile existed in
the metalimnion just above this depth. The diffusion of oxygen from
above into this stratum could easily have masked much of the actual
oxygen consumption. It also is possible that photosynthesis was occurring
in this stratum. At any rate this case does appear to be indirect evidence
for the existence of more respiration, or breakdown of organic matter, in
the area of the lake just below the trophogenic zone.

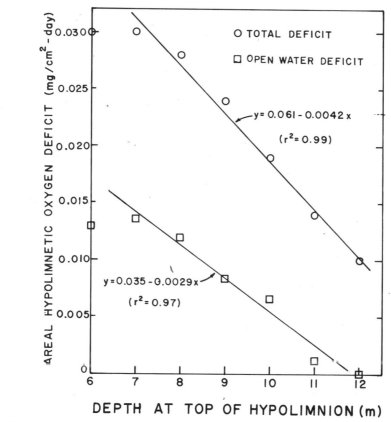

Figure 1. Areal hypolimnetic oxygen deficit of Lake Ozonia (Summer 1975) as a function of the depth chosen for the top of the hypolimnion. Lines represent least-squares fits to the data from 7 m and deeper.

Experience at Shagawa Lake, Minnesota (Larsen *et al.,* 1973) points out the potentially important contribution of water column decomposition and subsequent nutrient regeneration to algal population dynamics in a eutrophic lake. Having obtained reliable external (precipitation, runoff, point sources, outflow) nitrogen and phosphorus budgets for the lake, Larsen *et al.* (1973) attempted to model the seasonal chlorophyll *a* dynamics without including internal loadings such as water-column regeneration, nitrogen fixation or sediment nutrient release. Under these circumstances the calibrated model did not predict an observed large late-summer algal bloom, even with a total algal loss rate (death, grazing, sinking, etc.) as low as 0.025 day^{-1}. When calculated sediment nutrient release rates and estimated nitrogen-fixation rates were incorporated as additional nutrient sources, the model did predict a second chlorophyll *a* peak; however, the prediction underestimated both the magnitude and the duration of the bloom. One hypothesis for explaining the predicted phosphorus limitation in the second peak is the failure to include processes which recycle nutrients in the water column.

This author's field observations on a hypereutrophic, nitrogen-limited lake (Stone Lake, Michigan) provides further evidence for the occurrence of water column decomposition and nutrient regeneration (DePinto *et al.,* 1976). Phytoplankton blooms were found to be excellently correlated with the concentrations of dissolved oxygen and combined inorganic nitrogen at the bottom of the epilimnion. A plot of average algal biomass in the upper 4 m of the lake (average epilimnion depth) during the period of summer stratification (Figure 2) reveals two major blooms. The first peak occurred in late July and consisted of overlapping blooms; a bloom of *Aphanizomenon* and *Anabaena* (organisms capable of fixing atmospheric nitrogen) followed closely by a bloom of non-nitrogen-fixing phytoplankton consisting primarily of *Microcystis.* Note the rapid decline of these blue-green algal blooms. A mathematical model was used to test various hypotheses that can be used to explain this phenomenon, and the results suggested that water column decay and nutrient recycle were very important (see Modeling section). The second peak in the total crop (maximum in mid-October) could be attributed mainly to the bloom-forming green algae *Microspora.*

Figure 3 consists of plots of dissolved oxygen and combined inorganic nitrogen (NH_3 + NO_2 + NO_3) concentrations for the same time period and at a depth of 3.7 m. Water at this depth was generally above the thermocline but received less than 100 ft-candles (1,076 lux) light intensity. Note that the two minima in the dissolved oxygen (DO) curve correspond almost exactly with the two peaks in total algal crop (Figure 2), indicating that phytoplankton production in the euphotic zone creates an intense oxygen demand just below the trophogenic layer. It is also noteworthy that during the *Aphanizomenon-Anabaena* bloom a rather dramatic increase in soluble inorganic nitrogen occurred. Since the available nitrogen levels in the epilimnion were quite low at the start of the bloom ($< 100\ \mu g$-N l^{-1}), it is likely that these organisms were fixing atmospheric nitrogen. A corresponding high rate of decomposition and mineralization of these organisms could explain the increase in nitrogen which apparently fueled the *Microcystis* bloom in August. The tremendous rise in nitrogen in October was primarily the result of an increase in ammonia due to vertical transport from the very high NH_3-N concentration (1-3.7 mg-N l^{-1}) in the hypolimnion of the lake.

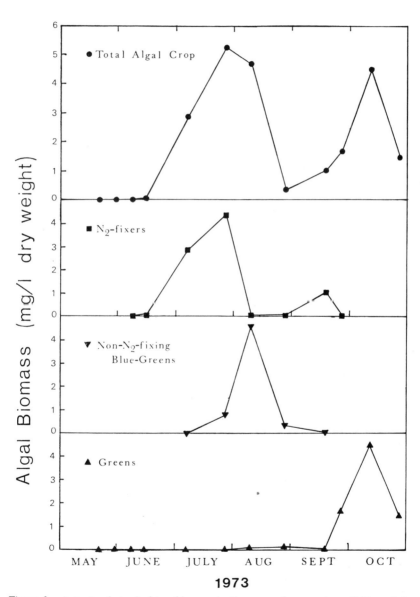

Figure 2. Average phytoplankton biomass in the upper four meters of Stone Lake, Michigan, during the summer stratification of 1973. Total crop is differentiated into the three prominent functional groups (from DePinto *et al.*, 1976).

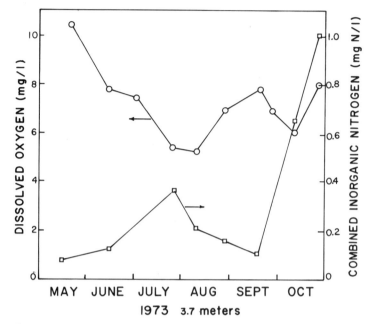

Figure 3. Dissolved oxygen and combined inorganic nitrogen at the bottom of the thermocline in Stone Lake, Michigan, during the summer stratification of 1973.

Additional evidence for nutrient regeneration in the water column can be found in results from sedimentation trap experiments. Kleerekoper (1953) concluded that most of the decomposition of sinking matter takes place in the region of the lake above the hypolimnion. He based this on the observation that the loss of ignition of epilimnetic and metalimnetic plankton in Lake Lauzon varied from 90.5-94.3% while the loss on ignition of detritus collected in sediment traps just below the thermocline and of surface bottom deposits ranged from 40-60%. About 70% of the nitrogen associated with plankton in the upper waters was mineralized by the time the detritus reached the bottom of the thermocline. Phosphorus regeneration was somewhat slower, while silica was actually enriched in the trapped detritus, presumably due to the resistance of diatom frustules to decay.

By combining a phosphorus budget for Canadarago Lake, New York, with phosphorus sedimentation measurements, Fuhs (1973) calculated that incoming phosphorus was recycled 10 to 11 times during the growing season. This remarkable result could not be due totally to sediment release; some contribution must come from solubilization and reuse of the phosphorus before it reaches the bottom muds. Charlton (1975) arrived at a similar conclusion by analyzing sedimentation data from traps installed in limnocorrals in the Bay of Quinte, Lake Ontario. From three enclosures, one a control, a second with phosphorus enrichment and a third with both nitrogen and phosphorus enrichments, he calculated that water column phosphorus would be recycled 12, 14 and 34 times per year, respectively. This result not only supports Fuhs' findings but suggests that regeneration rates are a function of lake trophic status.

Laboratory Studies

The impact of bacteria on the rate of algal decay in the dark has been demonstrated in studies using batch unialgal cultures. Uhlmann (1971) noted an exponential loss rate of *Scenedesmus obliquus* in stirred-batch experiments of 0.02 day^{-1} at 20°C and postulated the absence of specific algolytic microorganisms. By observing the chemical oxygen demand (COD) in bacteria-free darkened batch cultures of *Chlorella vulgaris* DePinto (1974) measured exponential loss rates, presumably due entirely to respiration, of 0.015 and 0.017 day^{-1}, in duplicate runs at 20°C. When a natural bacterial population from a eutrophic lake was present, decay rates of the culture varied from 0.02 to 0.09 day^{-1}. Also using COD as a measure of biomass, Jewell (1968) found that a 10-day-old axenic culture of *Chlorella pyrenoidosa* decayed at 0.022 day^{-1} (20°C, base e), while a comparably grown bacteria- and zooplankton-seeded culture decayed at 0.063 day^{-1}.

Previous laboratory work on nutrient regeneration from algal decay is not very extensive and has been confined to batch systems. These studies nevertheless have clearly demonstrated the importance of this phenomenon as well as of some of the parameters which govern its rate and extent. One of the earliest comprehensive studies was conducted by Grill and Richards (1964) using laboratory cultures of marine diatoms subjected to bacterial attack in a dark aerobic environment. At the end of five months, 64% of the initial total phosphorus was dissolved inorganic phosphorus, and 50% of the initial particulate nitrogen had been solubilized. Kamatani (1969) also studied diatom decomposition in laboratory cultures and obtained much quicker mineralization than Grill and Richards. At 30°C, about 70% of the phosphorus, nitrogen, carbon and silicon in the plankton was mineralized in the first 10 days, probably because his diatoms, a unialgal culture of *Skeletonema costatum,* had been cultured with a normal seawater bacterial population.

A number of authors have studied bacterial attack on dead algal cells (Otsuki and Hanya, 1972a,b; Fitzgerald, 1964; Varma and DiGiano, 1968; Golterman, 1964, 1973a). Golterman found that once the algae had undergone autolysis (whether natural or induced) mineralization of phosphate was very efficient with a 70-80% release in the first few days. These results in combination with those of DePinto and Verhoff (1977), suggest that once an algal cell dies virtually all excess cellular phosphate is liberated almost immediately with all but a small percentage of the remaining phosphorus available for mineralization by bacteria.

While few nutrient regeneration studies have been performed using live freshwater algae, data indicate that the process takes considerably longer when the initial conditions include viable algae. Foree *et al.* (1970) and Foree and Barrow (1970) studied the extent of nitrogen and phosphorus regeneration from the decomposition of various unialgal and mixed freshwater phytoplankton cultures. After six months to a year of decomposition, an average 50% of the initial particulate phosphorus and nitrogen had been regenerated with variations dependent on the initial nutrient content of the algal cells and the extent of organic decomposition. Green algae (*Chlorella* and *Scenedesmus*) tended to be more refractory than the blue-green alga *Anabaena* (also a conclusion of Gunnison and Alexander, 1975), with aerobic decomposition releasing more nutrients than anaerobic decay.

DePinto and Verhoff (1977), using smaller initial algal biomasses, found more rapid net phosphorus regeneration rates. An average of 74% of the initial cellular phosphorus was liberated from bacteria—inoculated cultures during dark incubation periods varying from 29 to 68 days (Figure 4). The first 10 to 20 days of the incubation period, however, represented a lag phase during which no net phosphorus release was observed. These results suggest that, for batch systems, as long as a certain fraction of the algae remains viable, any phosphorus that may be released by lysing cells is rapidly reincorporated into the algae and, as a result, not observed as net regeneration. We know that algae were dying during this lag phase because biomass measurements such as particulate COD and suspended solids decreased.

More recent laboratory studies on nutrient regeneration (Lee *et al.*, 1973; Mills and Alexander, 1974; DePinto and Verhoff, 1977) not only confirm its importance but point out the significance of several controlling parameters. An active bacterial population is apparently necessary to stimulate algal decomposition and net phosphorus regeneration. Note in Figure 4 that two *Chlorella* cultures not inoculated with bacteria remained viable for the duration of dark incubation and displayed no net phosphorus release. In addition to the importance of bacteria, DePinto and Verhoff (1977) observed that the rate and extent of phosphorus regeneration were dependent upon the phosphorus content of algal cells. Regeneration of cellular phosphorus was proportional to the degree of

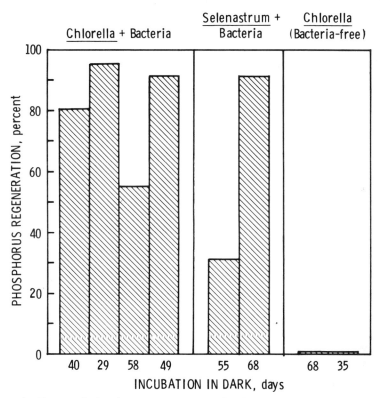

Figure 4. Extent of phosphorus regeneration in batch cultures of green algae incubated in the dark (from DePinto and Verhoff, 1977).

decomposition of the cellular organic matter, while excess phosphorus (due to luxury uptake) present in the cells as inorganic phosphorus was released rather rapidly following cell lysis. A plot of the ratio just before active decomposition of inorganic cellular phosphorus to organic cellular phosphorus versus the ultimate phosphorus release per unit dry weight of algae illustrates this point (Figure 5). Higher inorganic:organic phosphorus ratios are representative of algal systems in a state of luxury consumption because excess cellular phosphorus is stored in an inorganic form (Rhee, 1973).

Further evidence for the dependence of phosphorus regeneration on the cellular content can be derived from data on the rate of phosphorus regeneration as a function of cellular phosphorus. In the above experiments (DePinto and Verhoff, 1977) the two runs that had the lowest initial cellular phosphorus levels (0.48% and 0.38% P dry wt) produced

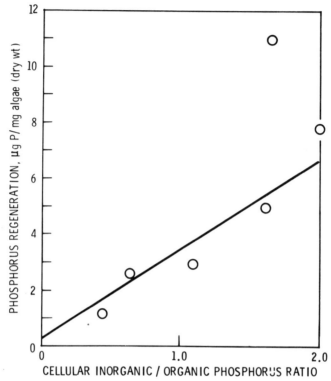

Figure 5. Extent of phosphorus regeneration in bacteria-inoculated batch green algae cultures as a function of cellular inorganic/organic phosphorus ratio (from DePinto and Verhoff, 1977).

phosphorus regeneration rates of 0.08 and 0.06 μg-P(mg algae dry wt)$^{-1}$ day^{-1}, respectively. Conversely, the rates for four runs where the algae appeared to be in a luxury state (cellular phosphorus ranging from 0.51-1.21%) ranged from 0.16 to 0.39 μg-P(mg algae dry wt)$^{-1}$ day^{-1}, with an average of 0.24 μg-P(mg algae dry wt)$^{-1}$ day^{-1}.

Nitrogen regeneration does not appear to behave in the same manner as phosphorus regeneration because the release of nitrogen from algal cells appears to occur simultaneously with organic decomposition (Figure 6) whether by respiration or active decomposition. Nitrogen regeneration was observed, although to a lesser extent, even in bacteria-free cultures.

The potential role of the metabiotic process of nitrogen-fixation, followed by nitrogen regeneration, and then by nitrification of the released ammonia has been demonstrated by monitoring batch cultures of *Anabaena flos-aquae*

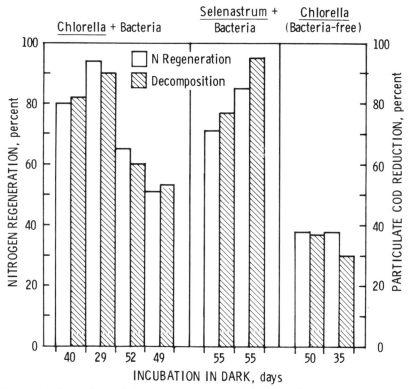

Figure 6. Comparison of the extent of decomposition and nitrogen regeneration (expressed as percent decrease in particulate organic nitrogen) in batch cultures of green algae (from DePinto and Verhoff, 1977).

(Rodgers and DePinto, in preparation). These cultures were grown under low-light conditions (110 ft-candle, 1,184 lux) with a 14-hr-light/10-hr-dark cycle and a temperature of 25 ± 2°C. No ionic nitrogen was supplied to any of the cultures. Culture a (Figure 7) contained an initial soluble phosphorus concentration of 68 μg-P 1^{-1} and received no inoculum of bacteria. Cultures b and c each received an inoculum of mixed bacteria subcultured from lakewater and had initial phosphorus concentrations of 69 and 205 μg-P 1^{-1}, respectively. Figure 7 is a plot of nitrate-nitrogen concentrations in the three cultures as a function of time of incubation. Note that the nitrate regeneration process has taken place in the bacteria-inoculated cultures, where peaks of 60 μg-N 1^{-1} and 80 μg-N 1^{-1} occurred on day 19 in the low and high phosphorus cultures, respectively. Very

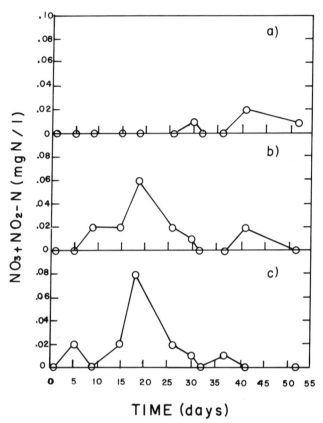

Figure 7. $NO_3 + NO_2 -$ nitrogen as a function of time in three batch cultures of *Anabaena* grown without the addition of any combined inorganic nitrogen to the medium: (a) no bacteria inoculum and initial phosphorus of 68 μg-P l^{-1}; (b) bacteria-inoculated and 68 μg-P l^{-1}; (c) bacteria-inoculated and 205 μg-P l^{-1}.

little nitrate was detected in the control. Also, the apparent reassimilation of nitrate by the algal population (reduction in soluble nitrate levels after the peak) suggests that gross nitrogen regeneration may actually have been greater than that observed in the batch system.

MODELING OF ALGAL DECAY AND NUTRIENT REGENERATION

Current efforts in ecological modeling of aquatic systems recognize the necessity of incorporating water column decomposition and nutrient regeneration in the system (*e.g.,* Bierman *et al.,* 1973, 1976; Cordiero *et al.,* 1973; Canale *et al.,* 1976; Fleming, 1975; Lehman *et al.,* 1975; Patten, 1968; Scavia *et al.,* 1976; Thomann *et al.,* 1975). In most cases the equation for the rate of change of phytoplankton concentration (A) as a function of time (t) for a completely mixed box of volume V takes the form:

$$V \frac{dA}{dt} = W - Q \cdot A + V \cdot A \cdot [G(T,L,N) - D(T) - ZG(Z,A) - S/\text{depth}]$$

where W = inputs of algae to the box
 G = growth rate as a function of temperature (T), light (L) and nutrients (N)
 D = algal death rate as a function of temperature
 ZG = zooplankton grazing rate as a function of zooplankton (Z) and phytoplankton biomass
 S = constant sinking rate
 Q = flow out of box

Most of the models developed by the authors above include a mechanism for zooplankton grazing on phytoplankton ranging from simple constant filtering rates to complex functions which depend on feedback from modeling zooplankton growth. The cell sinking rate is generally a species-specific constant but also has been approached at a more complex level (Scavia *et al.,* 1976).

Also, most models contain a term or a submodel framework that attempts to account for one or more of the biological losses of algal biomass: (1) algal respiration, (2) autolysis and (3) microbial attack. The purpose of this section is to discuss current approaches to modeling decomposition of algae and subsequent nutrient regeneration in view of the process description and experimental data presented in previous sections. The important considerations will be delineated along with the advantages and disadvantages of each basic approach.

The approaches to describing phytoplankton decay range from a very simple first-order rate coefficient, which is only a function of temperature (Thomann *et al.,* 1975; Canale *et al.,* 1976), to very sophisticated submodels, which include modeling bacteria explicitly as a state-variable (Cordiero *et al.,* 1973; Clesceri *et al.,* 1977). In view of the above discussions, the first-order coefficient probably only describes loss due to algal respiration and is unresponsive to such potentially important conditions as the physiological state of the algae and the size and influence of the decomposer population. Jewell (1968) and DePinto and Verhoff (1977) indicated that both of these effects could be important.

Attempts to correct the shortcomings of the first-order "death" rate include adjusting the first-order coefficient according to the value of certain system parameters (specific growth rate or length of time under stress) as well as according to forcing functions such as temperature and light. This approach is based on data that suggest that natural phytoplankton death and autolysis depend on some intrinsic property of the cell. Lehman *et al.* (1975) consider the death rate to be a function of the number of days of algal growth at suboptimal conditions in the following manner:

$$D = D_{max} [1 - \exp(-\ln 2 \cdot SG/K_D)]$$

where D = specific death rate
SG = number of days at suboptimal growth conditions (*i.e.*, μ/μ_{max} < 0.05 where μ and μ_{max} are the predicted and maximum growth rates, respectively)
K_D = number of days at suboptimal growth conditions until $D/D_{max} = 0.5$

Scavia *et al.* (1976) allowed phytoplankton death rate to vary as a function of temperature and of the physiological state of the cells. With regard to the physiological state parameter, the temperature-corrected maximum mortality rate was modified by multiplying it by the factor (1-u), where u was defined as the minimum of the individual nutrient or light limitation terms. In other words, if the growth rate happened to be limited by phosphorus to 75% of its temperature-dependent maximum rate, then the death rate would be (1-0.75) or 25% of its maximum value.

The adjusted first-order rates presented above probably represent actual conditions better than constant coefficients; however, at this time, there is insufficient quantitative validation of the functions. More experimental measurements of natural phytoplankton death rates as a function of the physiological state of the cells are needed. Furthermore, the effect of microbial attack has not been addressed in any of the previous formalisms.

One possible approach to simulating microbial effects on phytoplankton loss rates is to model bacteria implicitly. The implicit assumption is that bacterial biomass or activity is a dependent variable and exhibits a close correlation with phytoplankton biomass. This is the approach taken by Bierman (1976) when he uses a second-order algal death rate that is proportional to the product of the dry weight biomass of the phytoplankton group being modeled and the biomass of the entire assemblage. This term takes the form:

$$r_i = -KTA_iA_t$$

where r_i = change in biomass of ith algal group due to biological loss
K = decay constant
T = water temperature

A_i = biomass of ith algal group
A_t = biomass of total algal crop

The logic used here is that if, in fact, bacteria do respond quickly to algal growth, a second bloom species increasing shortly after a primary bloom would have to deal with the presence of an already large decomposer population.

The second-order mechanism employed by Bierman (1976) is rather stringent, but the often observed rapid crashes of phytoplankton blooms justify its use. In fact it was shown to be necessary, but not sufficient, to account for the rapid die-off of phytoplankton blooms observed in the hypereutrophic lake (Figure 2) discussed in the previous section (DePinto et al., 1976). An additional constraint had to be placed on individual bloom species during their domination. In order to simulate the timing of the bloom crashes, the specific growth rate of each group was constrained to be zero as that group reached its peak. While a zero growth rate is unlikely and this arbitrary constraint is not very elegant, it does point out that in certain cases current models are incomplete in their description of phytoplankton loss rates.

The effect of bacterial decomposition is generally implicit in nutrient regeneration submodels. This effect is often arrived at by placing dead algal biomass and its associated nutrients in an unavailable pool, which is independently caused to decompose, mineralize (by a first-order mechanism) and sink. This approach probably is not a bad estimate for nitrogen regeneration, assuming decomposition rates are known. Phosphorus, however, is a more difficult problem since this nutrient is known to be stored in algal cells in inorganic forms (Rhee, 1973). An intuitive approach to this problem, taken by Bierman (1974), is to release immediately all excess internal P to the available phosphorus pool upon the death of an algal cell and then add the remainder to the unavailable pool for normal mineralization. As presented in the previous section, some experimental evidence for this mechanism does exist, but more is needed. Golterman (1973a) reported rapid phosphorus release from induced autolysis of algae, but the results were not correlated with the size of the internal phosphorus fraction.

In a study of the Lake Huron-Saginaw Bay system, DiToro and Matystik (1976) concluded that a first-order rate of phosphorus recycling was inadequate to describe nutrient-phytoplankton dynamics in highly productive Saginaw Bay. A more general function was therefore postulated which again assumed a close correlation between phytoplankton biomass and bacterial activity. The construct chosen was an adjustment to the first-order recycling rate coefficient that made it a function of phytoplankton biomass in a Michaelis-Menten fashion:

$$R = \frac{R_{max}\ P}{(K_R + P)}$$

where   R   = recycle rate
        R_{max} = maximum recycle rate (~ 0.05 day^{-1})
        P   = phytoplankton biomass (μg-Chl a l^{-1})
        K_R = half-saturation constant (~ 5 μg-Chl a l^{-1})

If algae-related organic matter is the limiting substrate for heterotrophic bacteria in lakes, then their growth and subsequent recycle activity is quite possibly governed by Michaelis-Menten kinetics with respect to that limiting substrate (*i.e.*, phytoplankton). Much more field and laboratory work, however, is needed to validate this assumption. In the field, simultaneous observations of phytoplankton biomass and bacterial activity should be made on a short-term basis during a bloom.

Finally, the most complex approach to simulating phytoplankton-bacterial interactions is to model bacteria explicitly as a state variable. It then would be possible, for example, to simulate the death of a phytoplankton group by a second-order term which is a function of the phytoplankton biomass and the active bacteria biomass. It seems inevitable that further submodel development in this area will be needed, if only to elucidate mechanisms for phytoplankton death and nutrient recycling.

Recent modeling work in this area includes an attempt by Cordiero *et al.* (1973) to simulate nutrient recycling by actually modeling bacterial growth in a two-step manner (accumulation of limiting substrate and assimilation) based on an assumed cellular stoichiometry. They also appreciated the various ecological functions of bacteria and thus included three functional groups in the upper waters and three others in the often anaerobic hypolimnetic waters.

In addition to recognizing that the myriad of organic compounds in lakes leads to a diversified bacterial assemblage, Clesceri *et al.* (1977) developed a preliminary decomposer model which includes freely suspended decomposers as well as those attached to particulate organic matter. As discussed in the Process Description section, there is experimental evidence for this approach.

The current problem with developing decomposer community submodels and incorporating them into ecological models is one of having to add kinetic coefficients about which virtually no data exist. Most deterministic models of aquatic systems already include more state variables, processes and coefficients than the comparable available data, and making the models more complex without sufficient data will not solve the problem. In order to be able to improve ecological models, which include the decomposer community in nutrient cycles, more research in the following areas is needed:

1. further development of submodels that propose testable mechanisms for the impact of decomposers on phytoplankton decomposition and nutrient recycling;
2. process experimentation on the kinetics of bacterial growth using algae and their lysate as substrates;
3. experimentation that will elucidate the mechanism(s) of phytoplankton death due to microbial attack;
4. process experimentation on the stoichiometry and kinetics of nutrient regeneration from aerobic decomposition of algae and algae-related organic matter by microorganisms; and
5. monitoring of aquatic systems in such a manner as to allow the calibration of deterministic models that include bacteria as a state variable.

SUMMARY

Evidence has been presented of the importance of phytoplankton mortality and water-column decomposition in lakes. The regeneration of nutrients such as phosphorus and nitrogen commensurate with these processes can be an important internal nutrient source at certain times of the growing season in aquatic systems. It appears that these processes are more significant in highly productive eutrophic lakes. Whether this is a characteristic pattern or is simply more easily detectable in eutrophic systems remains to be seen.

Qualitative and quantitative research suggest that estimates of biological losses from phytoplankton biomass in lakes must include algal respiration, natural algal mortality, microbial-induced algal mortality, and grazing by higher trophic organisms. A number of parameters have been shown to have a significant influence on these processes. Among these are temperature; light; dissolved oxygen; species, nutrient status, cell age and stress condition of the phytoplankton; and interaction with the decomposer community. Further research is needed, however, to clarify and quantify the mechanisms involved.

Furthermore, the one aspect that is probably most important to decomposition and nutrient regeneration is the one which requires the most study, namely the microbial decomposer community. The importance of this component to a lake's trophic structure has been brought out in this chapter; however, until more information is obtained on the composition and activity of this trophic component, the processes in question cannot be modeled accurately. This is not to say that modelers should not attempt to construct and verify decomposer submodels until more experimental data are available. Experimental and modeling efforts should complement each other in an effort to complete the description of aquatic ecosystems.

Finally, in efforts to develop and test phytoplankton-bacteria mecha-nisms, the criterion for success or failure should not be the model's ability to describe field data. The reason for this is that most dynamic ecological models are overdetermined; often more than one set of coefficients will provide a reasonable fit to the data. The criterion for success or failure should depend on qualitative knowledge and direct measurement of the processes involved. A fundamentally sound submodel should be consistent with the process descriptions and the quantitative information reviewed in previous sections of this chapter. It is questionable whether a generally acceptable mathematical description of phytoplankton-bacteria interactions exists at this time. Until better, more testable submodels and more kinetics data are provided, we will not truly know.

REFERENCES

Belyaev, S. S. "Distribution of the Group Caulobacter in Volga-Don Reservoirs," *Mikrobiol.* 36:157-162 (1967).

Bere, R. "Numbers of Bacteria in Inland Lakes of Wisconsin as Shown by the Direct Microscopic Method," *Int. Rev. Ges. Hydrobiol. Hydrogr.* 29:248-263 (1933).

Bierman, V. J., Jr., F. H. Verhoff, T. L. Poulsen and M. W. Tenney. "Multi-Nutrient Dynamic Models of Algal Growth and Species Competi-tion in Eutrophic Lakes," in *Modeling the Eutrophication Process,* E. J. Middelbrooks, D. H. Falkenborg and T. E. Maloney, Eds. (Ann Arbor, MI: Ann Arbor Science Publishers, Inc., 1973), pp. 89-109.

Bierman, V. J., Jr. "Dynamic Mathematical Model of Algal Growth in Eutrophic Freshwater Lakes," Ph.D. Dissertation, University of Notre Dame, Notre Dame, IN (1974).

Bierman, V. J., Jr. "Mathematical Model of the Selective Enhancement of Blue-Green Algae by Nutrient Enrichment," in *Modeling Biochemical Processes in Aquatic Ecosystems,* R. P. Canale, Ed. (Ann Arbor, MI: Ann Arbor Science Publishers, Inc., 1976), pp. 1-32.

Canale, R. P., L. M. DePalma and A. H. Vogel. "A Plankton-Based Food Web Model for Lake Michigan," in *Modeling Biochemical Processes in Aquatic Ecosystems,* R. P. Canale, Ed. (Ann Arbor, MI: Ann Arbor Science Publishers, Inc., 1976), pp. 33-74.

Cangialosi, P. M. "A Phosphorus Budget and Lake Models for Lake Ozonia," Master's Thesis, Civil and Environmental Engr. Dept., Clarkson College of Technology, Potsdam, NY (1976).

Charlton, M. N. "Sedimentation: Measurements in Experimental Enclo-sures," *Verh. Int. Ver. Limnol.* 19:267-272 (1975).

Clesceri, L. S., R. A. Park and J. A. Bloomfield. "General Model of Microbial Growth and Decomposition in Aquatic Ecosystems," *Appl. Environ. Microbiol.* 33(5):1047-1058 (1977).

Collins, V. G. "Recent Studies of Bacterial Pathogens of Freshwater Fish," *Water Treatment Exam.* 19:3-31 (1970).

Cordiero, C. F., W. F. Echelberger, Jr. and F. H. Verhoff. "Rates of Carbon, Oxygen, Nitrogen and Phosphorus Cycling Through Microbial Populations in Stratified Lakes," in *Modeling the Eutrophication Process,* E. J. Middelbrooks, D. H. Falkenborg and T. E. Maloney, Eds. (Ann Arbor, MI: Ann Arbor Science Publishers, Inc., 1973), pp. 111-120.

Daft, M. J., and W. D. P. Stewart. "Light and Electron Microscope Observations on Algal Lysis by Bacterium CP-1," *New Phytol.* 72:799-806 (1973).

DePinto, J. V. "Studies on Phosphorus and Nitrogen Regeneration: The Effect of Aerobic Bacteria on Phytoplankton Decomposition and Succession in Freshwater Lakes," Ph.D. Dissertation, University of Notre Dame, Notre Dame, IN (1974).

DePinto, J. V., V. J. Bierman, Jr. and F. H. Verhoff. "Seasonal Phytoplankton Succession as a Function of Species Competition for Phosphorus and Nitrogen," in *Modeling Biochemical Processes in Aquatic Ecosystems,* R. P. Canale, Ed. (Ann Arbor, MI: Ann Arbor Science Publishers, Inc., 1976), pp. 141-170.

DePinto, J. V., and F. H. Verhoff. "Nutrient Regeneration from Aerobic Decomposition of Green Algae," *Env. Sci. Technol.* 11:371-377 (1977).

DiToro, D. M., and W. F. Matystik. "Phytoplankton Biomass Model of Lake Huron and Saginaw Bay," paper presented at 19th Conf. on Great Lakes Research, Guelph, Ontario, May 1976.

Drabkova, V. G. "Population Dynamics, Generation Time and Bacterial Production in Waters of Lake Krasnoe (Punnus-Yarvi)," *Mikrobiol.* 34(6):1063-1069 (1965).

Fitzgerald, G. P. "The Effect of Algae on BOD Measurements," *J. Water Poll. Control Fed.* 36:1524-1542 (1964).

Fleming, W. M. "A Model of the Phosphorus Cycle and Phytoplankton Growth in Skaha Lake, British Columbia, Canada," *Verh. Int. Ver. Limnol.* 19:229-249 (1975).

Fogg, G. E. "The Production of Extracellular Nitrogenous Substances by Blue-Green Algae," *Proc. Roy. Soc.* B139:372-397 (1952).

Fogg, G. E. "Extracellular Products of Algae in Freshwater," *Arch. Hydrobiol. Bih. Ergebn. Limnol.* 5:1-25 (1971).

Fogg, G. E., D. J. Eagle and M. E. Kinson. "The Occurrence of Glycolic Acid in Natural Waters," *Verh. Int. Ver. Limnol.* 17:480-484 (1969).

Foree, E. G., W. J. Jewell and P. L. McCarty. "The Extent of Nitrogen and Phosphorus Regeneration from Decomposing Algae," in *Advances in Water Pollution Research,* Vol. 2, S. H. Jenkins, Ed. (New York: Pergamon Press, 1970), pp. III-27/1-15.

Foree, E. G., and R. L. Barrow. "Algal Growth and Decomposition: Effects on Water Quality, Phase II. Nutrient Regeneration, Composition and Decomposition of Algae in Batch Culture," Office of Water Resources Research, Report No. 31, Kentucky Water Resources Institute, Lexington, KY (1970).

Fuhs, G. W. "Improved Device for the Collection of Sedimenting Matter," *Limnol. Oceanog.* 18(6):989-993 (1973).

Goldman, C. R. "Eutrophication of Lake Tahoe Emphasizing Water Quality," U.S. Environmental Protection Agency, Corvallis, Oregon, EPA-660/3-74-034 (1974).

Golterman, H. L. "Mineralization of Algae Under Sterile Condition or by Bacterial Breakdown," *Verh. Int. Ver. Limnol.* 15:544-548 (1964).

Golterman, H. L. "Natural Phosphate Sources in Relation to Phosphate Budgets: A Contribution to the Understanding of Eutrophication," *Water Res.,* 7:3-17 (1973a).

Golterman, H. L. "Vertical Movement of Phosphate in Freshwater," in *Environmental Phosphorus Handbook,* E. J. Griffith, A. Beeton, J. M. Spencer and D. T. Mitchell, Eds. (New York: John Wiley & Sons, Inc., 1973b), pp. 509-538.

Granhall, U., and B. Berg. "Antimicrobial Effects of *Cellvibrio* on Blue-Green Algae," *Arch. Mikrobiol.* 84:234-242 (1972).

Grill, E. V., and F. A. Richards. "Nutrient Regeneration from Phytoplankton Decomposing in Seawater," *J. Mar. Res.* 22:51-69 (1964).

Gromov, B. U., O. G. Ivanov, K. A. Mambaeva and I. A. Avilov. "Flexibacterium Lysing Blue-Green Algae," *Mikrobiol.* 41:1074-1079 (1972).

Gunnison, D., and M. Alexander. "Resistance and Susceptibility of Algae to Decomposition by Natural Microbial Communities," *Limnol. Oceanog.* 20(1):64-70 (1975).

Hargrave, B. T. "A Comparison of Sediment Oxygen Uptake, Hypolimnetic Oxygen Deficit and Primary Productivity in Lake Esrom, Denmark," *Verh. Int. Ver. Limnol.* 18:134-139 (1972).

Hellebust, J. A. "Excretion of Some Organic Compounds by Marine Phytoplankton," *Limnol. Oceanog.* 10:192-206 (1965).

Henrici, A. T. "The Distribution of Bacteria in Lakes," *Publ. Am. Assoc. Adv. Sci.* 10:39-54 (1940).

Ho, T. S., and M. Alexander. "The Feeding of Amoebae on Algae in Culture," *J. Phycol.* 10:95-100 (1974).

Jassby, A. D., and C. R. Goldman. "Loss Rates from a Lake Phytoplankton Community," *Limnol. Oceanog.* 19:618-627 (1974).

Jewell, W. J. "Aerobic Decomposition of Algae and Nutrient Regeneration," Ph.D. Dissertation, Stanford University, Stanford, CA (1968).

Jones, J. G. "Studies on Freshwater Bacteria: Factors Which Influence the Population and its Activity," *J. Ecol.* 59:593-613 (1971).

Kajak, Z., A. Hillbricht-Ilkowska and E. Pieczynska. "Production in Several Mazurian Lakes," prelim. papers for UNESCO-IBP Symposium on Productivity Problems of Fresh Waters, Kazimierz Dolny, Poland (1970), pp. 173-189.

Kalff, J., H. J. Kling, S. H. Holmgreen and H. E. Welch. "Phytoplankton, Phytoplankton Growth and Biomass Cycles in an Unpolluted and in a Polluted Polar Lake," *Verh. Int. Ver. Limnol.* 19:487-495 (1975).

Kamatani, A. "Regeneration of Inorganic Nutrients from Diatom Decomposition," *J. Oceanog. Soc.* (Japan) 25(2):63-74 (1969).

Kleerekoper, H. "The Mineralization of Plankton," *J. Fish. Res. Bd. Can.* 10(5):283-291 (1953).

Kuznetsov, S. I. "Determining the Intensity of Oxygen Uptake from the Water Mass of a Lake Due to Bacteriological Processes," *Trudy Limnol. Stantsii Kasine* 22:53-74 (1939).

Kuznetsov, S. I. "Application of Microbiological Methods for Studying Organic Matter in Lake Waters," *Mikrobiol.* 18(3):203-214 (1949).

Kuznetsov, S. I. "Recent Studies on the Role of Microorganisms in the Cycling of Substances in Lakes," *Limnol. Oceanog.* 13(2):211-223 (1968).

Kuznetsov, S. I. *The Microflora of Lakes and Its Geochemical Activity*, C. H. Oppenheimer, Ed. (Austin, TX: University of Texas Press, 1970), pp. 123-126.

Larsen, D. P., H. T. Mercier and K. W. Malueg. "Modeling Algal Growth Dynamics in Shagawa Lake, Minnesota," in *Modeling the Eutrophication Process*, E. J. Middlebrooks, D. H. Falkenborg and T. E. Maloney, Eds. (Ann Arbor, MI: Ann Arbor Science Publishers, Inc., 1973), pp. 15-32.

Lasenby, D. C. "Development of Oxygen Deficits in 14 Southern Ontario Lakes," *Limnol. Oceanog.* 20(6):993-999 (1975).

Lee, G. F., W. Cohen and N. Sridharon. "Algal Nutrient Availability and Limitation in Lake Ontario During IFYGL," *First Annual Reports of the EPA IFYGL Projects*, U.S. Environmental Protection Agency, Corvalus, Oregon, EPA 660/3-73-021 (1973), pp. 71-89.

Lehman, J. T., D. B. Botkin and G. E. Likens. "The Assumptions and Rationales of a Computer Model of Phytoplankton Population Dynamics," *Limnol. Oceanog.* 20(3):343-364 (1975).

Lepak, L. T. "Limiting Nutrient and Trophic Level Determination of Lake Ozonia by Algal Assay Procedure," Master's Thesis, Civil and Environmental Engr. Dept., Clarkson College of Technology, Potsdam, NY (1976).

Lund, J. W. G., F. J. H. Mackereth and C. H. Mortimer. "Changes in Depth and Time of Certain Chemical and Physical Conditions and of the Standing Crop of *Asterionella formosa* Hass. in the North Basin of Winderemere in 1947," *Phil. Trans. Roy. Soc. Lond. Ser.* B 246:255-290 (1963).

Menon, A. S., W. A. Glooschenbo and N. M. Burns. "Bacteria-Phytoplankton Relationships in Lake Erie," Proc. 15th Conf. Great Lakes Res., Intl. Assoc., Madison, WI, April 1972, pp. 94-101.

Mills, A. L., and M. Alexander. "Microbial Decomposition of Species of Freshwater Planktonic Algae," *J. Environ. Qual.* 3(4):423-428 (1974).

Nalewajko, C., and D. R. S. Lean. "Growth and Excretion in Planktonic Algae and Bacteria," *J. Phycol.* 8:361-366 (1972).

Niewolak, S. "Distribution of Microorganisms in the Waters of Kortowskie Lake," *Pol. Arch. Hydrobiol.* 21:315-333 (1974).

Otsuri, A., and T. Hanya. "Production of Dissolved Organic Matter from Dead Green Algal Cells I. Aerobic Microbial Decomposition," *Limnol. Oceanog.* 17(2):248-257 (1972a).

Otsuri, A., and T. Hanya. "Production of Dissolved Organic Matter from Dead Green Algal Cells II. Anaerobic Microbial Decomposition," *Limnol. Oceanog.* 17:258-264 (1972b).

Overback, J. "Distribution Pattern of Uptake Kinetic Responses in a Stratified Eutrophic Lake," *Verh. Int. Ver. Limnol.* 19:2600-2615 (1975).

Patten, B. C. "Mathematical Models of Plankton Production," *Int. Res. Ges. Hydrobiol.* 53:357-408 (1968).

Pearl, H. W. "The Regulation of Heterotrophic Activity by Environmental Factors in Lake Tahoe, California-Nevada," Ph.D. Thesis, University of California, Davis (1973).

Peters, R. H., and D. Lean. "The Characterization of Soluble Phosphorus Released by Limnetic Zooplankton," *Limnol. Oceanog.* 18(2):270-279 (1973).

Peters, R. H. "Phosphorus Regeneration by Natural Populations of Limnetic Zooplankton," *Verh. Int. Ver. Limnol.* 19:273-279 (1975).

Pomeroy, L. R., H. M. Mathews and S. M. Hong. "Excretion of Phosphate and Soluble Reactive Phosphorus Compounds by Zooplankton," *Limnol. Oceanog.* 8:50-55 (1963).

Rao, S. S., and A. A. Jurkovic. "Relationships Between Total Bacteria and Aerobic Heterotrophs in the Great Lakes," paper presented to the 20th Conf. on Great Lakes Research, Ann Arbor, MI, May 1977.

Razumov, A. S. "Microbial Plankton of Water," *Trudy Vsesoyuzhogo Gidrobiol. Obschchestva* 12:60-190 (1962).

Reim, R. L., M. S. Shane and R. E. Cannon. "The Characterization of a Bacillus Capable of Blue-Green Bactericidal Activity," *Can. J. Microbiol.* 20(7):981-989 (1974).

Rhee, G. Y. "A Continuous Culture Study of Phosphate Uptake Growth and Polyphosphate in *Scenedesmus* sp.," *J. Phycol.* 9:495-506 (1973).

Rodgers, P. R., and J. V. DePinto. "Algae-Bacteria Interaction in an Alternating Light-Dark Environment" (in preparation).

Scavia, D., B. J. Eadie and A. Robertson. "An Ecological Model for Lake Ontario Model Formulation, Calibration, and Preliminary Evaluation," NOAA Technical Report ERL 371–GLERL 12, Great Lakes Environmental Research Lab, Ann Arbor, MI (June 1976), 63 pp.

Sharp, J. H. "Excretion of Organic Matter by Marine Phytoplankton: Do Healthy Cells Do It?" *Limnol. Oceanog.* 22(3):381-399 (1977).

Shilo, M. "Lysis of Blue-Green Algae by a Myxobacter," *J. Bacteriol.* 104:453-461 (1970).

Shilo, M. "Biological Agents Which Cause Lysis of Blue-Green Algae," *Mitt. Int. Ver. Limnol.* 19:206-213 (1971).

Tanaka, M., M. Nakanishi and H. Kadota. "Nutrional Interrelation Between Bacteria and Phytoplankton in a Pelagic Ecosystem," in *Effect of the Ocean Environment on Microbial Activities,* R. Colwell and R. Morita, Eds. (Baltimore, MD: University Park Press, 1974), pp. 495-509.

Thomann, R. V., D. M. DiToro, R. P. Winfield and D. J. O'Connor. "Mathematical Modeling of Phytoplankton in Lake Ontario 1. Model Development and Verification," U.S. Environmental Protection Agency, Corvallis, Oregon, EPA-660/3-75-005 (1975), 177 pp.

Uhlmann, D. "Influence of Dilution, Sinking and Grazing Rate on Phytoplankton Populations of Hyperfertilized Ponds and Micro-Ecosystems," *Mitt. Int. Ver. Limnol.* 19:100-124 (1971).

Varma, M. M., and F. DiGiano. "Kinetics of Oxygen Uptake by Dead Algae," *J. Water Poll. Control Fed.* 40(4):613-626 (1968).

Waksman, S. A., J. L. Stokes and R. Butler. "Relations of Bacteria to Diatoms in Sea Water," *J. Mar. Biol. Assoc.* 22:359-373 (1937).

Watt, W. D. "Extracellular Release of Organic Matter from Two Freshwater Diatoms," *Ann. Bot.* 33:427-437 (1969).

Watt, W. D., and G. E. Fogg. "The Kinetics of Extracellular Gylcollate Production by *Chlorella pyrenoidosa, J. Exp. Bot.* 17:117-134 (1966).

Wright, R. T. "Studies on Glycolic Acid Metabolism by Freshwater Bacteria," *Limnol. Oceanog.* 20(4):626-633 (1975).

CHAPTER 3

ZOOPLANKTON GRAZING IN
SIMULATION MODELS: THE ROLE
OF VERTICAL MIGRATION

James A. Bowers*

Center for Great Lakes Studies
University of Wisconsin-Milwaukee
Milwaukee, Wisconsin 53201

INTRODUCTION

Most models of pelagic communities provide for herbivorous grazing upon phytoplankton. This trophic link is a crucial component in many simulation models (Dugdale, 1973). Unfortunately, zooplankton dynamics are generally one of the most poorly developed components in models (Steele and Mullin, 1977) because other functional relationships, especially those involving nutrients and phytoplankton, are better understood (Di-Toro et al., 1977). One complex problem encountered when modeling zooplankton grazing is diurnal vertical migration. These migrations have presented a challenge for both modelers and zooplankton ecologists. Vertical migration complicates the simulation of grazing pressure in time and space, and ecologists have yet to develop a comprehensive theory explaining the evolution, the controlling factors and the adaptive advantages of these migrations.

Often the modeler and the ecologist are suspicious of one another's approach to topics of common interest. The purpose of this paper is to present the approaches taken by these two groups regarding vertical migration. It will examine the role of vertical migration in models of planktonic

*Present Address: Great Lakes and Marine Waters Center, The University of Michigan, Ann Arbor, Michigan 48109

communities and offer recent research relevant to incorporation of migration in models. The ecologist and the modeler have much to offer one another in furthering our understanding of vertical migration.

GRAZING EQUATIONS IN SIMULATION MODELS

In most models, filtering rates are either constant or decrease with increasing phytoplankton concentrations. When utilizing constant filtering rates, total feeding is linearly dependent upon both phytoplankton and zooplankton biomass (Riley et al., 1949). In some recent models this feeding expression has been replaced by Ivlev (Ivlev, 1961; Parsons et al., 1967; O'Brien and Wroblewski, 1972) or Michaelis-Menten expressions (Steele, 1972). These formulations give saturation curves for ingestion rate vs algal density. The relationships have sometimes incorporated the concept of a threshold feeding concentration, a phytoplankton density below which grazing ceases. This phenomenon has been given considerable attention by zooplankton ecologists (Adams and Steele, 1966; Parsons et al., 1967; Marshall, 1973; Frost, 1974, 1975; Mullin et al., 1975). A better understanding of threshold feeding responses is needed since they have become important constants in some models (Dugdale, 1973; Steele, 1974). A comprehensive treatment of grazing functions and the role of threshold responses is given by Scavia (Chapter 6).

DIEL VERTICAL MIGRATION

The grazing functions discussed above are often part of a general energetic approach to simulation models. This approach, within obvious limits, may ignore zooplankton species composition but requires definition with regard to trophic and spatial structure (Vinogradov and Menshutkin, 1977), An obvious characteristic feature of spatial structure in planktonic communities is daily vertical migration of zooplankton, and many authors have reviewed the extensive literature describing migratory behavior (Rose, 1925; Russell, 1927; Kikuchi, 1930; Cushing, 1951; Bainbridge, 1961; Ringelberg, 1964; Hutchinson, 1967).

Several hypotheses have been proposed to explain the mechanism controlling migration. Although other stimuli have been implicated, illumination has been the principal variable considered (Hutchinson, 1967). An early hypothesis was that zooplankton migrated vertically to remain within an optimum light zone (Ewald, 1910; Rose, 1925). More recent discoveries have led to correlation of vertical movement with rates of change in light intensity. This relationship may be linear (Siebeck, 1960), logarithmic (McNaught and Hasler, 1964; Ringelberg, 1964) or both (Teraguchi et al., 1975).

Illumination also affects migration by serving as a synchronizer of endogenous rhythms which, in some zooplankton, regulate migration. These rhythms provide the animal with a temporal inertia when responding to the natural light cycle and disturbances in that cycle. Two excellent studies illustrate the existence of circadian rhythms that partially control vertical migration. In the first of these, Enright and Hamner (1967) observed a concentrated natural zooplankton assemblage in a large tank. Under an artificial light-dark cycle, most forms exhibited vertical migrations that were in phase with the tank's light-dark cycle even when this cycle was greatly out of phase with the cycle of the natural environment. A few species showed a true endogenous circadian rhythm by continuing such migrations after the light conditions were changed to a constant very dim level. Other species showed no evidence of such a rhythm and only descended from the surface when exposed to bright light. A third group appeared to possess an internal timing mechanism which was not self-sustaining. These forms rose to the surface after the light phase and then descended a few hours later. However, they then would not ascend again until a further light cycle had taken place. This suggested a non-rhythmic timing mechanism requiring daily resetting like an hourglass. Thus, a variety of physiological mechanisms were found to account for similar vertical movements. In a similar study illustrating circadian rhythms, LaRow (1976) observed *Chaoborus* larvae. These larvae emerge from the sediments at sunset, migrate upward into the water and remain there until morning. When captured at night and held under continuous darkness in the laboratory, the swimming activity of the larvae followed the natural light cycle. However, when the natural light cycle was reversed, swimming activity shifted 180° out of phase, indicating an exogenous regulation of migration. An endogenous rhythm regulated the vertical positioning of larvae in the sediments, where the animals were without visual stimuli. At sunset most of the population were found in the upper one centimeter of the sediment and thus were in a position to leave the sediments and migrate upwards. This circadian (endogenous) rhythm, which correctly positioned the population at the mud-water interface at sunset, was continually reset throughout the season. Thus, endogenous and exogenous factors regulated migratory behavior.

While much of the research on vertical migration has been directed toward determining controlling mechanisms, only a few hypotheses have been proposed to suggest any adaptive advantages in these migrations. Some of these proposals incorporate the animal's feeding habits. Worthington (1931) suggested that hunger is the principal reason for migration and, noting that phytoplankton production occurs in the surface layers, he emphasized the need for more information on the vertical distribution of nannoplankton. Hardy and Gunther (1935) believed that vertical migration aids in the horizontal disperson of the population and carries the herbivores into areas rich in food. Autrum (1960) thought that zooplankton are adapted ecologically in such as way as to place the animals at the depth of maximum photosynthesis. McLaren (1963) proposed that migration bestows an energy bonus for growth upon a migratory species, because optimum conditions of food and temperature never coincide in time and space for a zooplankter. The increased number of eggs gained by living in the deeper layers is offset by retarded development at colder temperatures. By feeding in the warmer surface waters at night, the animals feed at temperatures higher than those in the deeper waters which control their growth and development. Efficiencies are maximized for feeding at higher temperatures and for growth and development at lower temperatures. This results in an energy bonus for the migrating herbivore. Following Autrum's (1960) belief that a zooplankter's vertical position in the water column is crucial to feeding, Kerfoot (1970) viewed calanoid copepods as light-oriented consumers distributed in relation to the vertical distribution of phytoplankton, whose growth is also light-oriented. Depth profiles of potential energy (*i.e.,* food) were correlated with the migratory habits of *Calanus finmarchicus* in the Gulf of Maine. Within certain conditions, the biomass of copepods found in a particular range of light intensity was directly proportional to the daily primary production at that depth.

Enright (1977) sought to explain the selective advantages that compensate for a period of no feeding during the daylight hours. He proposed that, in early evening after the nonfeeding interval, feeding begins at a rate higher than the steady-state rate and some saturation of feeding occurs within a few hours. This behavior offers herbivorous zooplankton a greater energy gain for reproduction and growth compared to continuous grazing. Two critical assumptions of this hypothesis were that metabolic needs are reduced during the day when the animals are in the deeper colder layers and that primary production can result in significant increases in the plant biomass during daylight hours. The energy gain associated with this model offers a greater selective advantage than does avoidance of visual predation, a widely accepted concept to account for the evolution of diurnal migrations. A simple confirmation of Enright's hypothesis would be detection of a consistent migratory ascent before sunset. The results from attempts to detect such an ascent (Enright and Honegger, 1977) were inconclusive and suggested that, depending upon the period in the life cycle, either the energy gain from discontinuous feeding or that from predator avoidance could provide a greater selective advantage.

MODELS INCORPORATING VERTICAL MIGRATION

Nocturnal ascent and feeding is the most commonly observed pattern of vertical migration in the marine environment. Consequently, models accommodating vertical migration simulate only this type of diel grazing pressure. In this section we shall examine the techniques that have been used to incorporate nocturnal vertical ascent into simulation models. Vinogradov *et al.* (1972) simulated plankton succession in an upwelling zone in the western tropical Pacific. The trophic structure of the model was quite complex, including two size classes of herbivores and two of carnivores. Only the energetics of the large-sized herbivores will be considered here to illustrate how migration was simulated. The population was divided into two groups, one portion inhabiting the 0- to 50-m layer and one portion living in the 50- to 200-m layer. Equation 1 describes the daily biomass change in the large-sized herbivores:

$$f_{t+1} = f_t + U_f^{-1} (Z_{pf} + Z_{bf} + Z_{prf}) - N_f f_t - Z_{fsc} - Z_{flc} - U_f f_t \tag{1}$$

where

f	=	total large-sized herbivore biomass in both layers
U_f^{-1}	=	assimilation efficiency
Z_{pf}	=	ration of phytoplankton
Z_{bf}	=	ration of bacteria
Z_{prf}	=	ration of protozoans
N_f	=	coefficient of metabolic loss
Z_{fsc}	=	ration of f to small carnivores
Z_{flc}	=	ration of f to large carnivores
U_f	=	natural mortality

A proportion (k) of the total food (second term in Equation 1) of the population is consumed by the 50- to 200-m population. The proportion k is time-dependent according to Equation 2:

$$k = 0.02 + 0.0016t \qquad t \leqslant 50$$
$$= 0.1 \qquad\qquad t > 50 \tag{2}$$

With this technique, the deep-living portion of f was simulated as actually feeding in the 0- to 50-m layer, and thus it fed at the same rate as the 0- to 50-m population. The proportion increased linearly until day 50 and then remained constant until the last day of the simulation. This increase in k coincided with a rapid increase of phytoplankton biomass in the 0- to 50-m layer; this then made much of the increased biomass available to the migrating herbivores.

Walsh (1975) developed a simulation model of nutrient dynamics in the Peruvian upwelling system, where field data from this area indicated the occurrence of nocturnal grazing. The zooplankton in the model were positioned alternately at two depths depending on time; at 0-30 m between 1800 hr and 600 hr and at 30-60 m between 600 hr and 1800 hr. This alternation of level was expressed through a modification of the maximum grazing rate in a Michaelis-Menten equation. The maximum grazing rate (G) in this equation varied sinusoidally:

$$G = (G_m) (1.43) (\cos 0.2618t + 1.57) \quad \leqslant 30 \text{ m}$$
$$= (G_m) (1.43) (\sin 0.2618t) \quad > 30 \text{ m} \tag{3}$$

where t equals the cumulative time of day. The number 0.2618 sets the time periodicity of the sinusoidal function and (G_m) (1.43) is the absolute maximum rate. G is set equal to zero for the 0- to 30-m layer during the day and in the 30- to 60-m layer during the night.

Wroblewski and O'Brien's model (1976) was designed to examine critical length scales of phytoplankton patches which are also discussed by Kierstead and Slobodkin (1953). One set of simulations followed patch morphology under the stresses of turbulent diffusion, high nutrient concentrations, and a nocturnally migrating and feeding herbivore population. An Ivlev function was used to approximate grazing pressure. Vertical migration was introduced into the system by dividing the herbivore population into two groups. The first group inhabited the phytoplankton patch and was able to feed there continuously. The second group migrated downward out of the patch during the day and was unable to feed during this time. The vertical movement of the second group was centered about sunset and sunrise with 95% of the population completing an ascent or descent within four hours. Of the total herbivore population, 85% was removed from all system dynamics during the night.

The final expression of migration in all of the models is a spatial and diurnal variation in grazing pressure. Each of the above models offers a significantly different approach to simulating vertical migration. In the first example (Vinogradov et al., 1972), the food ration was redistributed over depth and time. Walsh (1975) varied the grazing rate sinusoidally on a diel cycle at two depths. In Wroblewski and O'Brien's (1976) model, a portion of the herbivore population was shuttled daily in and out of a phytoplankton patch.

We shall now inspect three exploratory models that specifically evaluate and illustrate the power and utility of the simulation approach to vertical migration. McAllister's (1971) model compared the effects of continuous and nocturnal grazing and the implications of these two grazing models on estimating secondary production. Using an Ivlev grazing function, McAllister found that grazing pressure was sensitive to the threshold phytoplankton density, maximum ingestion rate and the Ivlev constant chosen. The values chosen for these parameters were often more important in determining grazing pressure than the grazing scheme itself. Under certain initial conditions, nocturnal grazing resulted in higher rates of primary production and consequently higher rates of secondary production. Firm conclusions were unattainable because of inadequate field data for proper comparisons with model output. For example, simulation runs under steady-state restrictions pointed out the importance of using, for model calibration, phytoplankton densities that were measured at a specific time of day. Biomass densities estimated at different times of the day biased conclusions regarding grazing pressure.

Wroblewski and O'Brien (1976) demonstrated the widespread effects of including vertical migration in a model of a plankton community. When grazing was restricted to the night hours, grazing pressure was reduced by 50%. Lower rates of nutrient excretion in the phytoplankton patch reduced nitrogen levels 31%. Consequently, algal densities were 24% lower. Vertical migration had completely altered the biomass structure in the system.

Steele and Mullin (1977) applied a model patterned after the system of Steele and Henderson (1976) to simulate zooplankton dynamics in a shallow environment. Vertical migration was simulated in two ways. In one mode the zooplankton spent equal amounts of time at all depths. In the second mode, the amount of time at any depth was directly proportional to the phytoplankton concentration at that depth. The effects of these two types of migration and the effects of threshold densities were simultaneously investigated. Decreased vertical mixing rates in the water column had induced a distinct subsurface phytoplankton maximum in the model. Under conditions of food-independent migration and an assumed grazing threshold, the subsurface phytoplankton maximum was quite stable with regard to its magnitude and depth over a 300-day simulation. When the threshold level was removed and food-dependent migration introduced, the algal densities in the peak oscillated and the depth of the peak changed position over the 300 days.

How successful was the inclusion of vertical migration in these models? The models of Vinogradov et al. (1972) and Walsh (1975) were designed primarily to simulate energy and nutrient flow through a pelagic community, and the inclusion of migration was not specifically evaluated; however, the Vinogradov et al. model and a more recent version of this model (Vinogradov and Menshutken, 1977) realistically reproduced the vertical structure of the system. Migration played an important role in developing that structure.

The exploratory models offer a better measure of future successes. McAllister's (1971) results were inconclusive, but the model indicated the absence of the critical field data that would be required to evaluate properly model output. In this sense models are often excellent guides for defining future field and laboratory experiments. An example of a successful inclusion of migration is the model of Wroblewski and O'Brien (1976). Migration in their simulation induced zooplankton patchiness which in turn created patchiness in the phytoplankton through grazing. Although some models have required the inclusions of a phytoplankton refugium to prevent algal extinction, Wroblewski and O'Brien found that the addition of vertical migration made this unnecessary. Thus, migration was influential in creating the spatial and temporal realism that is often so difficult to achieve in system models (Steele, 1974; Vinogradov and Menshutkin, 1977).

Although I have emphasized the importance of migration and feeding, a small portion of the literature has suggested that migration may generate daily cycles in several other important biological processes. These observations illustrate the multiple effects of migration within a planktonic community. Dugdale (1967) suggested that excretion by zooplankton was the primary nitrogen source for phytoplankton in the surface waters of the central gyre of the North Pacific. Observing a diurnal periodicity in excretion rates in this same area, Eppley et al. (1967) indicated that a significant increase of nutrient loading into the mixed layer at night resulted from nocturnal migration. Thus, the migratory habits of the zooplankton played an important role in controlling the daily nutrient supply for the phytoplankton. Migration may also serve as a downward transport mechanism. Ferrante and Parker (1977) proposed that downward migration could cause the release of fecal pellets containing diatom frustules to occur in deep strata in Lake Michigan and thus increase the probability of the pellets reaching the bottom. Lowman et al. (1971) indicated that migration by zooplankton was responsible for a relatively small amount of downward transport of radionuclides. Koshov (1955a, b, 1960, 1963) indicated that the hunting behavior of the important planktivorous fish in Lake Baikal is closely coupled to the migratory behavior of herbivorous zooplankton. The planktivores fed only during the twilight hours when the zooplankton were ascending or descending and were relatively concentrated and visible. Future observations in these areas could further justify the investigation of vertical migration in models of pelagic communities.

VERTICAL DISTRIBUTION OF PHYTOPLANKTON

Migrating herbivores in thermally stratified lakes and oceans pass through depth strata where their nutritional environment changes considerably both quantitatively and qualitatively because the vertical distribution of phytoplankton is quite heterogeneous. This heterogeneity is most often observed as a subsurface maximum of chlorophyll (Parsons and Takahashi, 1973) and has been described in the Atlantic Ocean (Steele and Yentsch, 1960), the Pacific Ocean (Anderson, 1969), the Mediterranean Sea (Cahet et al., 1972), the Gulf of Mexico (Hobson and Lorenzen, 1972), Lake Tahoe (Kiefer et al., 1972), and smaller lakes (Fee, 1976). Several explanations have been given for these observations and some of these will be outlined below.

Riley et al. (1949), using pigment concentration (Harvey units) as a biomass estimator, noted subsurface maxima in the Sargasso Sea. Vertical turbulence and algal sinking were thought to be responsible for these subsurface peaks. In their model, these peaks occurred above the compensation depth; this contradicted most field results. Steele and Yentsch (1960) were the first to simulate subsurface maxima. Field data suggested that cells experienced a decrease in sinking rate at the bottom of the thermocline and an increase in buoyancy when entering darker nutrient-rich layers. An equation incorporating sinking rates, growth rates, cell respiration, zooplankton grazing and eddy diffusivity was derived to predict chlorophyll profiles. Anderson (1969) reported a high pigment content per cell in subsurface biomass peaks. This high pigment content was believed to be an adaptative response to the low-light environment. Anderson also suggested that these peaks were not unique to his study area and, in fact, may be a transpacific occurrence. In Lake Tahoe, Kiefer et al. (1972) observed subsurface peaks of cell volume and chlorophyll at 100 m. Unlike Anderson's results, the relationship between cell volume and chlorophyll was consistent with depth. The phytoplankton was composed of filamentous diatoms, Dinobryon and Elakatothrix; and qualitative changes over depth were not observed. The peaks were thought to have resulted from sinking out from the epilimnion.

Fee (1976) proposed a different explanation for subsurface peaks, based on comprehensive data from the Experimental Lakes Area, Ontario. Since the species present in the peaks were absent in the epilimnion, accumulation of sinking algae in the thermocline was not supported. The peaks could only have resulted from in situ growth. These hypolimnetic algae were adapted to dim light, low temperature and low turbulence. Primary production per unit of chlorophyll was very low, indicating a light-limited population which proliferated chlorophyll to trap as much light as possible. The high biomass resulted from a minimization of the losses due to respiration, grazing and sinking. The large mostly colonial cells in the peak probably had low respiratory losses per unit weight and were not easily fed upon by the dominant herbivore, a small calanoid copepod. Sinking was reduced since distinct peaks only occurred in sharp stable thermoclines.

The results from recent field work (Table I and Figure 1) illustrate the subsurface peaks during the summer in Lake Michigan. Mortonson (1977) and Ristic (1977) analyzed respectively the same water samples for chlorophyll and phytoplankton abundance at a deep station 27 km northeast of Milwaukee. Their combined results indicated dramatic changes in biomass, species composition and seasonal succession within subsurface maxima. In a June 12 sample, *Melosira* dominated the peak at 20 m, while at 30 m *Melosira* and *Rhodomonas* were codominant. At 20 m on July 14 *Dinobryon* composed almost half the total number of cells, while at 30 m *Synedra* and *Tabellaria* constituted most of the population. Within the

Table I. Percentage Abundance of Dominant Phytoplankton at Different Depths in the Subsurface Chlorophyll Peak in Lake Michigan (from Ristic, 1977)

12/6/76	14/7/76	26/7/76
20 m	**20 m**	**20 m**
68 *Melosira*	46 *Dinobryon*	31 *Rhodomonas*
11 *Rhodomonas*	14 "small green unicells"	13 *Dinobryon*
6 *Ankistrodesmus*	13 *Tabellaria*	13 "small green unicells"
30 m	**25 m**	**30 m**
38 *Melosira*	33 *Synedra*	21 *Fragilaria*
36 *Rhodomonas*	32 *Tabellaria*	17 *Rhodomonas*
9 "small green unicells"	9 *Dinobryon*	14 blue-green filaments
	30 m	**45 m**
	29 *Synedra*	24 *Melosira*
	16 *Tabellaria*	20 *Tabellaria*
	12 "small green unicells"	13 *Synedra*
	45 m	
	38 *Melosira*	
	21 *Tabellaria*	
	10 *Synedra*	

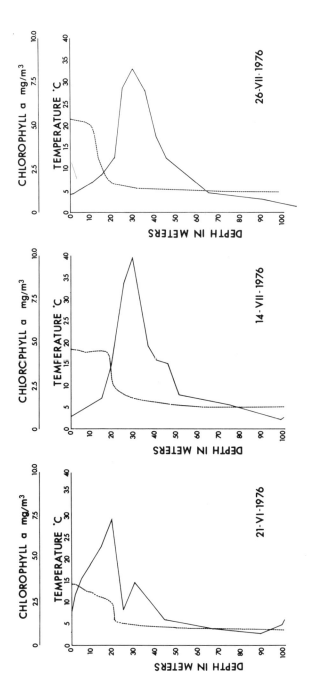

Figure 1. Vertical distribution of chlorophyll in Lake Michigan; solid line indicates chlorophyll, dashed line indicates temperature (redrawn from Mortonson, 1977).

hypolimnion at 45 m *Melosira* and *Tabellaria* were approximately 60% of the total. On July 26 the profile's dominant taxa also changed over depth.

Whether the subsurface phytoplankton maxima are created by *in situ* growth or by settling from the surface waters, they may have several important effects on herbivore grazing. Most importantly the peaks indicate how critical depth position may be for a herbivore during the summer months because vertical variations in cell concentration may alter ingestion rates. Although several studies have been done on the effects of cell density on grazing, few of the results have been related to vertical migratory behavior. Porter (1973, 1976) has shown that zooplankton differentially digest different types of phytoplankton. The feeding and assimilation rates of herbivores thus could also respond to the vertical changes in the quality of the phytoplankton. Also, migrating herbivores may feed at one depth dominated by small unicellular forms and at another depth dominated by large filamentous diatoms. If zooplankton are size-selective feeders (*e.g.,* Frost, 1977), how will ingestion rates change when moving from one depth to the next? Future studies of vertical migration must be paralleled by detailed sampling of the phytoplankton.

EFFECTS OF DEPTH POSITION AND ENDOGENOUS CYCLES

Based on observations of the subsurface chlorophyll maximum in Lake Michigan, Bowers (1977) examined the feeding habits of two herbivorous calanoid copepods, *Diaptomus ashlandi* and *Diaptomus sicilis,* with regard to depth location. Two experiments, part of a comprehensive seasonal study, illustrate how the two copepods respond to the vertical distribution of phytoplankton. Algal biomass during the summer months was estimated by measuring size-fractionated chlorophyll *a*, partitioning the phytoplankton into three size classes (Figure 2). Changes in concentration of chlorophyll *a* (Bowers, 1977) were used to estimate grazing in the same size fractions. *Diaptomus ashlandi,* the smaller copepod (1.1 mm), fed only on the smallest size class (Table II) and was unable to attack the larger net phytoplankton which constituted the larger portion of the subsurface chlorophyll peak. *Diaptomus sicilis,* a larger animal (1.6 mm), almost always fed on all three size fractions (Table III). This copepod had its highest ingestion rates and feeding rates in the metalimnetic chlorophyll peak (18 m) due to higher densities of the large phytoplankton there. These results indicate how differently two different-sized copepods can respond to fluctuations in the size composition of food over depth.

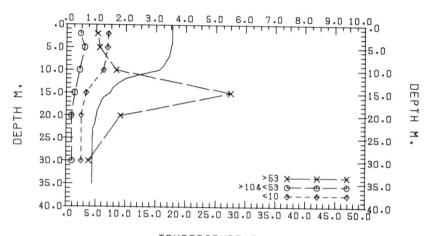

CHLØRØPHYLL A VS. DEPTH 28.AUG 1975
MG. CHLØRØPHYLL A/CUBIC M.

Figure 2. Vertical distribution of size-fractionated chlorophyll in Lake Michigan; solid line indicates temperature, legend indicates size fractions in microns.

Table II. Feeding of *Diaptomus ashlandi* in Lake Michigan
[D = depth (m); T = temperature (C°); SF = size fraction (μm); C = chlorophyll
concentration (mg/m³) at D; I = ingestion rate (mg/animal/day x 10^{-6});
F = filtering rate (ml/animal/day)]

D	T	SF	C	I	F
2	18	>53	1.27	0.00	0.00
		10-53	0.06	0.00	0.00
		<10	0.86	2.38	3.45
19	8	>53	1.45	0.00	0.00
		10-53	0.05	0.00	0.00
		<10	1.16	1.78	1.72
35	5	>53	1.34	0.00	0.00
		10-53	0.07	0.00	0.00
		<10	0.45	0.73	1.85

A separate study of *Mysis relicta* (Bowers and Grossnickle, 1978), using the same methods as Bowers (1977), attempted to establish possible relationships between the nocturnal migratory behavior (Beeton, 1960) and herbivorous habits of the mysid, and the vertical distribution of

Table III. Feeding of *Diaptomus sicilis* in Lake Michigan
[D = depth (m); T = temperature (C°); SF = size fraction (μm); C = chlorophyll
concentration (mg/m³) at D; I = ingestion rate (mg/animal/day x 10^{-6});
F = filtering rate (ml/animal/day)]

D	T	SF	C	I	F
		>53	0.92	2.43	3.34
2	14	10-53	0.06	0.00	0.00
		<10	0.71	0.27	1.53
		>53	3.38	3.72	1.20
18	7	10-53	0.46	1.26	3.49
		<10	1.57	0.49	1.11
		>53	1.82	2.63	1.61
30	5	10-53	0.21	0.34	1.81
		<10	0.79	0.54	0.72

phytoplankton in Lake Michigan. Laboratory experiments indicated *Mysis* has a strong preference for net phytoplankton. Using a high-frequency echosounder, the vertical migration of *Mysis* was observed at a shallow station near Milwaukee (Figure 3). On July 1-2 the population migrated off the bottom at 2125 hr. At 0125 hr the mysids reached the top of

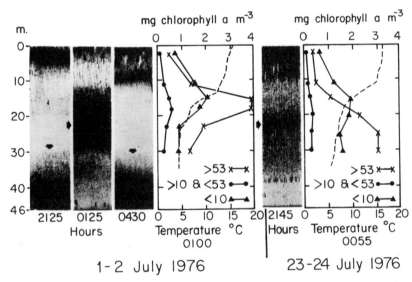

Figure 3. Echograms of vertical migration of *Mysis relicta* and the vertical distribution of size-fractionated chlorophyll; dashed line indicates temperature, arrows point to mysids (from Bowers and Grossnickle, 1978).

their ascent and were positioned within a subsurface chlorophyll peak. The echogram from the night of July 23-24 again revealed the mysids positioned within the subsurface peak. Grossnickle (1977) has recently demonstrated, with an *in situ* technique, that *Mysis* actively feeds at night at these depths. These observations suggest possible complex couplings among feeding habits, migration behavior and resource distributions. They also reveal our ignorance on how an individual zooplankter perceives and responds to changing food resources over short time scales during migratory movement. This behavior may be partially controlled by an endogenous component.

Haney (1973) has shown that increased grazing pressure at night in the epilimnion of Heart Lake, Ontario, resulted from both a migration of animals and an increase in individual filtering rates. The type of nocturnal feeding described by Haney had previously given the impression of a diurnal feeding cycle (Gauld, 1953). Gauld found the guts of zooplankton sampled at night from the surface layers were full, and those sampled during the day from deeper layers were almost empty. More recent evidence has also pointed to true endogenous feeding rhythms. Chisholm *et al.* (1975) described a bimodal feeding pattern in the cladoceran *Daphnia middendorffiana* from depression ponds in Alaska. A rhythm in temperature sensitivity (rate of change in ingestion in relation to rate of change in temperature) was approximately synchronous to the bimodal pattern of the ingestion rate. The rhythm was interpreted as an intrinsic cycle, a circadian clock and the bimodality resulted from the division of a unimodal circadian cycle of activity by the continuous light during the arctic summer. Two temporally distinct populations were identified with peak activities at 1400 hr and 2400 hr. The activity peaks coincided with optimum temperatures for the daphnid during the daily temperature cycle in a pond. Starkweather (1975) held *Daphnia pulex* on a light-dark 16:8 hr illumination cycle, and the resultant filtering rates followed a unimodal pattern with the maximum rate during the dark phase. After pretreatment in continuous darkness or light, a unimodal pattern was still discernible, but the periodicity was partially disrupted; consequently an endogenous component was proposed. However, the experiments were of insufficient length for an adequate determination. Duval and Geen (1976) reported a bimodal feeding rhythm for the zooplankton population in Eunice Lake, British Columbia. The periods of maximum activity in the rhythm occurred approximately two hours prior to sunrise and sunset. In Deer Lake these authors found that *Daphnia pulex* and *Cyclops scutifer* had bimodal feeding patterns after being kept two days without temperature or illumination cues, again implying an endogenous cycle. Mackas and Bohrer (1976) described unimodal and bimodal feeding cycles in several marine herbivores. *Balanus balanoides* nauplii, *Pseudocalanus minutus* and *Metridia lucens* had unimodal cycles with peak feeding activities during the night. *Centropages typicus* displayed a bimodal pattern with feeding maxima immediately after sunset and just prior to sunrise. The preceding results, though not conclusive, strongly suggest that endogenous feeding rhythms exist and are widespread amongst marine and freshwater zooplankton. This is important for both modelers and biologists. It may require reinterpretation of past experiments where the effects of temperature and light regimes on endogenous rhythms were not considered; thus reported grazing rates may be erroneous. Future experimental designs will have to consider the effects of such rhythms especially with regard to light environments. A fascinating aspect of these findings is the dual role of the light regimes as a synchronizers of both vertical migration and intrinsic feeding cycles.

CONCLUSIONS

Modelers and zooplankton ecologists have much to offer one another when dealing with vertical migration. Many of the assumptions and measurements made in developing hypotheses with regard to ecological problems will contain serious errors. McAllister (1971) discussed why models should be attempted in such cases. I concur with his answer: Modeling offers a reasonable approach to assess the implications of these errors where generalizations and conclusions are intuitively difficult. This philosophy is quite appropriate for zooplankton ecologists, who, in the future, will develop a general theory of vertical migration. A holistic approach will be necessary. It will not be sufficient to develop hypotheses based on one aspect of zooplankton life history. Future hypotheses must include, for example, nitrogen and phosphorus excretion, size selective feeding and migration as an avoidance behavior from vertebrate and invertebrate predators. This path will benefit most from the use of models as investigatory tools.

How may ecologists aid the modeler in a better accommodation of migration into large system models? Most past field studies of vertical migration dealt only with the description of migratory patterns. Future field studies should include data on phytoplankton distributions, nutrient profiles and diel movement of planktivorous fish over appropriate time scales. Laboratory feeding experiments should be designed to simulate the habitat changes experienced by a migratory herbivore with regard to temperature, light, phytoplankton quantity and quality, and endogenous rhythms.

ACKNOWLEDGMENTS

I wish to thank Dr. C. H. Mortimer for his helpful advice and encouragement. Dr. John Ferrante and an anonymous reviewer provided substantial improvements to the text. I am very grateful to Donald Scavia and Dr. Andrew Robertson for their patience in handling the paper. Becky Glover and Ratko J. Ristic drew the figures. Part of my research presented was supported by the Center for Great Lakes Studies, University of Wisconsin—Milwaukee.

REFERENCES

Adams, J. A., and J. H. Steele. In: *Some Contemporary Studies in Marine Science,* H. Barnes, Ed. (London: George Allen and Unwin Ltd., 1966), pp. 19-35.

Anderson, G. C. "Subsurface Chlorophyll Maximum in the Northeast Pacific Ocean," *Limnol. Oceanog.* 14:386-391 (1969).

Autrum, H. "Vergleichende Physiologie des Farbensehens," *For. Zool.* 12:176-205 (1960).

Bainbridge, R. "Migrations," in: *The Physiology of Crustacea,* T. H. Waterman, Ed. (New York: Academic Press, 1961), pp. 431-463.

Beeton, A. M. "The Vertical Migration of *Mysis relicta* in Lakes Huron and Michigan," *J. Fish. Res. Bd. Can.* 17:517-539 (1960).

Bowers, J. A. "The Feeding Habits of *Diaptomus ashlandi* and *Diaptomus sicilis* in Lake Michigan and the Seasonal Vertical Distribution of Chlorophyll at a Nearshore Station," Ph.D. Thesis, University of Wisconsin—Madison (1977).

Bowers, J. A., and N. E. Grossnickle. "The Herbivorous Habits of *Mysis relicta* in Lake Michigan," *Limnol. Oceanog.* 23:767-776(1978).

Cahet, G., M. Fiola, G. Jacques and M. Panouse. "Production Primaire au Niveau de la Thermocline en Zone Nesitique de Mediterranee Nord-Occidentale," *Mar. Biol.* 14:32-40 (1972).

Chisolm, S. W., R. G. Stross and P. A. Nobbs. "Environmental and Intrinsic Control of Filtering and Feeding Rates in Arctic *Daphnia*," *J. Fish. Res. Bd. Can.* 32:219-226 (1975).

Cushing, D. H. "The Vertical Migration of Plankton Crustacea," *Biol. Rev.* 26:158-192 (1951).

DiToro, D. M., R. V. Thomann, D. J. O'Connor and J. L. Macini. "Estuarine Phytoplankton Biomass Models—Verification Analyses and Preliminary Applications," in *The Sea Vol. 6 Marine Modeling,* E. D. Goldberg, I. W. McCave, J. J. O'Brien and J. H. Steele, Eds. (New York: John Wiley and Sons, Inc., 1977), pp. 969-1020.

Dugdale, R. C. "Nutrient Limitation in the Sea: Dynamics, Identification, and Significance," *Limnol. Oceanog.* 12:685-695 (1967).

Dugdale, R. C. "Biological Modeling I," in *Modeling of Marine Ecosystems,* J. Nihoul, Ed. (Amsterdam: Elsevier, 1973), pp. 187-205.

Dugdale, R. C., and J. J. MacIsaac. "A Computation Model for the Uptake of Nitrate in the Peru Upwelling Region," *Invest. Pesq.* 35:299-309 (1971).

Duval, W. S., and F. H. Geen. "Diel Feeding and Respiration Rhythms in Zooplankton," *Limnol. Oceanog.* 21:823-829 (1976).

Enright, J. T. "Diurnal Vertical Migration: Adaptive Significance and Timing. Part 1. Selective Advantage: A Metabolic Model," *Limnol. Oceanog.* 22:856-872 (1977).

Enright, J. T., and W. M. Hamner. "Vertical Diurnal Migration and Endogenous Rhythmicity," *Science* 157:937-941 (1967).

Enright, J. T., and H. W. Honegger. "Diurnal Vertical Migration: Adaptive Significance and Timing. Part 2. Test of the Model: Details of Timing," *Limnol. Oceanog.* 22:873-886 (1977).

Eppley, R. W., E. H. Renger, E. L. Venrick and M. M. Muller. "A Study of Plankton Dynamics and Nutrient Cycling in the Central Gyre of the North Pacific Ocean," *Limnol. Oceanog.* 12:685-695 (1967).

Ewald, W. F. "Uber Orientierung, Lokomotion und Lichtseaktionen einger Cladoceren und deren Bedentung fur die Theorie der Tropismen," *Biol. Zbl.* 30:1-399 (1910).

Fee, E. J. "The Vertical and Seasonal Distribution of Chlorophyll in Lakes of the Experimental Lakes Area, Northwestern Ontario: Implications for Primary Production Estimates," *Limnol. Oceanog.* 21:767-783 (1976).

Ferrante, J. G., and J. I. Parker. "Transport of Diatom Frustules by Copepod Fecal Pellets to the Sediments of Lake Michigan," *Limnol. Oceanog.* 22:92-98 (1977).

Frost, B. W. "Feeding Processes at Lower Trophic Levels in Pelagic Communities," in *The Biology of the Oceanic Pacific*, C. B. Miller, Ed. (Corvallis: Oregon State University, 1974), pp. 59-77.

Frost, B. W. "A Threshold Feeding Behavior in *Calanus pacificus,*" *Limnol. Oceanog.* 20:263-266 (1975).

Frost, B. W. "Feeding Behavior of *Calanus pacificus* in Mixtures of Food Particles," *Limnol. Oceanog.* 22:273-291 (1977).

Gauld, D. T. "Diurnal Variations in the Grazing Rate of Planktonic Copepods," *J. Mar. Biol. Assoc. (U.K.)* 31:461-474 (1953).

Grossnickle, N. E. Center of Great Lakes Studies, University of Wisconsin— Milwaukee. Personal communication (1977).

Haney, J. F. "An *in situ* Examination of the Grazing Activities of Natural Zooplankton Communities," *Arch. Hydrobiol.* 72:87-132 (1973).

Hardy, A. C., and E. R. Gunther. "The Plankton of the South Georgia Whaling Grounds and Adjacent Waters, 1926-27," *Dis Rept. (London)* 11:1-456 (1935).

Hobson, L. A., and C. J. Lorenzen. "Relationship of Chlorophyll Maxima to Density Structure in the Atlantic Ocean and Gulf of Mexico," *Deep-Sea Res.* 19:297-306 (1972).

Hutchinson, G. E. *A Treatise on Limnology Vol. II Introduction to Lake Biology and the Limnoplankton* (New York: John Wiley and Sons, Inc., 1967), 1,115 p.

Ivlev, V. S. *Experimental Ecology of the Feeding of Fishes.* D. Scott, Trans. (New Haven, CT: Yale University Press, 1961), 302 p.

Kerfoot, W. B. "Bioenergetics of Vertical Migration," *Am. Nat.* 104:529-546 (1970).

Kiefer, D. A., O. Holm-Hansen, C. R. Goldman, R. Richards and T. Berman. "Phytoplankton in Lake Tahoe: Deep-Living Populations," *Limnol. Oceanog.* 17:418-422 (1972).

Kierstead, H., and L. B. Slobodkin. "The Size of Water Masses Containing Plankton Blooms," *J. Mar. Res.* 12:141-147 (1953).

Kikuchi, K. "Diurnal Migration of Plankton Crustacea," *Quart. Rev. Biol.* 5:189-206 (1930).

Koshov, M. M. "Seasonal and Annual Changes in the Plankton of Lake Baikal," *Trudy USSR Hydriobiol. Soc.* 6:133-157 (1955a).

Koshov, M. M. "New Data on Life in the Mass of Water of Lake Baikal," *Zool. J.* 34:17-45 (1955b).

Koshov, M. M. "On Species Formation in Baikal Lake," *Bull. Mos. Nat. Soc.* 65:39-45 (1960).

Koshov, M. M. *Lake Baikal and Its Life* (The Hague, Netherlands: Dr. W. Junk, NV, Publishers, 1963), 344 p.

La Row, E. J. "Biorhythms and the Vertical Migration of Limnoplankton," in *Biological Rhythms in the Marine Environment*, P. J. De De Coursey, Ed. (Columbia, SC: University of South Carolina Press, 1976), pp. 225-238.

Lowman, F. G., T. P. Rice and F. A. Richards. "Accumulation and Redistribution of Radionuclides by Marine Organisms," in *Radioactivity in the Marine Environment* (Washington, DC: National Academy of Sciences, 1971), pp. 161-199.

Mackas, D., and R. Bohrer. "Fluorescence Analysis of Zooplankton Gut Contents and an Investigation of Diel Feeding Patterns," *J. Exp. Mar. Biol. Ecol.* 25:77-85 (1976).

Marshall, S. M. "Respiration and Feeding in Copepods," *Adv. Mar. Biol.* 11:57-120 (1973).

McAllister, C. D. "Some Aspects of Nocturnal and Continuous Grazing by Planktonic Herbivores in Relation to Production Studies," *Fish. Res. Bd. Can.* Tech. Rep. No. 248, Nanaimo, BC Biological Station (1971).

McLaren, I. A. "Effects of Temperature on Growth of Zooplankton, and the Adaptive Value of Vertical Migration," *J. Fish. Res. Bd. Can.* 20:685-722 (1963).

McNaught, D. C., and A. D. Hasler. "Rate of Movement of Populations of *Daphnia* in Relation to Changes in Light Intensity," *J. Fish. Res. Bd. Can.* 21:291-318 (1964).

Mortonson, J. A. "The Vertical Distribution of Chlorophyll *a* and Nutrients at a Deep Station in Lake Michigan," M.S. Thesis, University of Wisconsin—Milwaukee (1977).

Mullin, M. M., E. F. Stewart and F. J. Fuglister. "Ingestion by Plankton Grazers as a Function of Concentration of Food," *Limnol. Oceanog.* 20:259-262 (1975).

O'Brien, J. J., and J. S. Wroblewski. "An Ecological Model of the Lower Marine Trophic Levels on the Continental Shelf of West Florida," Geophysical Fluid Dynamics Institute, Florida State University, Tech. Rep. (Tallahasse, FL: Florida State University, 1972).

Parsons, T. R., and M. Takahashi, Eds. *Biological Oceanographic Processes* (New York: Pergamon Press, 1973), 186 p.

Parsons, T. R., R. J. LeBrasseur and J. D. Fulton. "Some Observations on the Dependence of Zooplankton Grazing on the Cell Size and Concentration of Phytoplankton Blooms," *J. Oceanog. Soc. (Japan)* 23:10-17 (1967).

Porter, K. G. "Selective Grazing and Differential Digestion of Algae by Zooplankton," *Nature* 244:179-180 (1973).

Porter, K. G. "Enhancement of Algal Growth and Productivity by Grazing Zooplankton," *Science* 192:1,332-1,334 (1976).

Riley, G. A., H. Stommel and D. F. Bumpus. "Quantitative Ecology of the Plankton of the Western North Atlantic," *Bull. Bingham Oceanog. Coll.* 12:1-169 (1949).

Ringelberg, J. "The Positively Phototactic Reaction of *Daphnia magna* Straus: A Contribution to the Understanding of Diurnal Vertical Migration," *Neth. J. Sea Res.* 2:319-406 (1964).

Ristic, J. "Seasonal and Vertical Distribution of Phytoplankton at an Offshore Station in Lake Michigan," M.S. Thesis, University of Wisconsin—Milwaukee (1977).

Rose, M. "Contribution a l'Etude de la Biologie du Plankton: Le Probleme des Migrations Verticales Journalieres," *Arch. Zool. Exper. Gen.* 64:387-542 (1925).

Russell, F. S. "The Vertical Distribution of Plankton in the Sea," *Biol. Rev.* 2:213-262 (1927).

Siebeck, O. "Untersuchungen uber die Vertikalwanderung planktishcer Crustaceen unter besonderer Berucksichtigung der Strahlungsverlialtnisse," *Int. Rev. Ges. Hydrobiol.* 45:381-454 (1960).

Starkweather, P. L. "Diel Patterns of Grazing in *Daphnia pulex* Leydig," *Verh. Int. Ver. Limnol.* 19:2851-2857 (1975).

Steele, J. H. In: *The Changing Chemistry of the Oceans: Nobel Symposium 20.* D. Dyrssen and D. Jaguer, Eds. (Stockholm: Almquist and Wiksell, 1972), pp. 209-221.

Steele, J. H. *The Structure of Marine Ecosystems* (Cambridge, MA: Harvard University Press, 1974), 128 p.

Steele, J. H., and E. Henderson. "Simulation of Vertical Structure in a Planktonic System," *Scot. Fish. Res. Rep.* 5:1-27 (1976).

Steele, J. H., and M. M. Mullin. "Zooplankton Dynamics," in *The Sea Vol. 6 Marine Modeling,* E. D. Goldberg, I. W. McCave, J. J. O'Brien and J. D. Steele, Eds. (New York: John Wiley and Sons, Inc., 1977), pp. 857-890.

Steele, J. H., and C. S. Yentsch. "The Vertical Distribution of Chlorophyll," *J. Mar. Biol. Assoc. (U.K.)* 39:217-226 (1960).

Teraguchi, M., A. D. Hasler and A. M. Beeton. "Seasonal Changes in the Response of *Mysis relicta* Loven to Illumination," *Verh. Int. Ver. Limnol.* 19:2,989-3,000 (1975).

Vinogradov, M. E., and V. V. Menshutkin. "The Modeling of Open-Sea Ecosystems," in *The Sea Vol. 6 Marine Modeling,* E. D. Goldberg, I. W. McCave, J. J. O'Brien and J. D. Steele, Eds. (New York: John Wiley and Sons, Inc., 1977), pp. 891-922.

Vinogradov, M. E., V. V. Menshutkin and E. A. Shushkina. "On Mathematical Simulation of a Pelagic Ecosystem in Tropical Waters of the Ocean," *Mar. Biol.* 16:261-268 (1972).

Walsh, J. J. "A Spatial Model of the Peru Upwelling System," *Deep-Sea Res.* 22:201-236 (1975).

Worthington, E. B. "Vertical Movements of Fresh-Water Macroplankton," *Int. Rev. Hydrobiol.* 25:394-436 (1931).

Wroblewski, J. S., and J. J. O'Brien. "A Spatial Model of Phytoplankton Patchiness," *Mar. Biol.* 35:161-175 (1976).

CHAPTER 4

MATHEMATICAL MODELING OF PHOSPHORUS DYNAMICS THROUGH INTEGRATION OF EXPERIMENTAL WORK AND SYSTEM THEORY

Efraim Halfon

Canada Centre for Inland Waters
Burlington, Ontario
Canada L7R 4A6

INTRODUCTION

Modeling an ecosystem requires knowledge of the system obtained from experiments, and an abstraction of the system within a mathematical framework. Systems methods can be used effectively in the latter phase of the modeling process. Indeed, when coupled with experimental work, these mathematical methods can help in the development of an ecologically realistic mathematical model. In this chapter, two commonly used systems methods—model order estimation and parameter estimation—are used to analyze phosphorus dynamics in freshwater and to suggest further research.

METHODS

Model order estimation techniques have been developed to find the model of minimal order capable of simulating a given process. Two methods are used: one in the time domain and the other in the frequency domain. The former uses a mathematical representation equivalent to a model in the form of a transfer function. This model describes the physical system as a black box with inputs and outputs. To determine the model order, a transfer function (G_M) in terms of the complex variable s of order n:

$$G_M(s) \;=\; \frac{b\,b_0 + b_1 s + \ldots + b_{n-2} s^{n-2} + b_{n-1} s^{n-1}}{a_0 + a_1 s + \ldots + a_{n-2} s^{n-2} + a_{n-1} s^{n-1} + s^n} \tag{1}$$

where a, b = constants

is simulated using observed inputs. The order of the transfer function is chosen large enough to ensure a good fit to the data; in this study it was set at nine. The order n is then reduced until the output of the simulation differs from the data by more than a specified error, here 1%. This indicates the model of lowest order needed to have a good simulation, given the error specified.

On the other hand, the frequency domain method for order checking transforms the data into responses at discrete frequencies (ω_i) by numerical Fourier transformation (Unbehauen, 1965). The order of the model is found by determining when the mean approximation error is of the order of magnitude of the measurement error. The mean approximation error decreases as the model order increases (Strobel, 1968). The error (ϵ_r) has the form:

$$\epsilon_r(j\omega_i) \;=\; \frac{G(j\omega_i) - G_M(j\omega_i)}{G(j\omega_i)} \tag{2}$$

and the problem is how a transfer function (Equation 1) with order n can be determined from $G(j\omega_i)$. This can be done by minimizing the criterion:

$$I \;=\; \sum_{i=1}^{N} \left\{ |\,\epsilon_r(j\omega_i)\,|^2 \, w_i^2 \right\} \;=\; Min \tag{3}$$

with respect to a_k and b_k. The weighting factor (w_i) is proportional to the inverse relative mean error of $G(j\omega_i)$. Should $\epsilon_r(j\omega_i)$ be considerably greater than the root mean square (rms) measurement error, then the model order must be increased. If the approximation error is of the order of magnitude of the measurement error, then the determined model order can be regarded as the true one. Mathematical details are given in Halfon et al. (1978).

The results of the two tests must agree in order to have confidence in the conclusions.

Parameter estimation techniques are used to estimate the values of the rate constants that describe the direction and the flux between compartments. Several techniques are now available. Here, an automated response-surface methodology (RSM) has been employed (Smith, 1976). The technique (Box and Wilson, 1951), practically an optimization method, is a combination of the statistical techniques of experimental design and regression analysis, and provides a sequential adaptive procedure for locating improved solutions. Mathematical details underlying the automated RSM program are given by Smith (1975a, b).

A coupling of experimental data with the results from this systems analysis can be used effectively in model development. These system techniques provide information about the system that is compatible with the results obtained experimentally. For example, the numbers and kinds of compartments also can be found with field and laboratory experiments and some fluxes between them can be measured or estimated. By comparing information from both laboratory studies and mathematical analysis, better models and an increased system understanding can be obtained.

RESULTS

Lean (1973b) proposed a four-compartment model (Figure 1) to explain the experimental observations of phosphorus movement in freshwater. The compartments are soluble phosphorus (PO_4), particulate

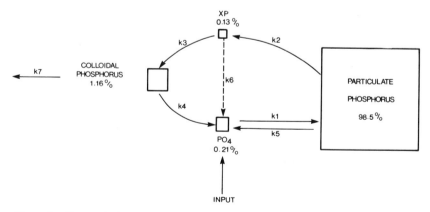

Figure 1. Four principal compartments in the phosphorus model. An exchange mechanism exists between phosphate and the particulate fraction. Moreover, an organic phosphorus compound (XP) is excreted that combines with a colloidal substance. This step involves the release of phosphate from the colloid. [Reprinted by permission, from Lean (1973b). Copyright 1973 by the American Association for the Advancement of Science.]

phosphorus, XP, and colloidal phosphorus; XP is a phosphorus compound with low molecular weight. Lean identified these chemical forms with gel filtration techniques. Also, by use of ^{32}P, he was able to estimate rates of transformation among these forms. This led to the development of a conceptual model with the purpose of describing the observed phosphorus behavior. When the model was tested against data it was not able to describe the phosphorus behavior in all of the experiments. The model could not be validated! This led to the systems analysis of the data described in this chapter. The purpose of the analysis was to estimate the rate coefficients of the four-compartment model and determine whether the model was correct or whether a higher or lower order model was more appropriate.

Two analyses were made. Data from Lean (1973a) were used to determine the model of lowest order capable of explaining the results, and the parameter values of Lean's four-compartment model were estimated by the optimization technique described above to obtain an estimate of parameter variability.

Six experiments (Table I) made from 1967 to 1970 were analyzed. The results of the model order estimation are presented in Table II. These results are consistent between time domain and frequency domain techniques. From Table II it is possible to deduce that a model with five or six compartments would be more appropriate for simulating and describing realistically the cycling dynamics of phosphorus. A larger model, with more compartments, would always be able to fit the data better because of the additional degrees of freedom, but since we are mostly interested in finding a realistic model structure, the problem is to find the lowest order model that can be used for prediction. The fact that the orders vary between three and six shows there is a strong dependence on the particular set of data, but the fact that some sets of data require a higher-order model than hypothesized by Lean indicates that there is a discrepancy between his conceptualization and reality.

Rate constants estimated experimentally and mathematically are provided in Table I. The weighted average values are the best that can be found mathematically for a four-compartment model. The parameter estimation technique indicates (Table II, last column) that the parameters associated with XP and colloidal phosphorus (k_2, k_3, k_4) have low variability. They are also similar to those measured experimentally by Lean (1973a). The other parameters measured by Lean (k_1, k_5) have high variability and differ substantially from his estimates. Thus, one may conclude that the inadequacy of the model to simulate equally well all the experiments may be due to the wrong formulation of the kinetics taking place between soluble and particulate phosphorus (Figure 1).

Table I.   Rate Constants [min^{-1}] of the Four-Compartment Phosphorus Model Obtained by Parameter Estimation and by Experimental Measurement[a]

Date Parameter	8/31/67 Estimated	5/29/69 Estimated	6/3/70 Estimated	6/11/70 Estimated	6/21/70 Estimated	7/27/70 Estimated	7/27/70 Measured	Weighted Average[b]	% SD/AV
k_1	0.37	0.12	0.34	0.31	0.41	0.80	0.90	0.21	24.06
k_2	0.009	0.030	0.030	0.031	0.011	0.037	0.022 0.	0.030	7.32
k_3	0.0014	0.0017	0.0017	0.0024	0.0013	0.0016	0.0017	0.0017	4.77
k_4	0.0023	0.0020	0.0020	0.0028	0.0016	0.0019	0.0019	0.0010	4.72
k_5	0.0017	0.00020	0.00020	0.00034	0.00045	0.00013	0.00003	0.00020	30.40
k_6	0.013	0.003	0.0	0.0	0.010	0.008	—	0.003	25.85
k_7	0.006	0.001	0.0	0.0	0.014	0.0	—	0.002	40.35
Normalized Weight[c]	0.016	0.634	0.130	0.122	0.065	0.034	—	—	—

[a] A weighted average of the parameter estimates is also presented. In the last column the percentage of the ratio standard deviation-mean gives an estimate of the confidence of the means.

[b] The mean estimate of the parameter values is obtained by multiplying the estimates by the normalized weight.

[c] The normalized weight indicates the relative confident of the parameter estimates. This is the inverse of the mean square error.

Table II. Model Order

Date	Order
August 21, 1967	5
May 29, 1969	3
June 3, 1970	5
June 11, 1970	5
June 21, 1970	3
July 27, 1970	6

The large variances associated with the rate constants imply that a four-compartment model is not sufficient to model the experimental results. A five-, or better, a six-compartment model seems more appropriate. These results agree with the model order estimation technique and with experimental results obtained after 1970 (Lean and Charlton, 1976). Since there is great uncertainty for the parameters relating soluble and particulate phosphorus, it seems appropriate to hypothesize that the model should have a different representation of the behavior of particulate phosphorus. This can be done in three ways, all in agreement with experimental results. Particulate phosphorus can be subdivided into two compartments. They might represent organisms with different phosphorus kinetics; for example, nannoplankton and netplankton. According to Lean (1977) and Sakshaug and Holm-Hansen (1977), storage polyphosphates play an important role in algal metabolism, so another option would be to add a phosphorus storage compartment. This second model, however, would be difficult to validate because of observability problems. A third hypothesis would be to include both models, that is, two sizes of particulate phosphorus, both with storage. Norman and Sager (1977) have suggested that the first model would be adequate; however, they lack data to support it.

DISCUSSION

A coupling of mathematical models, systems methods and experimental research is important for understanding phosphorus dynamics in freshwater. Mathematical models provide a means to test a hypothesis by simulation. Systems methods can be used to improve the models and to increase insight into the system. Experimental data provide a source for model validation. The researcher then puts data and model together to verify his ideas and see that they conform with the evidence. The jump from a four-compartment model to one with more compartments (higher order) implies a revision in the analysis of the experiments. In this instance, it seems appropriate that one additional compartment could be the storage of phosphorus within the cells (Figure 2). Another compartment (Figure 3) could result from the division of particulate phosphorus into two size fractions (< 30 μm and > 30 μm), which experiments indicate have

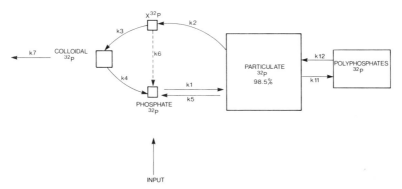

Figure 2. Five-compartment phosphorus model, option 1 (see Figure 1 and text for explanation).

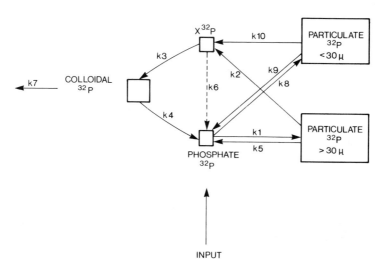

Figure 3. Five-compartment phosphorus model, option 2 (see Figure 1 and text for explanation).

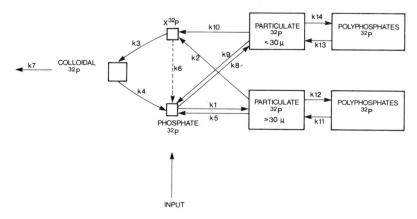

Figure 4. Seven-compartment phosphorus model (see Figure 1 and text for explanation).

different kinetics. If both forms have a storage compartment, then a 7th-order model (Figure 4) would be appropriate. This model would not be of minimal order but would be biologically realistic.

SUMMARY

System theory provides the basis for modeling and simulation; experimental work provides the data necessary to validate the model. Model order estimation and parameter estimation techniques have been used to develop a theoretical model of phosphorus dynamics in freshwater, based on work by Lean (1973a,b). Results indicate that a model less aggregated than that proposed by Lean (1973b) is more adequate. In particular, particulate phosphorus should be modeled with more than one compartment.

ACKNOWLEDGMENTS

Dr. D. R. S. Lean provided the data from his dissertation and helped in the analysis of the results. His collaboration is highly appreciated. Dr. H. Unbehauen kindly helped in the development and use of the model order estimation algorithm. Dr. A. El-Sharaawi gave helpful comments on the statistical aspects of the paper.

REFERENCES

Box, G. E. P., and K. B. Wilson. "On the Experimental Attainment of Optimum Conditions," *J. Roy. Stat. Soc. (Ser. B)* 13:1-45 (1951).

Halfon, E., H. Unbehauen and C. Schmid. "Model Order Estimation and System Identification, Theory and Application to the Modelling of ^{32}P Kinetics within the Trophogenic Zone of a Small Lake," *Ecol. Model.* (1978, in press).

Lean, D. R. S. "Phosphorus Compartments in Lake Water," Ph.D. thesis, University of Toronto (1973a), 192 pp.

Lean, D. R. S. "Phosphorus Dynamics in Lake Waters," *Science* 179: 678-680 (1973b).

Lean, D. R. S. Canadian Center for Inland Waters, Burlington, Ontario. Personal communication (1977).

Lean, D. R. S., and M. Charlton. "A Study of Phosphorus Kinetics in a Lake Ecosystem," in *Environmental Biogeochemistry*, J. O. Nriagu, Ed. (Ann Arbor, MI: Ann Arbor Science Publishers, Inc., 1976), pp. 283-294.

Norman, J. C., and P. E. Sager. "Differential Rates of ^{32}PO$_4$ Exchange between Soluble and Particulate Phosphate," Abstract, 40th Annual Meeting of the American Society of Limnology and Oceanography, East Lansing, Michigan, June 20-23 (1977).

Sakshaug, E., and O. Holm-Hansen. "Chemical Composition of *Skeletonema costatum* (Grev.) Cleve and *Pavlova (Monochrysis) lutheri* (Droop) Green as a Function of Nitrate-, Phosphate-, and Iron-Limited Growth," *J. Exp. Mar. Biol. Ecol.* 29:1-34 (1977)

Smith, D. A. "Automated Response Surface Methodology in Digital Computer Simulation, volume 1: Program Description and User's Guide," *Technical Report No. 101-1* (AD A016286) (Desmatics, Inc., 1975a), 98 pp.

Smith, D. A. "Automated Response Surface Methodology in Digital Computer Simulation, volume 2: Program Flowcharts and Listings," *Technical Report No. 101-2* (AD A016287) (Desmatics, Inc., 1975b), 85 pp.

Smith, D. A. "Automatic Optimum-Seeking Program for Digital Simulation," *Simulation* 27:29-31 (1976).

Strobel, H. *System Identification with Deterministic Test Signals* (Berlin: VEB-Verlag, 1968), in German.

Unbehauen, H. "Identification of the System Parameters and Development of the Mathematical Model of a Control System by Means of Measured Time Response," *IFAC-Symposium*, Tokyo (1965), pp. 193-201.

SECTION TWO

IDENTIFICATION OF RESEARCH NEEDS
THROUGH MODEL STUDIES

To build models that simulate significant portions of an aquatic eco-system, one must begin with a conceptual framework that delineates the major components and their interactions for the system of interest. How-ever, in order to test this hypothetical framework, one must specify equations for the component interactions and assign numerical values to the coefficients in these equations. Upon analysis of the resulting mathe-matical representation of the system, the initial conceptual framework can be reevaluated. Often this process leads to identification of areas that require further definition and thus further experimental research. The use of models to identify such research needs is obviously a valuable product of the modeling process, as the chapters in this section, which deals with identification, illustrate.

Chapter 4 by E. Halfon demonstrates the application of systems theory techniques to the identification of the appropriate structure for a model, in this case a model simulating phosphorus cycling. The model structure indicated from his analyses suggest that, in addition to the phosphorus components usually considered important, other phosphorus fractions play important roles in the dynamics of the phosphorus cycle. The results sug-gest further experimental work is needed to define and understand these fractions. In Chapter 5, R. A. Park, T. W. Groden and C. J. Desormeau discuss areas of research that are required before specific improvements to their ecosystem model can be made. Through analysis of their multi-compartment model, they have identified the most critical limitations on further development of this model—limitations that can only be relieved by further experimental work. In the final chapter of this section, D. Scavia discusses the major subdivisions of aquatic ecosystems in terms of the con-structs used to represent them in several ecological models and the experi-mental information used to support these constructs. Weaknesses in our understanding of these components, as identified by the state of present models, are identified and research priorities to remove these weaknesses are suggested.

These three chapters show, in separate ways, how the modeling process can be useful in identifying areas of weakness in our present understanding of ecosystem structure and function.

CHAPTER 5

MODIFICATIONS TO THE
MODEL CLEANER
REQUIRING FURTHER RESEARCH

Richard A. Park
Theresa W. Groden
Carol J. Desormeau
> Center for Ecological Modeling
> Rensselaer Polytechnic Institute
> Troy, New York   12181

INTRODUCTION

The trend in ecological modeling over the past several years has been to include more biological realism, resulting in more state variables and greater process resolution.   An intensive, multi-institutional effort, originally under the aegis of the International Biological Program and continuing over the past several years at Rensselaer Polytechnic Institute has resulted in MS. CLEANER, one of the more complex and biologically realistic ecosystem models.

The original model CLEAN was formulated by 25 investigators in the Eastern Deciduous Forest Biome (EDFB) section of the U.S. International Biological Program (Park et al., 1974) and incorporates a number of submodels for ecologic and physiologic processes (Bloomfield et al., 1973). The generalized EDFB version was implemented at Rensselaer Polytechnic Institute (Scavia et al., 1974).

Since then the model has been improved by adding state variables, by reformulating the decomposition submodels, and by including a subroutine to transform biomass values into environmental perception characteristics, resulting in CLEANER (Park et al., 1975; Scavia and Park, 1976; Clesceri et al., 1977).   The present version, MS. CLEANER, has been greatly expanded to include 40 state variables, much greater process resolution, and the capability for simulating as many as 10 vertical and horizontal segments simultaneously (Park, 1975; deCaprariis *et al.*, 1977).

Current research with MS. CLEANER, including verification in cooperation with several European hydrobiological laboratories, has suggested that continued model development is limited by the state of knowledge in aquatic ecology. Clearly, the time has come for modeling to advance beyond the synthesis stage and for modelers to assume responsibility for suggesting new directions for research. This is possible because the formalism of a mathematical model makes it easier to identify conceptual weaknesses and because modelers are forced to take a whole ecosystem perspective that is often missed by specialists.

Three levels of research are needed. These are: (1) process-level research, consisting of carefully devised experiments to aid in formulation of constructs representing environmental responses for specific processes; (2) detailed measurements to determine parameter values for both old and new constructs; and (3) comprehensive studies at experimental sites in order to provide data to verify the models under a range of natural and perturbed conditions.

Each of these levels of research is important. However, we will focus our attention on research needs relating to process responses because this is where we feel the impact on modeling will be the greatest. Furthermore, we will emphasize those research areas that are of particular concern to us in our current development of MS. CLEANER.

PHYTOPLANKTON

In an effort to make CLEANER more general, we undertook a detailed examination of the assumptions used in the model in light of current knowledge of algal physiology. We have concluded that by incorporating more of the complexities and feedback loops present in ecosystems (such as the ability of organisms to adapt to changing conditions by succession or by changes in their physiological characteristics) our models will eventually be both more realistic and more general, thus facilitating application to new or rapidly changing sites. However, because most process studies have been conducted with more restricted objectives than those of ecosystem modeling, there are gaps that must be bridged before we can realize the full benefit of such an approach.

In the following discussion we will consider the interactions among light, temperature and nutrients as well as the ability of algae to change their environmental requirements for optimal photosynthesis. We will also touch on the ways in which limitations of photosynthesis affect nonpredatory mortality and sinking rates

Light and Temperature Limitation

Photosynthesis consists of two processes—the temperature-independent photochemical process, which absorbs light energy and produces ATP and NADPH, and the temperature-dependent enzymatic process (Calvin cycle), which utilizes these products to produce organic compounds. The actual rate of photosynthesis is determined by the slowest process (Gabrielsen, 1948). At low light intensities the photochemical process determines the rate of photosynthesis because the products referred by the enzymatic process are insufficient for maximum rate of photosynthesis. At high light intensities the rate of the photochemical process exceeds that of the enzymatic process, making temperature, through its effects on the enzymatic process, the limiting factor. In exceedingly bright light a decrease in the rate of photosynthesis results from photooxidation of the enzymes participating in photosynthesis and inactivation of chlorophyll.

Currently available models of response of photosynthesis to light can be divided into the following two classes:

1. Those that represent the light limiting and light-saturation ranges. For such models the rate of photosynthesis initially increases linearly with increasing light intensities and ultimately reaches a saturation level where photosynthesis has a constant maximum value (for example Smith, 1936).

2. Those that attempt to represent photosynthesis as above, but also represent inhibition inhibition range where an increase in light intensity above a certain value produces a decrease in photosynthesis (for example, Steele, 1962).

In MS. CLEANER the effect of differing light intensities on photosynthesis is based on the formulations of both Smith (1936) and Steele (1962). Previously CLEANER used Steele's formulation for all light intensities (Scavia and Park, 1976). However, Steele's equation is not fully adequate to represent all conditions that can occur in nature (Straškraba, 1976) and is inaccurate for noninhibiting conditions because the photosynthetic **response** below the inhibition threshold is dependent upon the response above this threshold (Jassby and Platt, 1976). Therefore, the Smith equation is used for noninhibiting conditions.

It appears that many phytoplankters can adapt within a range of light intensities provided the duration of exposure is sufficiently long. Nielsen *et al.* (1962) have shown that algae collected near the surface in temperate waters are adapted to the light conditions of the previous one or two days. Tilzer (1972) has shown that phytoplankton in an Alpine lake can adapt during the course of a single day. Therefore, it is important that adaptation to light be included in our models.

The Smith equation incorporates the slope of the curve for photosynthesis vs light intensity. This slope is proportional to the light absorbed by the photosynthetic pigments and can be dependent upon the light condition to which the algae have adapted. Such adaptation is characteristic of those phytoplankters that can change their concentration of chlorophyll. For example, sun-adapted algae have smaller concentrations of chlorophyll than shade-adapted algae, a fact that can be represented by smaller values for the slope and higher values for the saturating light intensities (Ryther and Menzel, 1959).

Myers and Kratz (1955) give a linear relationship between the slope of the light-limiting portion of the photosynthesis vs light intensity curve and the chlorophyll *a* content for the blue-green alga *Anacystis nidulans*. Other researchers (Myers, 1946; Nielsen and Jørgensen, 1968) have shown that the concentration of chlorophyll increases exponentially as algae become adapted to lower light intensities. By combining these two relationships one obtains the following equation for the slope (*a*) as a function of the light intensity (I) to which the algae, other than diatoms, have adapted.

$$a = k_1 \ln(I) + k_2$$

Jørgensen (1964) has shown that the diatom *Cyclotella meneghinian* can adapt to increasing light intensities by changing the concentration of photosynthetic enzymes. This produces an increase in the light-saturated rate of photosynthesis, but the slope of the curve for photosynthesis vs light intensity is not affected because the chlorophyll content of the diatoms remains constant. Future research should determine if this change in the light-saturated photosynthetic rate is characteristic of all diatoms and, if so, provide sufficient data to develop and calibrate a function relating the light-saturated photosynthetic rate of diatoms to the light intensity.

For light intensities great enough to cause photoinhibition, Steele's (1962) formula is used in MS. CLEANER. The onset of light inhibition is dependent on both the light condition to which the algae have adapted and the ambient temperature (Sorokin and Krauss, 1958; Nielsen, 1962). Unfortunately, it is not clear from published data how we should determine the onset of photoinhibition; this needs to be investigated more fully for all groups of phytoplankton.

In the light-saturation range the rate of the enzymatic process is slower than the rate of the photochemical process; therefore, temperature affects the maximum photosynthetic rate. The relationship between the relative rate of photosynthesis at the light-saturated intensity and temperature is illustrated by Figure 1 from Aruga (1965b). Although the empirical equation given in Table I successfully simulates this relationship and replaces the TEMP function of Park *et al.* (1974), additional research is needed to calibrate such an equation. Specifically, data such as those of Sheldon and Boylen (1975) for mixed natural assemblages are needed. Also, the absolute light-saturated rate of photosynthesis (as opposed to the relative rate) measured at the optimal temperature and at 0°C is required for different phytoplankton groups if this equation is to be used.

The ability of algae to optimize the light-saturated rate of photosynthesis based on water temperature is best illustrated by Straškraba's (1976) replotting of Aruga's (1965a) data on seasonal adaptation (Figure 2). This adaptive capability is also characteristic of thermophilic algae (Brock, 1970); however, one cannot assume that the algae can adapt to all possible temperature extremes. For example, it has been shown for algal and bacterial assemblages from diverse temperature habitats, where environmental temperatures cycle on a seasonal basis from 0°C to 20-30°C, that community metabolic optima do not drop below 20°C (Boylen and Brock, 1973, 1974). Therefore, in such habitats these organisms are only optimally adapted to their enviornment during the summer but do not become adapted to colder temperatures.

Evidence from natural assemblages suggests that adaptation to temperature is a genotypic and not a phenotypic response. If an organism is optimally adapted to the temperature of its environment, its composite enzyme systems are usually optimally active at that temperature as well. An organism that adapts to higher or lower temperatures generally possesses

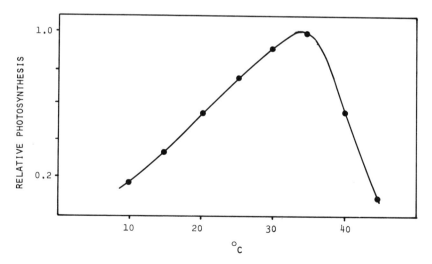

Figure 1. Relative photosynthesis as a function of temperature in *Chlorella*. Data from Aruga (1965b).

Table I. Empirical Temperature Function used to Evaluate the Light-Saturated Rate of Photosynthesis

$$R(T) = EXP\ [k \cdot (aT^2 - bT^c - 1)]$$

where

$R(T)$ is the light-saturated rate of photosynthesis at temperature T.

The coefficients a, b, c and k are dependent on the optimum temperature (TOPT), the maximum temperature (TMAX) and the value of $R(T)$ at $0°C$ and the optimum temperature. These coefficients are evaluated using the following equations:

$$k = -\ln[R(0)]$$

$$a = b(TMAX)^{c-2}$$

$$b = \frac{1 + \ln\ [R(TOPT)]/k}{(TMAX)^{c-2}\ (TOPT)^a - TOPT^c}$$

$$\frac{2^{\frac{1}{c-2}}}{c} = \frac{TOPT}{TMAX}$$

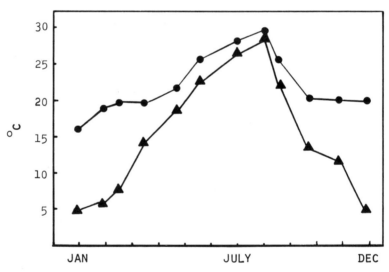

Figure 2. Relationship between optimum temperature of phytoplankton (circles) and ambient temperature (triangles). After Straskraba (1976) using monthly data from Aruga (1965a).

isozymes (enzymes carrying out similar functions but coded for at different genetic loci) or allozymes (enzymes produced by altered loci that have different temperature optima from the original enzymes). It is our hypothesis that the cold winter is not sufficiently long to allow enough generations to produce the genetic changes required for adaptation.

Future research should determine the maximum and minimum temperatures to which mixed algal assemblages can adapt, should indicate the genetic basis for the adaptations, and should provide sufficient data to improve, or at least to calibrate, the function given by Straškraba (1976), which relates the optimum temperature (TOPT) to the ambient water temperature (T):

$$TOPT = T + 28 \; EXP \; (-0.1155T)$$

Nutrient Limitation

We rejected the idea of including excess intracellular nutrients when developing CLEAN (Bloomfield et al., 1973). However, we have encountered a time lag between modeled and observed phytoplankton peaks in both eutrophic and ultraoligotrophic systems; it has been suggested that this is because we have coupled growth to external nutrient supplies. Therefore, we have recently turned our attention to development of a submodel for intracellular nutrients for CLEANER. Several modelers have developed formulations for simulating luxury consumption of nutrients. Koonce's (1972) empirical approach was probably the first attempt. Apparently independent of one another, Grenney et al. (1973), Bierman (1974) and Nyholm (1975, 1977) developed intracellular nutrient submodels based on principles of cellular physiology. However, these fail to take into account some of the previous process research and are especially inadequate in light of recent developments.

Kinetic studies of phosphorus uptake (Ketchum, 1939; Fuhs, 1969; Healey, 1973; Rhee, 1973; Lehman, 1976), nitrogen uptake (MacIsaac and Dugdale, 1969; Caperon and Meyers, 1972; Eppley and Renger, 1974; Underhill, 1977), and silica uptake (Kilham, 1975; Guillard et al., 1973) provide a good basis for modeling. But ever since Ketchum (1939) recognized that phosphate uptake increased in the presence of nitrate, it has been felt that nutrient interactions should be considered; this has proven difficult for modeling and has resulted in several diverse constructs for multiple nutrient limitations (cf. Chen, 1970; Larsen et al., 1973; Bloomfield et al., 1973; Scavia and Park, 1976).

Rhee (1974, 1978) investigating the interaction of nitrogen and phosphorus in Scenedesmus spp., has shown that only one nutrient is limiting at a time and that there is storage of the other nutrient in excess of what is needed for growth. Therefore, it appears that Liebig's law of the minimum has been validated for nutrient limitation! Rhee (1978) found that a species-specific atomic ratio of stored nitrogen to phosphorus, which he called the "threshold ratio," describes the point at which primary productivity is limited by a single nutrient. Titman (1976) and Kilham and Kilham (in press) have suggested species-specific "switching points" for silica to phosphorus ratios of some freshwater diatoms. Switching points are based on external concentrations of nutrients and represent species-optimum points for resource competition.

Since an algal cell is capable of storing excess nutrients, it is necessary to model the intracellular levels of nutrients and to use threshold ratios in order to predict competition advantage and succession. At this time, additional research is needed to determine the threshold ratios for more algal species, as well as functional groups. Model predictions of algal succession should be verified by growing algal assemblages in media of various nutrient ratios to confirm the existence of competitive advantages that reflect differing threshold ratios.

The effects of temperature and light intensity must be considered when simulating the uptake of nutrients. The uptake rate for nitrogen generally increases with increasing light intensity (Eppley and Coatswork, 1968; Morris, 1974; Bates, 1976). Davis (1976) has shown that nitrogen, phosphorus and silica uptake ratios are light-dependent in *Skeletonema costatum*. Other authors have found the rate of phosphorus uptake to increase with increasing temperature (Blum, 1966; Fuhs *et al.*, 1972).

Maddux and Jones (1964) demonstrated a change in the optimum temperature for growth under varying levels of light and nutrients and a similar effect on optimum light intensity under low- and high-nutrient concentrations. For example, it was evident that at lower concentrations of nitrate and phosphate their organisms were particularly susceptible to increasing light. Cloern (1977) has shown that there are interactive effects of light intensity and temperature on growth and uptake rates in *Cryptomonas ovata*. However, much more work is needed on a variety of species before the complexity of interactions will be understood adequately. Experiments to determine the effects of light, temperature and nutrients should be carried out both independently and with combinations of control variables in order to evaluate the interactive influences.

Mortality

Nonpredatory mortality is an important aspect of the dynamics of phytoplankton populations. Although many models do not explicitly account for nonpredatory mortality, DePinto *et al.* (1976) and Bierman (1976) found that it was necessary to include a density-dependent mortality term to simulate the decline of blue-green algae.

In developing CLEANER we have assumed that increased mortality occurs when conditions become limiting, analogous to the occurrence of osmotic fragility in bacteria (Mitchell and Moyle, 1959). A similar mechanism, developed of a "phosphorus debt," has been postulated to account for mass mortality of phytoplankton in the prairie ponds of Manitoba (Barica, 1975). Whatever the physiological basis, phytoplankton definitely exhibit a pattern of decreasing viability, accompanied by an increase in colonization by decomposers and leading to the eventual loss of cell contents (Jones, 1976). It has been demonstrated that in the absence of decomposers, *Chlorella* and *Selenastrum* cells do not lyse (DePinto and Verhoff, 1977); however, it is unlikely that decomposers can kill healthy cells [see later section on *Colonization* and DePinto (Chapter 2) for further discussions].

It is evident that more observations should be made under controlled conditions. In particular, it would be helpful if the relationship of mortality to limitation of growth could be quantified.

Sinking

The sinking rate of phytoplankton, detritus and decomposers has been represented by Stoke's law in several models, including the present version of MS. CLEANER. The calibration of the construct has proven difficult for both unstratified and stratified lakes and may represent a significant source of error in the simulations. It is evident that we must not be content to represent sinking as a simple application of Stoke's law, even with provision for deviations in shape and specific gravity (Fogg, 1975).

It has been shown that, with decreasing growth rate, there is an increase in sinking rate of phytoplankton (Eppley *et al.*, 1967; Smayda, 1970; Titman and Kilham, 1976); this implies that the construct for sinking should include a term related to limitation of growth. Scavia *et al.* (1976) have suggested an approximate formulation that is based on Stoke's law but which takes physiological state and shape into account. This formulation could be quantified concurrently with the investigation of mortality (see above).

A more serious problem is that, under turbulent conditions commonly found in nature, the free-falling motion represented by Stoke's law may be completely counteracted by advective transport. Difficulty in calibrating the sinking term in MS. CLEANER suggests that the hydrodynamics may exert an important control over the maintenance of phytoplankton in the euphotic zone.

In the unstratified lakes and shallow-water areas the shear velocity of wind-driven currents and shallow-water waves can be taken into consideration in determining if sedimentation of phytoplankton and detritus will occur. However, there is a need for innovative experiments, such as those performed by the Freshwater Biological Association Staff on Blenham Tarn, England, involving marine diatomites and other markers (Pennington, 1974), to determine actual patterns of sedimentation.

For stratified lakes the approach of Stefan *et al.* (1976) may be the key to simulating the transfer of phytoplankton from the euphotic zone. They assume that turbulence in the mixed layer effectively prevents sinking and that transfer occurs only when calm periods prevail and the depth of the mixed layer decreases. However, this approach requires simulation of daily changes in mixing depth and should be investigated more thoroughly before being routinely adopted. Titman and Kilham (1976), in an interesting theoretical discussion of the effects of Langmuir circulation on maintenance of phytoplankton in the epilimnion, suggest that transfer can occur during turbulent conditions; we thoroughly agree with their conclusion that much more experimental and theoretical work needs to be done.

DECOMPOSERS

A recent paper (Clesceri *et al.*, 1977) described the progress that has been made in modeling microbial growth and activity in MS. CLEANER. *disc.* However, there is still a pressing need for additional work to quantify the relationships of decomposition to other ecosystem functions.

Colonization

Bacteria seem to colonize phytoplankton cells readily toward the end of the cells' active growth phase (Jones, 1976). However, it is not clear what the actual relationship is to the viability of the cell (see Mortality above). Colonization may be stimulated by increased loss of photosynthate (Bell and Mitchell, 1972), or it may be increased due to a change in the electrostatic charge of the cell (Oppenheimer and Vance, 1960). There are also differences in degree of colonization on different types of algae (Jones, 1972).

The effectiveness of colonization and the rapidity with which algal cells are decomposed are important to understanding both nutrient cycling (Mills and Alexander, 1974; DePinto and Verhoff, 1977) and trophic interactions. As an ecosystem goes from oligotrophy to eutrophy, there is a less effective functioning of the plankton communities. This is a result of more direct utilization of primary production by zooplankton in oligotrophic lakes, while in eutrophic lakes decomposers quite often intervene between the producer and primary consumer trophic levels (Gliwicz and Hillbricht-Ilkowska, 1972). The proportion of nannoplanktonic algae consumed directly is thereby greatly reduced in eutrophic lakes (Hillbricht-Ilkowska *et al.*, 1970). In order to model these relationships explicitly, it is necessary to determine the actual controls on colonization, preferably under laboratory conditions. [See DePinto (Chapter 2) for further discussion of the role of bacterial colonization on phytoplankton dynamics.]

Uptake of Organic and Inorganic Materials

Dissolved organic matter (DOM) serves an important function in providing organic substrates for heterotrophic metabolism in freshwater environments. Its composition is highly variable, but in an ecosystem context it is most generally separated into labile and refractory fractions. The former is rapidly taken up by heterotrophic bacteria while the latter is very slowly acted upon. In both cases the DOM is partly assimilated to generate bacterial biomass and partly remineralized and excreted.

The sensitivity of decomposers to environmental controls affects the dynamics of remineralization. The relative utilization of different types of organic matter by the decomposers serves both to drive and to stabilize the aquatic ecosystem (Clesceri *et al.*, 1977). Experimentation with MS. CLEANER has shown that the explicit modeling of labile and refractory components imparts stability to the simulation as well. However, with the exception of a few papers (*cf.* Foree and McCarty, 1970; Otsuki and Hanya, 1972; Mills and Alexander, 1974) data on these two fractions are sparse. Additional data are needed on decomposition of different algal types and of allochthonous material under varying conditions.

ZOOPLANKTON

CLEANER contains a very complex construct for predator-prey interaction (Park *et al.*, 1974; Scavia and Park, 1976), and one might expect it to be quite realistic. However, we have had a persistent problem of the predicted zooplankton biomass not decreasing sufficiently in the fall and winter (*cf.* Figure 3). Analysis of these results in consultation with limnologists in several European laboratories has led us to the conclusion that two additional processes should be considered in the model: starvation and the formation of resting stages. Furthermore, excretion should be considered if we are to model zooplankton dynamics realistically.

Starvation and Resting Stages

In productive lakes it is not uncommon for the zooplankton to exceed the carrying capacity of the system, either through their own rapid development or due to a crash of the phytoplankton (see Mortality above). The

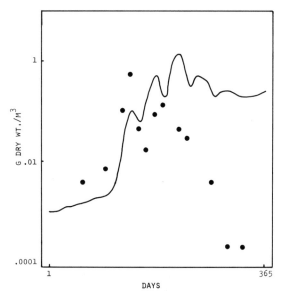

Figure 3. Predicted (line) and observed (circles) biomass levels of omnivorous zooplankton in epilimnion of Slapy Reservoir, Czechoslovakia. From Youngberg (1977), with observed values from Straškraba and Hrbáček (1966).

result can be starvation of large numbers of zooplankters. Food availability is particularly important in controlling the mortality of adult zooplankton (Argentesi *et al.*, 1977).

Unfortunately, few quantitative studies have been conducted on the effects of low food levels on zooplankton production (Bottrell *et al.*, 1976). There are some observations on the loss of fat and changing carbon: nitrogen ratios under austere conditions (*cf.* Conover, 1964; Omori, 1970) from which one might estimate threshold values for starvation. However, more data are needed from controlled experiments if the process of starvation is to be modeled realistically.

In order to minimize the effects of low food levels and adverse environmental conditions, many zooplankton species have developed either resting juvenile (copepodite) stages or resting eggs. The production and hatching of resting eggs in cladocerans is relatively well understood and should be amenable to modeling. The production of fertilized resting eggs may be in response to any one of several adverse factors (Hutchinson, 1967). The cues for hatching may include agitation (Larsson, 1977) or combinations of light, temperature, O_2 and CO_2 changes (Stross, 1971). There may or may not be an increase in sensitivity to environmental change with increasing age of the eggs. The result is to ensure synchroneity of production and hatching of resting eggs.

The controls on resting copepodite stages do not seem to be well documented. Because the encapsulated copepodites sink to the bottom sediments (Wierzbicka, 1962; Elgmork, 1967), they are effectively removed from the plankton and the sampled population declines rapidly. Upon ending diapause, the copepodites return to the plankton and may contribute to an apparently sudden increase in standing crop (as suggested in Figure 3). Although the literature is contradictory (*cf.* Smyly, 1973), it seems that, in general, diapause occurs when the carrying capacity is exceeded. Thus, both diapause and starvation probably should be modeled responses to consumption rate. The cues for ending diapause in copepodites may be similar to those for resting eggs. However, there seems to be a definite need for controlled experiments similar to those that have been conducted on resting eggs.

Excretion

One of the effects of modeling excess internal nutrients in the phytoplankton is that variable stoichiometry must be taken into account in considering the mass balance of nutrients in the food chain. At the present time MS. CLEANER accounts for two types of excretion: the release of phosphorus, nitrogen and CO_2 when biomass is metabolized for energy, and the release of phosphorus and nitrogen, which is in the food in excess of that required to maintain the average stoichiometry of the zooplankton. This approach is appealing because there may be as much as a tenfold difference in phosphorus and a threefold difference in nitrogen storage in phytoplankton, reflecting the nutrient balance of the lake. This could help to account for the large differences in excretion noted by different authors (*cf.* Rigler, 1961; Ganf and Blazka, 1974; LaRow *et al.,* 1975).

However, zooplankton do not maintain a constant stoichiometry; there seems to be less excretion in proportion to consumption and higher values of body phosphorus and nitrogen under conditions of abundant phytoplankton in contrast to increased excretion and lower values for body nitrogen and phosphorus under conditions of low food availability (Martin, 1968). Storage of phosphorus and nitrogen represents accumulation in lipids (Conover, 1964; Omori, 1970) and invalidates our assumption that excretion is directly related to the stoichiometry of prey. Unfortunately, we do not know the actual stoichiometry of the prey (phytoplankton and yeast) that have been fed to the zooplankton by investigators studying excretion, nor do we know how rapidly the stored phosphorus and nitrogen are lost from the zooplankton. Until these types of data are obtained, it is difficult to assess how much error is promulgated by our simplifying assumption of constant stoichiometry in the zooplankton.

FISH

CLEANER has been applied in simulating fish for several years (Park *et al.*, 1974, 1975); however, we have been dissatisfied because we felt that, in using single compartments for dominant species, important aspects of population dynamics and trophic interactions were being ignored. It would be tempting to model year classes of dominant species as Ursin (1976) does in his very complex ecosystem model of the North Sea, but the computational load is felt to be prohibitive.

Chen *et al.* (1975) have suggested that a compromise approach can be used (see also Chapter 11), with major growth stages being differentiated; that approach is currently being incorporated in MS. CLEANER. However, each growth stage has to be characterized by a different set of parameters for factors such as optimal growth rate, optimal temperature, prey preferences, and habitat preference. Unfortunately, these data for larval and immature stages of many important species seem to be missing from the literature; perhaps what is needed is for ecosystem modelers to make a conscious effort to communicate with fishery biologists.

Adaptive Prey Preference

It is difficult to obtain adequate data on prey preferences because one must not only determine gut contents, but also estimate the biomass of potential prey types at the time of feeding. However, it is apparent that prey preference also entails preconditioning and that a fish population may switch preference to an abundant prey (as any successful fisherman can attest). This adaptive preference should be taken into account because it functions as a density-dependent feedback and can be a significant stabilizing factor in the ecosystem. Unfortunately, data suitable for determining adaptive preference are almost nonexistent. Perhaps the simplest way to obtain the necessary data would be to vary systematically the proportions of prey types in a fish tank, while periodically sacrificing a representative sample of fish in order to determine gut contents (because individual fish will behave differently, each sample will have to be sufficiently large to ensure statistical significance).

Migration

MS. CLEANER contains a function to determine the migration of both zooplankton and fish between modeled segments. This function includes responses to dissolved oxygen levels, food availability and temperature, and we plan to add light intensity (*cf.* Scherer, 1971); average swimming rates are taken into consideration in order to constrain short-term (*e.g.*, diurnal) migration. However, it would be helpful if there were data on directed movement under controlled conditions. In particular, one might establish gradients of temperature, light and possibly dissolved oxygen and food in various combinations in order to determine optima and relative importance.

SEDIMENTS

The problem of sedimentation of phytoplankton, decomposers and detritus has been discussed above (see Sinking). However, a related problem in shallow lakes is the resuspension of these components, along with inorganic sediments. Resuspension of sediments can greatly affect the turbidity of the water, thereby controlling photosynthesis; it can also keep the sediments from becoming anaerobic, maintain phytoplankton and decomposers in the pelagic zone, and prevent macrophytes from becoming established.

There are numerous references that provide a basis for modeling the occurrence of resuspension. However, we have not found any study that relates the quantity of suspended lake sediments to the shear velocities. In other words, we have no way of predicting how much material can be resuspended by a storm event. What seems to be needed is continuous, electronic monitoring of turbidity with depth during a series of storms.

SUMMARY AND CONCLUSIONS

Additional process-level research is needed if the development of more realistic and general models is to be continued. In particular, improvement of MS. CLEANER would be facilitated if studies were performed to determine:

Improvements of both

1. the maximum light intensity to which algae can adapt and the factors controlling the onset of photoinhibition,
2. the absolute light-saturated rate of photosynthesis at the optimal temperature and at $0°C$,
3. the relationship between optimal and ambient temperature and the values of the maximum and minimum temperatures to which natural assemblages of algae can adapt, as well as the rate of adaptation,

4. threshold ratios of nitrogen, phosphorus and silica necessary to model resource competition among major taxa and functional groups of algae,

5. the nature of the interacting effects of light, temperature and nutrient kinetics on photosynthesis in pure and mixed cultures,

6. the relationships of phytoplankton mortality and sinking rates to limitation of growth,

7. sinking rates under turbulent conditions in both unstratified and stratified lakes,

8. the colonization, solubilization and rates of uptake by decomposers acting on materials characteristic of natural aquatic systems,

9. threshold values of starvation in zooplankton,

10. the environmental cues for the onset and end of diapause in copepods and for the production and hatching of resting eggs in cladocerans,

11. the relationship between phytoplankton and zooplankton stoichiometries and the effect on nutrient excretion rates,

12. the parameters necessary to model fish growth stages as separate compartments,

13. the adaptive-preference response of fish to abundant prey types,

14. the quantitative relationships between fish migration and dissolved oxygen, food, temperature and light,

15. the amount of sediments that are suspended during storm events of known hydrodynamic characteristics.

ACKNOWLEDGMENTS

We wish to thank the other past and present members of our research group, including James Albanese, Charles Boylen, Catharine Collins, Pascal de Caprariis, Shirley Gully, Robert Haimes, Donna Leung and Barbara Youngberg, for their help in formulating the concepts and in preparing this chapter. We also appreciate the many ideas that were advanced by our European colleagues, including Anthony Bailey-Watts, Gernot Bretschko, Karl Iver Dahl-Madsen, Ricardo de Bernardi, Helmut Forstner, Eivind Gargas, Glen George, Sandor Herodek, Sven Erik Jørgensen, Petter Larsson, Jorma Niemi, Niels Nyholm, Ian Smith, Duncan Stewart, Milan Straškraba and John Thorpe. The manuscript was much improved thanks to the critical reviews of Joseph DePinto, Don Scavia and an anonymous reviewer.

This project has been financed with federal funds from the Environmental Protection Agency under Grants No. RH05047010 and No. 68032142; from the National Science Foundation under Grant No. DEB 75-141168 A01; and from the Eastern Deciduous Forest Biome, U.S. International Biological Program, funded by the National Science Foundation under Interagency Agreement AG-199, BMS 76-00761 with the Energy Research and Development Administration—Oak Ridge National Laboratory.

REFERENCES

Argentesi, F., R. DeBernardi and G. DiCola. "Some Mathematical Methods for the Study of Population Dynamics," First World Conference on Mathematics at the Service of Man, Barcelona, Spain (1977).

Aruga, Y. "Ecological Studies of Photosynthesis and Matter Production of Phytoplankton I. Seasonal Changes in Photosynthesis of Natural Phytoplankton," *Bot. Mag. (Tokyo)* 78:28-288 (1965a).

Aruga, Y. "Ecological Studies of Photosynthesis and Matter Production of Phytoplankton II. Photosynthesis of Algae in Relation to Light Intensity and Temperature," *Bot. Mag. (Tokyo)* 78:360-365 (1965b).

Barica, J. "Collapses of Algal Blooms in Prairie Pothold Lakes: Their Mechanism and Ecological Impact," *Verh. Int. Verein. Limnol.* 19: 606-615 (1975).

Bates, S. "Effects of Light and Ammonium on Nitrate Uptake by Two Species of Estuarine Phytoplankton," *Limnol. Oceanog.* 21:212-218 (1976).

Bell, W., and R. Mitchell. "Chemostatic and Growth Responses of Marine Bacteria to Algal Extracellular Products," *Biol. Bull.* 143:265-277 (1972).

Bierman, V. J., Jr. "Dynamic Mathematical Model of Algal Growth," Ph.D. Thesis, University of Notre Dame, South Bend, IN (1974).

Bierman, V. J., Jr. "Mathematical Model of the Selective Enhancement of Blue-Green Algae by Nutrient Enrichment," in *Modeling Biochemical Processes in Aquatic Ecosystems*, R. P. Canale, Ed. (Ann Arbor, MI: Ann Arbor Science Publishers, Inc., 1976), pp. 1-31.

Bloomfield, J. A., R. A. Park, D. Scavia and C. S. Zahorcak. "Aquatic Modeling in the Eastern Deciduous Forest Biome, U.S. International Biological Program," in *Modeling the Eutrophication Process*, E. J. Middlebrooks, D. H. Falkenborg and T. E. Maloney, Eds. (Ann Arbor, MI: Ann Arbor Science Publishers, Inc., 1973), pp. 139-158.

Blum, J. J. "Phosphate Uptake by Phosphate-Starved *Euglena*," *J. Gen. Physiol.* 49:1125-1137 (1966).

Bottrell, H. H., A. Duncan, Z. M. Gliwicz, E. Grygiereck, A. Herzig, A. Hillbricht-Ilkowska, H. Kurasawa, P. Larsson and T. Waglenska. "A Review of Some Problems in Zooplankton Production Studies," *Norw. J. Zool.* 24:419-456 (1976).

Boylen, C. W., and T. D. Brock. "Effects of Thermal Additions from the Yellowstone Geyser Basins on the Benthic Algae of the Firehold River," *Ecology* 54(6):1282-1291 (1973).

Boylen, C. W., and T. D. Brock. "A Seasonal Diatom in a Frozen Wisconsin Lake," *J. Physiol.* 10(2):210-213 (1974).

Brock, T. D. "High Temperature Systems," *Ann. Rev. Ecol. System* 1:191-220 (1970).

Caperon, J., and J. Meyers. "Nitrogen-Limited Growth of Marine Phytoplankton II. Uptake Kinetics and Their Role in Nutrient-Limited Growth of Phytoplankton," *Deep-Sea Res.* 19:619-632 (1972).

Chen, C. W. "Concepts and Utilities of Ecological Model," *J. San. Eng. Div., Proc. Am. Soc. Civil Eng.* 96(SA5):1085-1097 (1970).

Chen, C. W., M. Lorenzen and D. J. Smith. *A Comprehensive Water Quality-Ecological Model for Lake Ontario* (Lafayette, CA: Tetra Tech, Inc., 1975).

Clesceri, L. S., R. A. Park and J. A. Bloomfield. "A General Model of Microbial Growth and Decomposition in Aquatic Ecosystems," *Appl. Environ. Microbiol.* 33:1047-1058 (1977).

Cloern, J. E. "Effect of Light Intensity and Temperature on *Cryptomonas ovata* (Cryptophyceae) Growth and Nutrient Uptake Rates," *J. Phycol.* 13:389-395 (1977).

Conover, R. J. "Food Relations and Nutrition of Zooplankton," *Occasional Publ., Proc. Sym., Exp. Mar. Ecol.* 2:81-91 (1964).

Davis, C. O. "Continuous Culture of Marine Diatoms Under Silicate Limitation. II. Effect of Light Intensity on Growth and Nutrient Uptake of *Skeletonema costatum*," *J. Phycol.* 12:291-300 (1976).

deCaprariis, P., R. A. Park, R. Haimes, J. Albanese, C. Collins, C. Desormeau, T. Groden, D. Leung and B. Youngberg. "Utility of the Complex Ecosystem Model MS. CLEANER," in *Proceedings of the International Conference on Cybernetics and Society* (Washington, DC, 1977), pp. 87-89.

DePinto, J. V., V. J. Bierman, Jr. and F. H. Verhoff. "Seasonal Phytoplankton Succession as a Function of Species Competition for Phosphorus and Nitrogen," in *Modeling Biochemical Processes in Aquatic Ecosystems*, R. P. Canale, Ed. (Ann Arbor, MI: Ann Arbor Science Publishers, Inc., 1976), pp. 141-170.

DePinto, J. V., and F. H. Verhoff. "Nutrient Regeneration from Aerobic Decomposition of Green Algae," *Environ. Sci. Technol.* 11:371-377 (1977).

Elgmork, K. "Ecological Aspects of Diapause in Copepods," *Proc. Symposium Crustacea* 3:947-954 (1967).

Eppley, R. W., and J. L. Coatsworth. "Uptake of Nitrate and Nitrite by *Ditylum brightwellii*—Kinetics and Mechanisms," *J. Phycol.* 4:151-156 (1968).

Eppley, R. W., R. W. Holmes and J. D. H. Strickland. "Sinking Rates of Marine Phytoplankton Measured with a Fluorometer," *J. Exp. Mar. Biol. Ecol.* 1:191-208 (1967).

Eppley, R. W., and E. H. Renger. "Nitrogen Assimilation of an Oceanic Diatom in Nitrogen-Limited Continuous Culture," *J. Phycol.* 10:15-23 (1974).

Fogg, G. E. *Algal Cultures and Phytoplankton Ecology* (Madison, WI: University of Wisconsin Press, 1975).

Foree, E. G., and P. L. McCarty. "Anaerobic Decomposition of Algae," *Environ. Sci. Technol.* 4:842-849 (1970).

Fuhs, G. W. "Phosphorus Content and Rate of Growth in the Diatoms *Cyclotella nana* and *Thalassiorsira fluviatilis*," *J. Phycol.* 5:312-321 (1969).

Fuhs, G. W., S. D. Demmerle, E. Canelli and M. Chen. "Characterization of Phosphorus-Limited Plankton Algae (With Reflections on the Limiting Nutrient Concept)," in *Nutrients and Eutrophication*, G. Likens, Ed. (*Limnology and Oceanography*, Spec. Pub., 1972), pp. 113-132.

Gabrielsen, E. K. "Effects of Different Chlorophyll Concentrations on Photosynthesis in Foliage Leaves," *Physiol. Plantarum* 1:5-37 (1948).

Ganf, G. G., and P. Blazka. "Oxygen Uptake, Ammonia and Phosphate Excretion by Zooplankton of a Shallow Equatorial Lake (Lake George, Uganda)," *Limnol. Oceanog.* 19:313-325 (1974).

Gliwics, Z. M., and A. Hillbricht-Ilkowska. "Efficiency of the Utilization of Nannoplankton Primary Production by Communities of Filter-Feeding Animals Measured *in situ*," *Verh. Int. Verein. Limnol.* 18:197-203 (1972).

Grenney, W. J., D. A. Bella and H. C. Curl, Jr. "A Mathematical Model of the Nutrient Dynamics of Phytoplankton in a Nitrate-Limited Environment," *Biotech. Bioeng.* 15:331-358 (1973).

Guillard, R., P. Kilham and T. Jackson. "Kinetics of Silicon-Limited Growth in the Marine Diatom *Thalassiosira pseudonana* Hasle and Heimdal (= *Cyclotella nana Hustedt*)," *J. Phycol.* 9:233-237 (1973).

Healey, F. P. "Characteristics of Phosphorus Deficiency in *Anabaena*," *J. Phycol.* 9:383-394 (1973).

Hillbricht-Ilkowska, A., I. Spodniewska, T. Weglenska and A. Karabin. "The Variation of Some Ecological Efficiencies and Production Rates in Plankton Community of Several Lakes of Different Trophy," *Proceedings IBP-UNESCO Symposium on Productivity Problems of Freshwaters*, Kazimierz Dolny, Poland (June 12, 1970).

Hutchinson, G. E. *A Treatise on Limnology, II. Introduction to Lake Biology and the Limnoplankton* (New York: John Wiley and Sons, Inc., 1967).

Jassby, A. D., and T. Platt. "Mathematical Formulation of the Relationship Between Photosynthesis and Light for Phytoplankton," *Limnol. Oceanog.* 21:540-547 (1976).

Jones, J. G. "Studies on Freshwater Bacteria: Association with Algae and Alkaline Phosphatase Activity," *J. Ecol.* 69:59-75 (1972).

Jones, J. G. "The Microbiology and Decomposition of Seston in Open Water and Experimental Enclosures in a Productive Lake," *J. Ecol.* 64:241-278 (1976).

Jørgensen, E. G. "Adaptation to Different Light Intensities in the Diatom *Cyclotella meneghiniana* Kütz," *Physiol. Plantarum* 17:136-145 (1964).

Ketchum, B. H. "The Absorption of Phosphate and Nitrate by Illuminated Cultures of *Nitzschia closterium*," *J. Am. Bot.* 26:399-407 (1939).

Kilham, S. S. "Kinetics of Silicon-Limited Growth in the Freshwater Diatom *Asterionella formosa*," *J. Phycol.* 11:396-399 (1975).

Kilham, S. S., and P. Kilham. "Natural Community Bioassays: Predictions of Results Based on Nutrient Physiology and Competition," *Verh. Int. Verein. Limnol.* (in press).

Koonce, J. F. "Seasonal Succession of Phytoplankton and a Model of the Dynamics of Phytoplankton Growth and Nutrient Uptake," Ph.D. Thesis, University of Wisconsin, Madison, WI (1972).

LaRow, E. J., J. W. Wilkinson and D. K. Kumar. "The Effect of Food Concentration and Temperature on Respiration and Excretion in Herbivorous Zooplankton," *Verh. Int. Verein. Limnol.* 19:966-973 (1975).

Larsen, D. P., H. T. Mercier and K. W. Malueg. "Modeling Algal Growth Dynamics in Shagawa Lake, Minnesota, with Comments Concerning Projected Restoration of the Lake," in *Modeling the Eutrophication*

Process, E. J. Middlebrooks, D. H. Falkenborg and T. E. Maloney, Eds. (Ann Arbor, MI: Ann Arbor Science Publishers, Inc., 1974), pp. 15-32.

Larsson, P. Zoological Museum, Oslo, Norway. Personal communication (1977).

Lehman, J. T. "Photosynthetic Capacity and Luxury Uptake of Carbon During Phosphate Limitation in *Pediastrum duplex* (Chlorophyceae)," *J. Phycol.* 12:190-193 (1976).

MacIsaac, J. J., and R. C. Dugdale. "The Kinetics of Nitrate and Ammonia Uptake by Natural Populations of Marine Phytoplankton," *Deep-Sea Res.* 16:45-57 (1969).

Maddux, W. S., and R. F. Jones. "Some Interactions of Temperature, Light Intensity, and Nutrient Concentration During the Continuous Culture of *Nitzschia closterium* and *Tetraselmis* sp.," *Limnol. Oceanog.* 9:79-86 (1964).

Martin, J. H. "Phytoplankton-Zooplankton Relationships in Narragansett Bay III. Seasonal Changes in Zooplankton Excretion Rates in Relation to Phytoplankton Abundance," *Limnol. Oceanog.* 13:63-71 (1968).

Mills, A., and J. Alexander. "Microbial Decomposition of Species of Freshwater Planktonic Algae," *J. Environ. Quality* 3:423-428 (1974).

Mitchell, P., and J. Moyle. "Autolytic Release and Osmotic Properties of 'Protoplasts' from *Staphylococeus aureus*," *J. Gen. Microbiol.* 16:184-194 (1959).

Morris, I. "Nitrogen Assimilation and Protein Synthesis," in *Algal Physiology and Biochemistry*, W. D. F. Stewart, Ed. (Oxford: Blackwell Press, 1974), pp. 583-609.

Myers, J. "Culture Conditions and the Development of the Photosynthetic Mechanism. III. Influence of Light Intensity on Cellular Characteristics of *Chlorella*," *J. Gen. Physiol.* 26:419-427 (1946).

Myers, J., and W. A. Kratz. "Relations between Pigment Content and Photosynthetic Characteristics in a Blue-Green Alga," *J. Gen. Physiol.* 39:11-22 (1955).

Nielsen, E. S. "Inactivation of the Photochemical Mechanism in Photosynthesis as a Means to Protect the Cells Against Too High Light Intensities," *Physiol. Plantarum* 15:161-171 (1962).

Nielsen, E. S., V. K. Hansen and E. G. Jørgensen. "The Adaptation to Different Light Intensities in *Chlorella vulgaris* and the Time Dependence on Transfer to a New Light Intensity," *Physiol. Plantarum* 15:505-517 (1962).

Nielsen, E. S., and E. G. Jørgensen. "The Adaptation of Plankton Algae I. General Part," *Physiol. Plantarum* 21:401-413 (1968).

Nyholm, N. "Kinetic Studies of Phosphate-Limited Algal Growth," Ph.D. Thesis, Technical University of Denmark, Department of Applied Biochemistry (1975), in Danish.

Nyholm. N. "Kinetic Studies of Phosphate-Limited Algal Growth," *Biotech. Bioeng.* 14:467-492 (1977).

Omori, M. "Variations of Length, Weight, Respiratory Rate, and Chemical Composition of *Calanus cristatus* in Relation to Its Food and Feedings," in *Marine Food Chains*, J. H. Steele, Ed. (Berkeley: University of California Press, 1970), pp. 113-126.

Oppenheimer, C. H., and M. H. Vance. "Attachment of Bacteria to the Surfaces of Living and Dead Microorganisms in Marine Sediments," *Z. Allg. Mikro. Biol.* 1:47-52 (1960).

Otsuki, A., and T. Hanya. "Production of Dissolved Organic Matter from Dead Green Algal Cells II. Anaerobic Microbial Decomposition," *Limnol. Oceanog.* 17:258-264 (1972).

Park, R. A., R. V. O'Neill, J. A. Bloomfield, H. H, Shugart, R. S. Booth, R. A. Goldstein, J. B. Mankin, J. F. Koonce, D. Scavia, M. S. Adams, L. S. Clesceri, E. M. Colon, E. H. Dettmann, J. Hoopes, D. D. Huff, S. Katz, J. F. Kitchell, R. C. Kohberger, E. J. LaRow, D. C. McNaught, J. Peterson, J. Titus, P. R. Weiler, J. W. Wilkinson and C. S. Zahorcak. "A Generalized Model for Simulating Lake Ecosystmes," *Simulation* 23(2):33-50 (1974).

Park, R. A. "Ecological Modeling and Estimation of Stress," *2nd Joint US/USSR Symposium on the Comprehensive Analysis of the Environment* (Washington, DC: U.S. Environmental Protection Agency, 1975), pp. 119-126.

Park, R. A., D. Scavia and N. L. Clesceri. "CLEANER, The Lake George Model," in *Ecological Modeling in a Management Context*, C. S. Russell, Ed. (Washington, DC: Resources for the Future, Inc., 1975), pp. 49-81.

Pennington, W. "Seston and Sediment Formation in Five Lake District Lakes," *J. Ecol.* 62:215-251 (1974).

Rhee, G. "A Continuous Cuture Study of Phosphate Uptake, Growth Rate and Polyphosphate in *Scenedesmus* sp.," *J. Phycol.* 9:495-506 (1973).

Rhee, G. "Phosphate Uptake Under Nitrate Limitation by *Schenedesmus* sp. and Its Ecological Implications," *J. Phycol.* 10:470-475 (1974).

Rhee, G. "Effects of N:P Atomic Ratios and Nitrate Limitation on Algal Growth, Cell Composition, and Nitrate Uptake," *Limnol. Oceanog.* 23:10-25 (1978).

Rigler, R. H. "The Uptake and Release of Inorganic Phosphorous by *Daphnia magna* Straus," *Limnol. Oceanog.* 8'239-250 (1961).

Ryther, J. H., and D. W. Menzel. "Light Adaptation by Marine Phytoplankton," *Limnol. Oceanog.* 4:492-497 (1959).

Scavia, D., J. A. Bloomfield, J. S. Fisher, J. Nagy and R. A. Park. "Documentation of CLEANX: A Generalized Model for Simulating the Open-Water Ecosystems of Lakes," *Simulation* 23(2):51-56 (1974).

Scavia, D., and R. A Park. "Documentaiton of Selected Constructs and Parameter Values in the Aquatic Model CLEANER," *Ecol. Model.* 2:33-58 (1976).

Scherer, E. "Effects of Oxygen Depletion and of Carbon Dioxide Buildup on the Photic Behavior of the Walleye (*Stizostedion vitreum vitreium*)," *J. Fish. Res. Bd. Can.* 28:1303-1307 (1971).

Sheldon, R. B., and C. W. Boylen. "Factors Affecting the Contribution of Epiphytic Algae to the Productivity of an Oligotrophic Freshwater Lake," *Appl. Microbiol.* 30:657-667 (1975).

Smayda, J. J. "The Suspension and Sinking of Phytoplankton in the Sea," *Oceanog. Mar. Biol. Ann. Rev.* 8:353-414 (1970).

Smith, E. L. "Photosynthesis in Relation to Light and Carbon Dioxide," *Proc. Nat. Acad. Sci.* 22:504-511 (1936).

Smyly, W. J. P. "Bionomics of *Cyclops strenuus abyssorum* Sars (Copepoda:Cyclopoida)," *Oecologia* 11:163-186 (1973).

Sorokin, C., and R. W. Krauss. "The Effects of Light Intensity on the Growth Rates of Green Algae," *Plant Physiol.* 33:109-113 (1958).

Steele, J. H. "Environmental Control of Photosynthesis in the Sea," *Limnol. Oceanog.* 7:137-150 (1962).

Stefan, H., T. Skoglund and R. Megard. "Wind Control of Algae Growth in Eutrophic Lakes," *J. Environ. Eng. Div., Proc. Am. Soc. Civil Eng.* 102:1201-1213 (1976).

Straškraba, M. "Development of an Analytical Phytoplankton Model with Parameters Empirically Related to Dominant Controlling Variables," in *Umweltbiophysik*, R. Glaser, K. Unger, and M. Koch, Eds. (Berlin, G.D.R.: Akademic Verlag, 1976), pp. 33-65.

Straškraba, M., and J. Hrbáček. "Net-Plankton Cycle in Slapy Reservoir During 1959-1960," in *Hydrobiological Studies I.*, J. Hrbáček, Ed. (Prague: Academia Publishing House of the Czechoslovak Academy of Sciences, 1966), pp. 113-154.

Stross, R. G. "Photoperiod Control of Diapause in *Daphnia*. IV. Light and CO_2-Sensitive Phases Within the Cycle of Activation," *Biol. Bull.* 140:137-155 (1971).

Tilzer, M. "Dynamics and Productivity of Phytoplankton and Pelagic Bacteria in a High-Mountain Lake," *Arch. Hydrobiol.* 40(3):201-273 (1972), in German.

Titman, D. "Ecological Competition Between Algae: Experimental Confirmation of Resource-Based Competition Theory," *Science* 192:463-465 (1976).

Titman, D., and P. Kilham. "Sinking in Freshwater Phytoplankton: Some Ecological Implications of Cell Nutrient Status and Physical Mixing Processes," *Limnol. Oceanog.* 21(3):409-417 (1976).

Underhill, P. A. "Nitrate Uptake Kinetics and Clonal Variability in the Neritic Diatom *Biddulphia aurita*," *J. Phycol.* 13:170-179 (1977).

Ursin, E. Danish Institute for Fishery and Marine Research, Charlottenlund, Denmark. Personal communication (1976).

Youngberg, B. A. "Application of the Aquatic Model CLEANER to a Stratified Reservoir System," Master's Thesis, Rensselaer Polytechnic Institute, reprinted by Rensselaer Center for Ecological Modeling, Report 1 (1977).

Wierzbicka, M. "On Resting Stage and Mode of Life of Some Species of Cyclopoida," *Poll. Arch. Hydrobiol.* 10:215-229 (1962).

CHAPTER 6

THE USE OF ECOLOGICAL MODELS OF LAKES IN SYNTHESIZING AVAILABLE INFORMATION AND IDENTIFYING RESEARCH NEEDS*

Donald Scavia

U.S. Department of Commerce
National Oceanic and Atmospheric Administration
Environmental Research Laboratories
Great Lakes Environmental Research Laboratory
Ann Arbor, Michigan 48104

INTRODUCTION

The development of aquatic ecological modeling programs has been partially justified in the past as a means of identifying and setting research priorities for advancing our understanding of aquatic ecosystems (Chen, 1970; Park *et al.,* 1974; Canale *et al.,* 1976; Chen and Orlob, 1975; Bierman, 1976; Scavia *et al.,* 1976a). This purpose has often been lost, however, during the formidable tasks of model development, calibration, verification and application. Quite often modelers are so pleased with accomplishments (or so discouraged with failures) that major deficiencies in available information that were uncovered throughout analyses go unreported. Even where research needs have been identified, they usually are buried in reams of model jargon and simulation output. The purpose of this chapter is to focus specifically on research areas that, at present, represent blocks to further model development.

The models reviewed in this chapter are restricted mainly to those

---

*GLERL Contribution No. 132

developed over the past decade that address the trophic ecology of the aquatic system (*i.e.*, models simulating food chains or webs and nutrient cycles). This restricted review eliminates several categories of models that focus mainly on dissolved oxygen, biological oxygen demand (BOD) and certain specific nutrients alone and that have been developed generally for application in management contexts. The emphasis here is on non-linear models because in this paper the interactions between experimental results and model studies are emphasized and most experimental work has led to nonlinear process equations. Most of the models included have been developed and used for freshwater environments. The emphasis is also oriented somewhat toward the Great Lakes, although much of the process work has been marine.

One might ask why an ecological model can be useful in identifying weak points in our knowledge of aquatic systems. Steele (1974) addressed this point succinctly in describing his model:

> It does not in any sense produce new facts, but merely permits the evaluation of laboratory experiments carried out on different components in isolation. By forcing one to produce formulas to define each process and put numbers to the coefficients, it reveals the lacunae in one's knowledge. Although the output of the model can be tested against existing field observations and experimental results, the main aim is to determine where the model breaks down and use it to suggest further field or experimental work.

Riley (1946) introduced this concept of assuming mathematical formulations for individual biological processes, and then synthesizing the formulations in a model. He suggested:

> . . . developing the mathematical relationships on theoretical grounds and then testing them statistically by applying them to observed cases of growth in the natural environment. At present this can be done only tentatively, with over-simplification of theory and without the preciseness of mathematical treatment that might be desired. It is not expected that any marine biologist, including the writer, would fully believe all the arbitrary assumptions that will be introduced. However, the purpose of the paper is not to arrive at exact results but rather to describe promising techniques that warrant further study and development.

Mortimer (1975) also has recently emphasized the "two-pronged approach" to systems studies, which involves an interaction between modeling efforts and experimental investigations.

The rationale behind this use of models is based on two assumptions: (1) that one can actually represent isolated biological and chemical processes or events with mathematical equations (process constructs) and

(2) that combining these expressions results in a whole system model that represents the actual structure of the real world. The quantification of processes by Riley *et al.,* (1949) and the formulation of more recent process constructs indicate the general acceptance of the first assumption (*e.g.,* for nutrient uptake and phytoplankton growth—Munk and Riley, 1952; Droop, 1968; Eppley and Thomas, 1969; Fuhs, 1969; Dugdale, 1975, 1977; for photosynthetic light limitation—Steele, 1962; Vollenweider, 1965; Jassby and Platt, 1976a,b; Platt *et al.,* 1977; for zooplankton and fish processes—McAllister, 1970; Mullin *et al.,* 1975; Steel and Mullin, 1977; Ivlev 1945; Ursin, 1967). The validity of the second assumption has only begun to be investigated for ecological models. To prove that combinations of process equations can indeed simulate the dynamics of ecological systems, model output must be compared to data from the real world. This has been done for a wide range of models differing in orientation, complexity and utility. One example, consisting of one phytoplankton group, one zooplankton group and total inorganic nitrogen, is the model of DiToro *et al.* (1971), which was applied to a river system to assess nutrient loads for management purposes. Since that time, other models have been developed that simulate variations in several phytoplankton and zooplankton groups as well as cycles of nitrogen, phosphorus, silicon, carbon and oxygen in lakes for research purposes. There also exists a spectrum of models, both engineering- and research-oriented, with varying degrees of complexity. A number of recent examples are identified in Table I. Most of these models have been reasonably successful in describing the dynamics of the systems for which they were built and therefore lend credence to the second assumption; however, final acceptance of this assumption can be brought about only by continued success in modeling efforts. It is important to note here that validation of both assumptions is not a sufficient criterion for model application for predictive purposes. This validation only demonstrates the model's capability of describing (*i.e.,* simulating) important biological and chemical processes simultaneously. The verification procedures necessary before proceeding with predictions are far more vigorous (Orlob, 1975; O'Connor *et al.,* 1975) and not the subject of this chapter.

If one accepts the two assumptions, then the models can be considered to be mathematical representations of existing knowledge concerning the modeled processes and interactions. Therefore, systematic investigation of the model can be used to identify obstacles to developing better models and to obtain a clearer understanding of the system and thus to determine specific research areas needing further study.

In the following sections, seven components of the aquatic system will be discussed with respect to current modeling concepts and supporting

Table I. Comparison of Model Complexity

Reference	System	Primary Producers	Herbivores	Carnivores	Nutrient Cycles	Physical Segments	Compartments per Segment
DiToro et al., 1971	San Joaquin River	1	1	0	1	1	3
Bloomfield et al., 1973	Lake Wingra, WI	2	2	2	0	1	8
Larson et al., 1973	Shagawa Lake, MN	1	0	0	2	1	3
Thomann et al., 1974	Potomac Estuary	1	1	0	3	38	9
MacCormick et al., 1974	Lake Wingra, WI	1	1	2	0	1	8
Park et al., 1974	Lake George, NY	2	4	4	1	1	28
Scavia et al., 1974	Lake George, NY	2	2	3	1	1	11
Steele, 1974	North Sea	1	1	0	1	1	4
Thomann et al., 1975	Lake Ontario	1	1	3	2	2	10
Chen and Orlob, 1975	Lake Washington	2	1	0	3	33	17
Chen and Orlob, 1975	San Francisco Bay	2	1	0	3	67	17
DiToro et al., 1975	Lake Erie	1	1	0	2	7	7
Lehman et al., 1975	Linsley Pond, CT	4	0	0	2	1	6
Thoman and Winfield, 1976	Lake Ontario	1	1	3	2	67	10
Bierman, 1976	Saginaw Bay, Lake Huron	3	2	1	3	1	14
Canale et al., 1976	Lake Michigan	4	6	3	3	2	25
DePinto et al., 1976	Stone Lake, MI	3	2	1	2	1	11
DiToro and Matystick, 1976	Saginaw Bay, Lake Huron	1	1	2	2	5	9
Scavia et al., 1976a	Lake Ontario	4	5	1	3	3	17
McNaught and Scavia, 1976	Lakes Ontario and Michigan	2	3	0	0	1	6
Steele and Mullin, 1977	None	1	1	0	1	12	3
Walsh, 1977	Several Marine Systems	1	2	0	3	?	9
Scavia, 1978a,b	Lake Ontario	5	6	1	4	2	20

experimental evidence: nutrients, phytoplankton, zooplankton, fish, bacteria, sediments and hydrodynamics. The discussion of each component will set the stage for consideration of the research needed before further model development, and thus system understanding, can be accomplished.

Since the purpose of this chapter is to indicate research needs by outlining model developments in terms of available information, a thorough review of each specific field will not be given. Where possible, previous reviews will be cited, with only specific pertinent details discussed here. At the conclusion, a summary of research needs and a hierarchical priority system will be suggested to ensure that research on the important problems progresses evenly on all fronts.

NUTRIENTS

The importance of simulating nutrient cycles as an integral part of ecosystem models rather than inputting specified ambient nutrient concentrations was realized early in the development of model building. Nutrient simulation is necessary because nutrients are the components controlling much of the system behavior, and specifying future nutrient conditions is extremely difficult.

The first major question concerning the inclusion of nutrients in a model is: How many nutrients should be included and which are most important? Modelers have included as few as one nutrient (e.g., nitrogen—DiToro et al., 1971) and as many as four nutrients (e.g., carbon, nitrogen, phosphorus and silica—Scavia, 1978a) in simulation models. If the model is to be used to investigate the control of phytoplankton production under varying environmental conditions, then cycles of all the potentially limiting nutrients must be included, because under future conditions it is not always clear which nutrients will control phytoplankton growth. If, however, one is interested in the effects of a particular nutrient on the system, it is not necessary to include calculations for other nutrients.

One important aspect of nutrient-phytoplankton dynamics and control of the process of eutrophication is nutrient availability. Nutrients used by phytoplankton are supplied to lakes in forms ranging from largely undegraded terrestrial litter, which undergoes decomposition releasing nutrients, to dissolved inorganic ions from rainwater, which are immediately available for algal assimilation. Upon reaching the lake, these elements enter complex cycles. The relative amounts of material in the compartments of the nutrient cycles and the control on the rates of transfer among the compartments are not completely understood. More research, as outlined below, is needed to better identify the nature of the cycles and quantify the availability of various nutrient forms to the phytoplankton.

Phosphorus

Some recent works (Herbes, 1974; Cowen and Lee, 1976) have indicated that certain fractions of soluble unreactive phosphorus (SUP) are available for phytoplankton growth. It generally is assumed that this availability is of little consequence when supplies of soluble reactive phosphorus (SRP) are plentiful. However, under conditions of SRP depletion the ratio SUP: SRP may become much greater than 1.0. This can be readily seen by making simple calculations with the Lake Ontario data in Stadelman and Fraser (1974). It is, therefore, important to determine the ability of phytoplankton to use various components of SUP. The significance of the ability of phytoplankton to use SUP would take on most meaning if this ability, or utilization rate, was determined and compared to the transfer rate of SUP to SRP. Results from such studies would help determine whether, in models, the two forms of phosphorus should be (1) combined, (2) separated with preferential uptake constants or (3) separated and SUP made unavailable. Canale *et al.* (1976), Thomann *et al.* (1975) and Bierman (1976) use the third assumption. Scavia *et al.* (1976a) model SRP but assume the SUP pool is relatively constant and that transfer through it is fast; therefore, SUP is not modeled explicitly.

Other phosphorus components exist in the environment; yet less work has been done to identify their composition and to determine their input and output transfer rates. Recent investigations (Lean, 1973; Halfon, Chapter 4) have shown that compartments other than SUP and SRP also may be necessary for describing the phosphorus cycle adequately. These studies identify organic phosphorus compartments other than sestonic as being important. More study is also needed to measure the amount of phosphate adsorbed to particles and the interactions of physical and biological desorption rates versus particle sedimentation.

Nitrogen

Most phytoplankton models include separate compartments for ammonium (AM) and nitrate plus nitrite (NI), and use various constructs to simulate AM preference over NI. (This preference has been demonstrated experimentally by Harvey, 1955; Walsh and Dugdale, 1972; Bates, 1976). Each model contains a construct that represents the amount of ammonium taken up during phytoplankton growth as: $\alpha * k * PROD$, where PROD is primary production (*e.g.*, mgC/mgC/day) and k is ratio N:C in algal cells. The functional form of α, the preference factor, varies in different models:

$$\alpha = \frac{AM}{AM + NI} \qquad \text{(Thomman et al., 1975)} \qquad \text{(1a)}$$

$$\alpha = \frac{\beta(AM)}{\beta(AM) + NI} \qquad \text{(Scavia } et\ al.,\ 1976a) \qquad (1b)$$

$$\alpha = \frac{AM}{\beta + AM} \qquad \text{(DiToro } et\ al.,\ 1975) \qquad (1c)$$

$$\alpha = \frac{\beta(AM)}{\beta(AM) + (1 - \beta)\ NI} \qquad \text{(Canale } et\ al.,\ 1976) \qquad (1d)$$

where β = an empirical constant

Further investigation into this preferential uptake will allow a more unified formulation for this process and a better estimation of appropriate coefficients.

This preference of phytoplankton for ammonia rather than nitrate as a nitrogen source is complicated by nitrifying bacteria. Competition between bacteria and phytoplankton for ammonium may be significant in controlling phytoplankton growth in some circumstances. In these cases, it may become necessary to include explicit formulations for bacterial kinetics, or at least their effects, in models, and therefore additional studies will be necessary to measure simultaneously rates of ammonium assimilation by phytoplankton and rates of nitrification by bacteria.

The use of dissolved molecular nitrogen (nitrogen fixation) by N-fixing phytoplankton or bacteria can be important in eutrophic environments, and it can be a significant factor in the nitrogen budget of a lake (Horne and Goldman, 1972; Vanderhoef et al., 1974). The organisms most often associated with this phenomenon, filamentous blue-green algae with heterocysts, typically form nuisance-blooms in eutrophic lakes. They also serve as "bottle-necks" in the food web because most of these species are inedible and generally have low sinking rates, thus remaining in the eplimnion for a longer period of time. In models addressing the ecology of aquatic systems, especially in models involving eutrophication processes, N-fixing algae may be necessary; however, they often have not been included.

When nitrogen fixation has been included in models, it has been described either as an "on-off" process (DePinto et al., 1976) or as a continuous function of ambient ammonia and nitrate concentrations (Scavia, 1978a). It is probably best described as a switch at the onset of fixation (mimicking the beginning of nitrogenase production) and some continuous function thereafter, related to both ambient nitrogen concentrations and the recent history of the organism. This nitrogen fixation construct also should allow simultaneous uptake of ammonia, nitrate and molecular nitrogen as demonstrated by Dugdale and Dugdale (1965). High phosphorus concentrations are required to promote nitrogen fixation

(Vanderhoef *et al.,* 1974), and it generally, but not always (Brezonik, 1972) occurs when ammonia and nitrate reach limiting levels.

Carbon

Models usually assume that carbon cannot limit algal growth, although some models (Chen, 1970; Chen and Orlob, 1975) do allow for potential limitation by including Michaelis-Menten expressions for carbon. Recent studies dealing with carbon-limited algal cultures raise questions regarding the form of carbon needed for photosynthesis by different algae. Goldman *et al.* (1974) contend that two species of green algae use total inorganic carbon, while King and Novak (1974), using the same data, conclude that those plankters respond to dissolved CO_2 and H_2CO_3 only. It has also been suggested that the relative abilities of green and blue-green algae to use various forms of inorganic carbon could partially explain the succession from greens to blue-greens in enriched natural waters (King, 1972; Shapiro, 1973). These as well as other developments (see references in Goldman *et al.,* 1974) indicate that control of primary production by carbon limitation could be important in certain lakes. If this is true, then appropriate mechanisms must be included in multiple-group phytoplankton models, and this will require adding carbonate species calculations (*e.g.,* Chen and Orlob 1975; DiToro, 1976; Scavia *et al.,* 1976a). Further study of the control of competition by carbon limitation is necessary before these processes can be modeled explicitly.

Carbon is also a convenient element upon which to base food web dynamics (*cf.* Robertson and Scavia, Chapter 11). Models following the flow of carbon in this way are amenable to having oxygen calculations included. For example, the rates of decay of detrital carbon, respiration rates and production rates can all be related to oxygen by standard stoichiometric constants.

The importance of rate process measurements as part of calibration and verification data sets becomes apparent in terms of carbon and oxygen calculations. Even though there are questions pertaining to the interpretation of measurements of ^{14}C primary production (cf. Strickland, 1960; Schindler, 1972; Stadelman *et al.,* 1974) and BOD, such measurements can be useful for verifying the adequacy of model constructs and coefficients. The fact that these measurements are relatively easy to make may compensate for the uncertainties associated with their estimation. Further, BOD measurements approximate community respiration rates, but it also would be useful to determine the respiration rates of phytoplankton, zooplankton and bacteria separately.

Silicon

Diatoms use dissolved reactive silicon (mainly as $Si(OH)_4$) to build a frustule that surrounds the cell. Silicon in frustules is not available to other diatoms and therefore ambient concentrations of available silicon can become low enough to limit further growth of these algae. In a series of papers concerned with silicon dynamics in Lake Michigan, Conway et al. (1977) and Parker et al. (1977a,b) demonstrated that 80-100% of the silica in the water column is recycled annually. It appears that even though most of the silicon is incorporated into diatom frustules, a large fraction of it is recycled. This recycling is accomplished primarily through the effects of zooplankton grazing (i.e., the breaking of frustules that assists dissolution), although evidence also exists indicating that appreciable amounts of silicon may be dissolving from actively growing diatoms (Nelson et al., 1976). The siliceous frustules, alone or in fecal pellets, eventually are deposited in the sediments, although quite often they are partially dissolved by the time they reach the bottom (Ferrante and Parker, 1977). Once in the surficial sediments, the frustules undergo resuspension and additional dissolution (Parker et al., 1977a). Little is known quantitatively about other mechanisms of silicon cycling in lakes; more work is needed to identify and quantify the silicon cycle because this element undoubtedly plays an important role in the seasonal and long-term succession of phytoplankton in certain lakes (Kilham, 1971; Tarapchak and Stoermer, 1976; Schelske and Stoermer, 1972).

Nutrient Recycling

At least two modeling studies on large lake systems (Thomann et al., 1975; Scavia and Park, 1976) have suggested that seasonal nutrient dynamics are almost completely controlled by nutrient cycling rather than by external loads. (Loads do play a larger role in many smaller lakes.) Recycling has been suggested as the major reason why large lakes do not respond immediately to alterations in inputs. Thus, it is important to model nutrient cycles with special emphasis on recycling mechanisms and rates. Most ecosystem models include at least some elements of the recycling process; however, more research on the specific mechanisms is needed. For example, it is interesting to note that in the models of Thomann et al. (1975) and Scavia et al. (1976a), recycling through plant and animal excretion is more important than decomposer remineralization. Scavia and Park (1976), however, show such remineralization to be more important. In the latter model, decomposers are modeled explicitly, whereas in the other models remineralization is described as simple first-order decay. Further studies into the feasibility and importance of explicit decomposer submodels are discussed below and by DePinto (Chapter 2).

Most models couple inorganic portions of nutrient cycles to biological portions (*i.e.*, phytoplankton, zooplankton, decomposers, fish and detritus) with constant stoichiometric ratios. These models are usually based on the relative atomic proportions of C:N:P for phytoplankton established by Redfield *et al.* (1963) at 106:16:1 and then substantiated by many authors as a general average ratio. This ratio, however, varies from species to species (Strickland, 1960) and as a function of environmental conditions, as discussed by Parsons and Takahashi (1973) and Schelske (1975). Therefore, some recent models allow for luxury consumption of nutrients (*i.e.*, simulation of internal phytoplankton nutrient stores as well as external nutrient concentrations), one of the potential causes of variable nutrient ratios. This mechanism relaxes the dependence of phytoplankton growth on ambient nutrient concentrations and may have serious implications in terms of nutrient cycling (see Chapter 2). The implications of luxury consumption for phytoplankton dynamics are discussed in more detail below.

Recently, several field and laboratory studies have addressed individual processes affecting nutrient recycling within the water column [see Ferrante, 1976, and references therein concerning zooplankton excretion; see Foree *et al.* (1970), Bell *et al.* (1974), Nelson *et al.* (1976), Nalewajko and Schindler (1976), Nalewajko and Lean (1972) concerning phytoplankton release and phytoplankton-bacteria interactions]. Such studies are critical to understanding nutrient cycles, and there is a special need for work in the following areas: (1) the release of nutrients by both actively growing and senescent phytoplankton; (2) the causes of cell lysis and the magnitude of nutrient releases; (3) the bacteria-mediated transformations from particulate organic matter to dissolved inorganic nutrients through dissolved organic phases, and (4) the nature of the nutrients excreted by animals. In studies in these areas, it would be very valuable to determine the nutrient release rates as functions of other metabolic processes [*e.g.*, nitrogen and phosphorus excretion as a function of oxygen uptake in zooplankton (Ganf and Blazka, 1974)] in order to develop mechanistic constructs. Also, synthesis papers organizing these studies into more complete cycles [*e.g.*, Brezonik (1972) for nitrogen; Kerr *et al.* (1972) for carbon and Rigler (1973) for phosphorus] are necessary to determine the relative significance of the individual compartments and transformations.

Both modelers [Welch *et al.* (1973), Larson *et al.* (1973), Scavia *et al.* (1976b), Lam and Jaquet (1976), Scavia and Chapra (1977)] and experimentalists [see Mortimer's 1941 and 1942 classic investigations and Lee (1970) for literature review] have cited interactions between nutrients in the water column and those in the sedimentary interstitial water as critical to a description of the control of the whole system. The effects of anoxia on phosphorus release from sediments have been shown to be important in both the laboratory (see Mortimer, 1971) and the field (Burns and Ross, 1972). Mechanistic submodels of this process, amenable to incorporation in ecosystem models, have not been developed as yet. For that reason, this very important component of the system is discussed in greater detail below.

PHYTOPLANKTON

Functional Groups

The development of phytoplankton models has proceeded along two lines. The first group of models simulate phytoplankton as one group represented by chlorophyll *a* (DiToro *et al.*, 1971; Thomann *et al.*, 1975; DiToro *et al.*, 1975; Larsen *et al.*, 1973; Canale *et al.*, 1974; DiToro and Matystik, 1976). The second group simulates several phytoplankton groups, differentiating the phytoplankton according to size (Park *et al.*, 1974; Bloomfield *et al.*, 1973), physiology (Bierman *et al.*, 1973; Bierman, 1976; DePinto *et al.*, 1976), or, more recently, size and physiology (Scavia *et al.* 1976a; Canale *et al.*, 1976; Scavia, 1978a).

In the past, the major reasons for representing all phytoplankton as a single group were that quantitative estimates of the species composition of the phytoplankton communities were often not available; and that chlorophyll *a*, which was more commonly measured routinely, was considered to be a good indicator of total phytoplankton biomass. Also, the context of many modeling works (*e.g.*, DiToro *et al.*, 1971; Thomann *et al.*, 1975; O'Connor *et al.*, 1975) was such that the interest lay primarily with gross parameters like chlorophyll *a*.

There are two basic reasons for dividing the phytoplankton community into two or more groups. One reason is that questions have been raised recently concerning the reliability of chlorophyll *a* as a measure of total algal biomass (*cf.* Dolan *et al.*, in press). More important though is that certain problems by their very nature dictate the separation of the phytoplankton. For example, the dominance of particular algal species causes taste and/or odor problems in drinking water supplies. Other algae form surface mats and cause filter clogging as well as aesthetic problems. To use models to address problems like these, one must include more than one algal group in order to simulate or predict the dominance of the particular algal type of interest.

Mortimer (1975) has also discussed the importance of allowing "new actors to come onto the stage" when conditions in a lake being modeled change, and Scavia *et al.* (1976b) have demonstrated in a modeling exercise for the Great Lakes that, in some circumstances, it is probably important to allow both long-term and seasonal succession of phytoplankton groups to occur in predictive models. To do this, multiple group phytoplankton (MGP) models must be implemented. Fortunately, many recent integrated research programs, such as the International Biological Program (IBP) and the International Field Year for the Great Lakes (IFYGL), have included detailed analyses of phytoplankton succession, thus allowing modelers the opportunity to develop and calibrate MGP models.

Physiology

A major concern in many current modeling efforts is the development of criteria for segregating the phytoplankton community. The most obvious distinction from a physiological standpoint is the difference in nutrient requirements of certain phytoplankton groups. All phytoplankters require carbon, nitrogen and phosphorus, whereas only diatoms require silicon to any significant degree. Certain species of blue-green algae are able to use dissolved nitrogen gas (nitrogen-fixation) and are, therefore, not limited to ammonium and nitrate as nitrogen sources. So, an initial segregation could lead to these subdivisions: (1) diatoms, (2) nitrogen-fixing blue-green algae and (3) all other algae. These subdivisions have also been useful categories to group species based on other growth-related functions (*e.g.*, light requirements—Ryther, 1956; specific growth rates and temperature effects—Canale and Vogel, 1974). The metabolic requirements associated with motility in flagellates (Parsons and Takahashi, 1973), as well as their special attributes with regard to searching for optimal light and nutrients and their defenses against sinking, may make it necessary to treat this group separately.

Size

A second way to distinguish groups is the use of phytoplankton size. Algal size is generally a determining factor in causing filter clogging and similar problems at water intakes. Some modeling studies, using this distinction, have attempted to separate two size categories based on zooplankton feeding preferences. Some work has been done in an effort to quantify this selectivity by zooplankton, and the results will be described in more detail with respect to zooplankton dynamics. There are, however, other reasons for size distinctions. In a review of the suspension of marine phytoplankton, Smayda (1970) provides a thorough discussion of the effects effects of size and shape on sinking rates. Generally, larger cells sink faster than smaller ones. The effects of cell surface area to volume ratios (S:V) on sinking rates also follow a general pattern, which is inversely related to the effect of cell diameter (Smayda, 1970). There are, of course, many complicating factors, such as colony formation, cell age or physiological condition, "health," protuberances, motility and a host of other factors that affect sinking rates (Hutchinson, 1967).

Several recent works have explored the effects of cell size on photosynthesis and respiration. Banse (1976), in a review of previously published data, reports that weight-specific growth, respiration and gross photosynthesis rates are lower for larger species (using cellular carbon, volume and S:V ratios as measure of cell size). He also indicates that chemical composition and many biochemical processes are size-dependent, as might be expected from general size dependencies in other plants and animals. An inverse relationship between either photosynthetic activity or growth rate and cell size has also been shown by others (Eppley *et al.*, 1969; Gutel'makher, 1974; and references cited in both). Eppley *et al.* (1969), for example, has shown a direct relationship between cell sizes and half-saturation constants for nitrogen uptake. Taguchi (1976) presents similar direct relationships between cell size and both photosynthesis and nutrient uptake by marine diatoms, and he relates these processes to S:V ratios.

The idea that phytoplankton succession may be partially controlled by S:V ratios becomes reasonable when one considers even the few size-related processes described above. In fact, Lewis (1976) has suggested that one might expect very similar S:V ratios among autotrophs found in a specific environment. S:V ratios may be the best criterion for subdividing the primary physiological groups for modeling purposes; more studies exploring the relation of S:V ratios to phytoplankton succession are needed.

By segmenting the phytoplankton community into several groups, the problem of coefficient estimation becomes formidable. One can no longer use an average value of the range of values reported in the literature for phytoplankton in general because, in MGP models, distinctions within the phytoplankton community become critical. Some of the coefficients that vary among the physiological groups have been discussed above; others have not yet been as thoroughly studied. Coefficients that seem to be functions of cell size include: (1) maximum growth rates, (2) half-saturation constants for nutrient uptake and phytoplankton growth, (3) maximum respiration rates, and (4) sinking rates. Synthesis work similar to that of Smayda (1970) and Banse (1976) should be done to quantify relationships between these coefficients and cell size (or S:V ratio). This particular form of quantification is not new [Zeuthen (1947, 1953) summarized size-specific process rates in animals], but it has seldom been done for algae. Although these relationships classify rates in a continuum rather than as separate functional groups, they can be used to evaluate average size-dependent rates within given size ranges.

It is important to note that in aggregating rates to determine specific group functions one should use an expression corresponding to an envelope encompassing the individual rate functions rather than an expression for the average values. Under a given set of environmental conditions, the species within a group best suited to those conditions, rather than the average species, will dominate (Steele, 1974, p. 62).

Process Constructs

Uptake/Growth

The actual formulations used to describe various phytoplankton processes have evolved substantially since the first phytoplankton models. Some constructs have remained the same over the past 10 years, but others have been refined continuously. Dugdale (1977) and Platt et al. (1977) provide reviews with respect to nutrient kinetics and light limitation. Although the effects of temperature on various processes were modeled linearly at first (DiToro et al., 1971), most models now consider this relationship to be exponential. Only a few models consider rate reductions above some optimum temperature for a given group of organisms (Bloomfield et al., 1973; Lassiter and Kearns, 1973; Scavia et al., 1976a). Except for a few cases (e.g., Jassby and Platt, 1976a), most investigators use a depth- and time-integrated form of Steele's (1965) formulation for light-limited photosynthesis, the derivation of which is shown in DiToro et al. (1971). Park et al. (Chapter 5) discuss adaptation formulations for both light and temperature.

Most investigators use a Michaelis-Menten formulation to relate phytoplankton growth (or nutrient uptake) rate, μ, to a single external nutrient concentration, N, and the maximum rate, μ_{max}:

$$R = \frac{\mu}{\mu_{max}} = \frac{N}{K+N} \qquad (2)$$

where   K = the half-saturation constant
        R = the limitation term

The effect of multiple nutrient limitation on phytoplankton growth, however, has not been modeled consistently. Several constructs have been proposed to link the various individual nutrient reduction terms, R_i:

$$R = R_1 \cdot R_2 \cdot R_3 \cdot \ldots R_i \qquad \text{(Chen, 1970; DiToro et al., 1971; Thomann et al., 1975)} \qquad (3)$$

$$R = \text{Min} (R_1, R_2, \ldots R_i) \qquad \text{(Larsen et al., 1973; Scavia et al., 1976a, Bierman, 1976)} \qquad (4)$$

$$R = \frac{n}{\sum_{i=1}^{n} (1/R_i)} \qquad \text{(Bloomfield et al., 1973; Park et al., 1974)} \qquad (5)$$

$$R = \frac{1}{n} \sum_{i=1}^{n} R_i \qquad \text{(Patten } et\ al.,\ 1975) \qquad (6)$$

The first three constructs produce similar values of R if one nutrient is significantly more limiting than others and the number of potentially limiting nutrients is low. Large discrepancies arise when several nutrient concentrations are close to limiting levels. For example, given four individual nutrient reduction terms:

$$R_1 = 0.8,\ R_2 = 0.7,\ R_3 = 0.6,\ \text{and}\ R_4 = 0.5$$

the overall reductions calculated from Equations 3 through 6 are 0.17, 0.50, 0.63 and 0.65, respectively. These large discrepancies in R may result in different standing crop estimates, depending on the timing of other processes controlling phytoplankton dynamics.

DiToro et al. (1971), using phosphorus uptake rates as a function of phosphorus and nitrogen concentrations (Ketchum, 1939), produced a good fit to uptake rates with Equation 3. Rhee (1978), however, found Equation 4 to be better than Equation 3 in relating the growth rate of Scenedesmus spp. to intracellular concentrations of nitrogen and phosphorus. (The question of relating growth rates, uptake rates, intracellular nutrients and extracellular nutrients will be discussed below.) More work is needed to determine the best construct for describing multiple nutrient limitation because both of the above studies were conducted with individual species, whereas many recent models include groups of species. One must use a construct that describes competition *among* the groups modeled, while allowing implicit adaptation and/or succession of species *within* each group.

Respiration

Respiration is an important loss term in phytoplankton calculations. Also, in models that do not include luxury consumption of nutrients, mass conservation is assured by relating nutrient excretion (release of dissolved organic matter and inorganic nutrient) stoichiometrically to respiration rates. Since both phytoplankton calculations and nutrient cycle calculations are dependent on respiration, it is important to use a proper formulation for this process.

Phytoplankton respiration usually is modeled as a function of temperature only (DiToro et al., 1971; Thomann et al., 1975; Canale et al., 1976; DiToro et al., 1975; Scavia et al., 1976a). Nihei et al. (1954) have shown that respiration rates also are a function of the physiological state of the algae. Platt et al. (1977) also cite studies with higher plants indicating that respiration is a function of photosynthesis (or light), and they suggest that this should be true for phytoplankton as well. More work is needed to quantify relationships between respiration, excretion, photosynthesis and expressions of physiological state.

Sinking

The sinking rates of phytoplankters have been discussed above with respect to cell size; however, environmental conditions also affect these rates (Smayda, 1970; Smayda and Boleyn, 1966a, b; Smayda, 1974). Models have considered sinking rates to be: (1) constant (Chen and Orlob, 1975; Thomann et al., 1975), (2) a linear function of temperature (Scavia and Park, 1976), (3) a function of the water column temperature gradient (Bloomfield et al., 1973), (4) the values measured in situ (Canale et al., 1976), or (5) functions of temperature and cell "health" (Scavia et al., 1976a).

Sinking rates can be estimated with knowledge of the particle size and shape and the effect of temperature on the density and viscosity of water (Hutchinson, 1967; Smayda, 1974). The quantitative effect of physiological state on sinking, however, is not well known. Smayda (1974) suggests that physiological state can influence the production of gas vacuoles in blue-green algae, the ionic charge of the cells, the production of gelatinous sheaths and other mechanisms affecting phytoplankton buoyancy. Titman and Kilham (1976) demonstrated the importance of physiological conditions on sinking rates in algae. They found a fourfold increase in sinking rates of four diatoms in the stationary growth phase over the sinking rates of the same populations in the exponential growth phase. Changes in sinking rates also were observed as the culture medium varied from nutrient-poor to nutrient-rich.

Scavia et al. (1976a) parameterized physiological effects on sinking rates by relating these rates to the overall growth-limitation term, R (Equation 4). The best test data available at the time were in situ sinking rates of particulate organic carbon (Burns and Pashley, 1974) that did not discriminate among algal types. Due to the restricted suitability of these data and also because sinking rates estimated from quiescent laboratory studies are of questionable pertinence (Titman and Kilham, 1976), the true test of this algorithm must await more quantitative and realistic studies. Work should be done to determine sinking rates and the factors controlling these rates under conditions found in the field.

Luxury Consumption

In the above discussion of several constructs, reference was made to modeling luxury consumption of nutrients by phytoplankton, that is the process by which phytoplankton assimilate nutrients and create internal stores for future use. The phenomena often cited in support of including luxury consumption, or internal pool, models are: (1) a time lag between the decrease in ambient nutrient concentrations and phytoplankton growth, (2) significant phytoplankton growth during severe nutrient depletion, and (3) seasonal succession caused by competition for available nutrients. The underlying support for the importance of luxury consumption comes from a rather large body of experimental evidence (e.g., references cited in Bierman, 1976) indicating that nutrient uptake and phytoplankton growth are not always in steady-state; that is, the sizes of nutrient pools in phytoplankton are, in fact, variable.

The first phenomenon, the time lag, has been measured as well as simulated to be about seven days in phosphorus kinetic studies (Koonce, 1972). Bierman (1974) ran hypothetical competition simulations and observed a similar time lag between phosphorus removal and the growth of green and blue-green algae. Unfortunately, field data rarely are obtained on a timeframe suitable for observing short-term lags in lakes.

There is a question whether or not storage due to luxury consumption is the main source of nutrients that permit significant growth during periods of nutrient depletion. In typical Lake Ontario model simulations (Scavia, unpublished) that match 1972 data (Stadelman and Fraser, 1974; Stoermer et al., 1975), phytoplankton populations are moderately high during the period July-September, while simulated phosphate concentrations remain in the range of 1 to 2 μg-P l^{-1}. During this time, the primary production rates are sufficient to combat simulated losses due to respiration, sinking, natural mortality and grazing (Figure 1). One might expect that with these low nutrient concentrations, the phytoplankton productivity would rapidly drive the concentrations even lower so that they become severely limiting to further algal growth. While phosphorus is controlling growth, the algae are not severely limited because long-term limitation is not controlled solely by instantaneous nutrient concentration, but also by the supply of nutrients to the system (recycled as well as externally supplied). This control of growth by the supply of nutrients, rather than by ambient concentrations, also has been emphasized in discussions on eutrophication control (Shapiro, 1970).

Finally, with regard to the third phenomenon, although experimental studies have indicated luxury consumption may influence phytoplankton succession, the need for explicit functions of luxury consumption in models has not been demonstrated. Koonce's (1972) constructs for luxury consumption were incorporated in an ecosystem model for Lake Wingra, Wisconsin (Huff et al., 1973), which was capable of simulating the dynamics of some of the major ecological compartments of the system. However, another model for the same lake that did not include luxury uptake

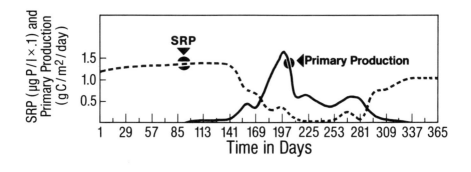

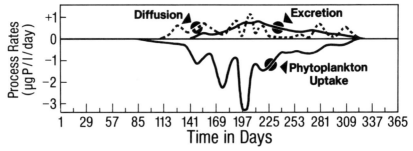

Figure 1. Simulated seasonal dynamics of soluble reactive phosphorus (SRP) and primary production (top); and contributions to SRP pool (bottom) in Lake Ontario.

(Bloomfield *et al.*, 1973) produced results of similar quality. More recently, Bierman (1974) derived a two-step model of nutrient uptake, which was based on carrier-mediated transport of nutrients, and growth which was dependent on internal stores of nitrogen and phosphorus. Bierman's uptake and growth formulations have been incorporated in a model of Saginaw Bay, Lake Huron (Bierman, 1976); the results of this model compare favorably with observed measurements of chlorophyll *a*, dissolved phosphate, inorganic nitrogen and dissolved silica. DiToro and Matystik (1976) also have developed a model of Saginaw Bay (coupled to Lake Huron). Their simulations agree with actual measurements as well as Bierman's, yet the model does not include luxury consumption of nutrients by phytoplankton.

At this point in the development of phytoplankton models, it is not clear that explicit formulations for luxury uptake are necessary for predicting seasonal phytoplankton succession. There are, however, other reasons to support the explicit formulation of this process. If long-term storage of various nutrients does occur, and models simulate this, then the nutrient status or physiological state of the population can be determined at any particular point in time. The internal nutrient status (*i.e.*, the actual amount of nutrients in the cell relative to the minimum needed for growth) can then be used to aid in simulating the regulation of respiration, sinking and cell death (*cf.* Lehman *et al.*, 1975).

The controversy surrounding the concept of modeling internal nutrient pools can be summarized with two statements. The first is that nutrient uptake and phytoplankton growth are indeed two separate processes. A number of laboratory studies have shown this (*e.g.*, Fuhs, 1969; Hallman and Stiller, 1974; Caperon and Meyer, 1972a,b). If, however, internal and external nutrient concentrations are at steady state (Gavis, 1976), which implies that growth rate is proportional to uptake rate, then growth kinetic constants can be related to external nutrient concentrations (e.g., Eppley and Thomas, 1969; Tilman and Kilham, 1976). The frequency of occurrence of this steady-state condition in the natural environment is unclear at this time.

The second summary statement addresses this assumption of steady state. Although luxury uptake (and its influence on growth kinetics) has been documented in the laboratory, it has not been shown to occur significantly in the field. That is, algae do have the capacity to build internal stores of nutrients, but whether they actually store these nutrients over long enough time periods (or in enough quantity) to affect their dynamics in the field is unclear.

The effects of luxury consumption become even less clear in models that simulate phytoplankton *groups* (as any ecosystem model must do) rather than individual *species*. By grouping many species (even similar ones) together, one may be creating "biological integrators" that tend to smooth out short-term variations. That is, adaptation and succession within each modeled group are not seen. Only the response of the entire assemblage is simulated. This analysis of various modeling strategies indicates research is needed to assess the actual extent and influence of luxury uptake on phytoplankton assemblages in the field. The primary point of reference should be to compare nutrient recycling rates with long-term phytoplankton stores of those nutrients.

ZOOPLANKTON

The effects of zooplankton on phytoplankton populations have been discussed since the mid-1930s (Harvey *et al.*, 1935; Fuller, 1937; Fleming, 1939) and have been included in most mathematical models since Riley (1946). The importance of the zooplankton community on plant nutrient recycling has been discussed previously (see Nutrient Section) and is accounted for in most recent ecosystem models.

Interactions among zooplankton groups (*i.e.,* carnivory) have been included in some models (Thomann *et al.,* 1975; Canale *et al.,* 1976; Scavia and Park, 1976; Scavia *et al.,* 1976a), whereas the dynamic relationships between zooplankton and fish have rarely been included in whole ecosystem models (but see MacCormick *et al.,* 1974; Park *et al.,* 1974). Most models assume that loss to these predators can be modeled as a constant fraction of prey biomass. As discussed below and by Park *et al.* (Chapter 5), model formulations at this end of the food web obviously need improvement.

Functional Groups

Simulating zooplankton dynamics as part of an ecosystem model requires consideration of how to form meaningful species aggregations. Riley (1947) included only one zooplankton group in his work, as did most early modelers. This approach treats the entire zooplankton community as a single entity and assumes that average process rates adequately represent the community.

The development of a logical basis for separating the zooplankton into function groups began with a consideration of feeding strategies. One method of separation recognized two categories, carnivorous and herbivorous zooplankton (Thomann *et al.,* 1975). A further separation has been made by dividing the herbivorous group into cladocerans and copepods (Bloomfield *et al.,* 1973), based on different feeding strategies and susceptibilities to predation. The most detailed segregation of the zooplankton community is part of a Lake Michigan model developed by Canale *et al.* (1976). The conceptualization includes two raptors, five selective filter feeders and two nonselective filter feeders. Within each of these three feeding-strategy groups, distinctions are based on the apparently variable susceptibilities to predation by higher trophic levels. The model thus includes nine functional groups: predators, large omnivores, small omnivores, large herbivorous cladocerans, small herbivorous cladocerans, small herbivorous copepods and three naupliar groups.

A major component missing from this model, as well as from most other models, is the rotifer group. Because rotifers are generally small they are often considered unimportant in the food web and thus omitted from models. Hutchinson (1967) refers to this group as "the most important soft-bodied invertebrates in the freshwater plankton," and, in spite of their small size, their numerical abundance and high production rates suggest an integral role in both nutrient cycling and the food web. The importance of rotifers in food webs should be investigated. However, more work needs to be done to quantify various aspects of rotifer dynamics (*e.g.*, feeding strategies and rates, respiration rates, and assimilation efficiencies as functions of environmental conditions).

Process Constructs

Two of the most important processes included in zooplankton models are consumption and respiration. Consumption rate is an important mechanism in controlling both phytoplankton and zooplankton population sizes, and respiration is an important mechanism in controlling zooplankton population sizes and in modeling nutrient cycles. Both processes have been handled differently by different models and each will be discussed below. Feeding will be discussed in two contexts: (1) consumption as a function of total food and (2) selective feeding.

Consumption

The development of mathematical expressions for zooplankton filter-feeding has come not from modelers, but rather from experimentalists. Prior to 1960, most researchers (*e.g.*, Fuller, 1937; Fleming, 1939; Marshall and Orr, 1956) accepted the view that filtering rate is independent of food concentration and that therefore the rate of ingestion is directly proportional to food concentration (Figure 2a). Since that time, most researchers (Rigler, 1961; McMahon and Rigler, 1963, 1965; Reeve, 1963; Mullin, 1963; Richman, 1966; Kersting and van der Leeuw-Leegwater, 1976; Kersting and van der Leeuw, 1976) have found suppressed filtering rates and constant ingestion rates at high food concentrations (Figure 2b). Several other investigators (Burns and Rigler, 1967; Parsons *et al.*, 1967; Maly, 1969; McQueen, 1970; Sushchenya, 1970; McAllister, 1970; Frost, 1972; Monakov, 1972; Semenova, 1974; Gaudy, 1974; Chisholm *et al.*, 1075) have reported a curvilinear relationship (Figure 2c). The relationship shown in Figure 2d has been reported occasionally (Mullin, 1963; Conover, 1966); however, Rigler (1971) explained that the curve is incorrect and due mainly to errors in calculating feeding rates.

A few studies have pointed out that the actual feeding process at low food concentrations may not yet be known. Parsons *et al.* (1967) observed " . . . that grazing occurred down to some low prey density . . . and then ceased" in four marine species of copepods (Figure 2e). McAllister (1970) also observed this for *Calanus pacificus.*

Frost (1975) performed feeding experiments at low food concentrations to examine this phenomenon more closely. His data show that below a certain threshold concentration of food the filtering rate decreases, but does not cease. In his feeding experiments with *C. pacificus* feeding on algae of four sizes, he found the threshold to be $\sim$20 μg-C 1^{-1} for each food even though that level represented a range of 4-200 cells ml^{-1}. This food level corresponded to the concentration at which the ingestion rate is 15% of the maximal hourly carbon intake rate. He concluded that, although his data do not show a cessation of feeding, they do indicate a significantly suppressed rate below this threshold value. Two theoretical studies that treated the feeding process for the marine *Calanus pacificus* (Lam and Frost, 1976) and for filter feeders in general (Lehman, 1976) as an optimization of energy consumption and expenditure obtained curves supporting Frost's observations (Figure 2f).

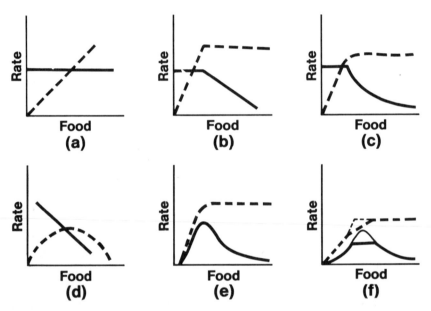

Figure 2. Possible relationships of food consumption rate (broken line) and filtering rate (solid line) with total food concentration.

Recent studies on the effects of food concentration on carnivorous raptorial feeders (Ambler and Frost, 1974; Dodson, 1975) have indicated that curves similar to Figure 1c or 1d are acceptable. Special problems arise in these studies on raptors however. The method most used to determine ingestion rates is counting prey in controls and in experimental chambers after feeding. This measure is most indicative of the kill rate since these raptors do not always eat the whole body of the prey (Ambler and Frost, 1974). Dagg (1974) has shown this waste of food increases with both food size and concentration for a marine amphipod. This also occurs in some predatory freshwater copepods: *Cyclops vernalis* (Nalepa, 1977), *Mesocyclops edax* and *Cyclops vicinus* (Brandl and Fernando, 1975). It is important to conduct more work on quantifying ingestion efficiencies for these animals.

The three expressions most commonly used to describe ingestion rate as a function of food are the rectilinear form (Figure 2b), represented simply as two straight lines with different slopes above and below a specified food concentration; the curvilinear forms, which are generally represented by Ivlev (1966) and Michaelis-Menten equations (Figure 2c); or those same equations modified for the threshold concentration (Figure 2e):

$$I = I_{max} (1 - e^{-kP}) \qquad \text{(Ivlev)} \qquad (7)$$

$$I = I_{max} (1 - e^{-k(P - P_o)}) \qquad \text{(Ivlev, modified)} \qquad (8)$$

$$I = \frac{I_{max} P}{k + P} \qquad \text{(Michaelis-Menten)} \qquad (9)$$

$$I = \frac{I_{max}(P - P_o)}{k + (P - P_o)} \qquad \text{(Michaelis-Menten, modified)} \qquad (10)$$

where I_{max} = maximum ingestion (grazing) rate
k = a constant
P = food concentration
P_o = threshold P
I = ingestion (grazing) rate (mg mg^{-1} day^{-1})

Mullin *et al.* (1975) could demonstrate no preference statistically for Equations 8, 10 or the rectilinear model using filtering data from *C. pacificus* (Frost, 1972). They do indicate, however, that differences in the models are more important at low food concentrations.

The incorporation of various feeding mechanisms into ecosystem models has generally lagged the experimental studies by several years. Fleming (1939) and Riley and Bumpus (1946) used time-dependent specified grazing losses for describing the changes in algal populations. Riley (1946) later used a more mechanistic model by replacing the specified grazing rate with a constant times the size of the ambient zooplankton population. By supplying zooplankton biomass data to his model, Riley was able to simulate the general dynamics of the phytoplankton population as well as many models can today!

The first explicit differential equation for zooplankton dynamics was written by Riley (1947). He still formulated zooplankton assimilation as a constant times the phytoplankton population; however, realizing there is an upper limit to the rate at which an animal can assimilate food, he truncated the rate at an upper limit. Therefore, in their classical paper coupling phytoplankton and zooplankton differential equations, Riley *et al.* (1949) made a distinction between the grazing loss from the phytoplankton (ingestion) I, and the grazing input to the zooplankton (assimilation) A:

$$I = FZP \text{ (mg l}^{-1} \text{ day}^{-1}) \tag{11}$$

$$A = FZ(P - P') \tag{12}$$

where   F = zooplankton filtering rate
          p = phytoplankton concentration
          Z = zooplankton concentration
          P' = amount of phytoplankton consumed, but not assimilated

Davidson and Clymer (1966) and Parker (1968) subsequently used formulations similar to Equations 11 and 12, but without P'.

This basic approach was modified by DiToro *et al.* (1971) who made the assimilation efficiency a continuous function of food concentration, rather than applying an empirical threshold as imposed by Riley (1946). The newer model proposed:

$$I = FZP \tag{13}$$

$$A = FZPa(1 - \frac{P}{K + P}) = FZaK \frac{P}{K + P} \tag{14}$$

where   a = maximum assimilation efficiency
          k = half-saturation constant
          other variables are as before

Both Riley and DiToro *et al.* assumed grazing always remains linearly proportional to food concentration, but in fact, as food becomes more abundant, more is wasted. Many models have subsequently used this form of plant-animal interaction (Thomann *et al.*, 1975; DiToro *et al.*, 1975; Chen and Orlob, 1975; Canale *et al.*, 1974).

A second line of models uses the ingestion rate vs food expression (Equation 9) directly, reflecting the experimental evidence that, with increasing food, filtering rate decreases and grazing rate, I, approaches a maximum asymptotically. Zooplankton assimilation is then modeled as a constant fraction of the ingested food; the remaining fraction is then defecated or otherwise lost. Several models have used this approach (MacCormick *et al.*, 1974; Park *et al.*, 1974; Canale *et al.*, 1976; Scavia *et al.*, 1976a). Bloomfield *et al.*, (1973) and Scavia *et al.*, (1976a) have used the modified version that includes the minimum food level (Equation 10).

The main difference between the two groups of models above is that the first group assumes a constant filtering rate and variable assimilation efficiency, whereas the second group assumes a constant efficiency and variable filtering rate. As discussed above, the filtering rate does vary with food concentration; however, feeding or assimilation efficiency also varies with food concentration for raptors (see above) and filter feeders (Infante, 1973). So, possibly both mechanisms should be included.

Models have begun to be formulated that include the present level of understanding for feeding constructs (see Steele and Mullin, 1977). The particular issues that still remain in question are: (1) the existence and impact of a lower threshold food concentration, (2) the variability of feeding and assimilation efficiencies, and (3) the specific values for the constants to be used in the model equations. The first two issues are presently receiving considerable attention; however, studies to determine values for controlling coefficients are not being conducted. Feeding studies continue to be done in terms of food cell numbers, a meaningless unit from a systems standpoint. McMahon and Rigler (1963) indicated that reduced filtering rates were caused by an animal's gut being filled rather than by the concentration of cells in the medium. McMahon and Rigler (1965) subsequently demonstrated this by feeding *Daphnia magna* food of four sizes. In their study, the concentration of cells at which the filtering rate began to decrease was different for various cell sizes; however, the maximum volumes of food ingested per animal per unit time were similar. This was also true for *Artemia* (Reeve, 1963). Mullin (1963) suggested that "a more realistic calculation of grazing rate is based on the removal of biomass of food." It is important that investigators report results in terms of biovolume, or biomass, instead of cell numbers. This generally has not been the case, and it has often been difficult to compare studies.

It would be valuable to conduct investigations, similar to the studies in phycology, where certain control parameters (*e.g.,* half-saturation constant for feeding, maximum ingestion rates, minimum threshold values) are measured in relation to other variables (*e.g.,* animal size, taxonomic group, trophic position) for many species of zooplankton. In this way, functional groups, and therefore compartments amenable to systems analysis, can be described and compared.

Selective Feeding

Another concept discussed in early works (Harvey, 1937) and investigated in detail both experimentally (*e.g.,* Mullin, 1963; Wilson, 1973) and theoretically (*e.g.,* Lam and Frost, 1976) is the apparent size-selective nature of some zooplankton feeding. This concept of size selection can be confused in a modeling context and an important distinction should be made in terminology. All aquatic invertebrates are selective with regard to the size of their food. Raptoral feeders can seize large prey and tear them apart before eating (Ambler and Frost, 1974; Brandl and Fernando, 1975), but there are limits to the minimum and maximum size of prey they can capture. Filter-feeders are limited physically on both ends of the size spectrum by the dimensions of their sieve-like filtering apparatus. The animals filter water and retain only material large enough to be caught by the seive and yet small enough to fit through the filter chamber (Parsons and Takahashi; 1973). These physical limits can be incorporated simply into models by not allowing specific zooplankton groups to feed on prey whose size lies outside this range. True selection occurs when food of a certain size that is within these limits is eaten disproportionately with respect to its relative availability. "Passive selection" (Frost, 1977) can result from an indiscriminant feeder filtering with an appendage acting as a leaky sieve. That is, the sieve retains certain size (shape) particles more effectively than others. "Active selection" results when an animal searches and captures its food, or is capable of adjusting the characteristics of its filter. Many investigators have identified specific feeding preferences of zooplankton based not only on food size but on its shape and chemical composition as well (e.g., Parsons et al., 1967; McQueen, 1970; Ambler and Frost, 1974; Frost, 1972; Burns, 1969; Bogdan and McNaught, 1975; Berman and Richman, 1974).

A mathematical construct useful in describing this selection was first proposed by O'Neill (1969):

$$X_i = \frac{W_i P_i}{\Sigma W_i P_i} \tag{15}$$

where X_i = fraction of total food eaten that came from category i
              P_i = available food in category i
              W_i = preference or selection coefficient

This approach has been adopted in most multiple group phytoplankton models (Bloomfield et al., 1973; Park et al., 1974; Canale et al., 1976; Scavia et al., 1976a), and it can be represented in the feeding term (Equation 10) as:

$$I_i = I_{max} \frac{W_i (P_i - P_{io})}{K + \Sigma W_i (P_i - P_{io})} \tag{16}$$

where I_i = grazing on prey group i (total feeding is then the sum of all I_i)

The most appropriate way to calculate values for the relative preference coefficients from experimental data is:

$$W_i = \frac{F_i}{\Sigma F_i} \tag{17}$$

where F_i = the filtering rate on a specific food category

However, W_i may also be calculated from other commonly reported data, grazing rates, electivity coefficients, and initial and final particle abundance spectra (Vanderploeg and Scavia, 1978).

In selectivity experiments, efforts should be made to include a spectrum of food sizes similar to that available in nature. Previous studies have looked at subsets of this spectrum (*e.g.*, Burns, 1969, 1-30μm; Hargrave and Geen, 1970, 1-15μm; McQueen, 1970, 2-15μm; Kersting and Holterman, 1973, 1-5μm; Berman and Richman, 1974, 2-16μm). Other studies have used wider ranges, but several performed independent experiments with pure cultures rather than mixtures of various size categories (McMahon and Rigler, 1965; Frost, 1972). Inferences drawn from studies based on filtering or grazing rates for pure cultures are not necessarily applicable to field conditions, where actual food choices are available to the animal. Some studies have used size spectra approaching those available in nature (Mullin, 1963, 6-55μm; Parsons *et al.*, 1967, 1-100μm; Wilson, 1973, 1-60μm); however few studies (*e.g.*, Bogdan and McNaught, 1975; Bowers, Chapter 3) have used an entire natural assemblage.

Information gained by conducting experiments that segment a wide spectrum of prey into many categories will improve our knowledge of the selection processes. These data can be applied to model studies only if they are presented in such a way as to allow a coarser segmentation of food types or sizes. That is, for the entire food spectrum, the amount (weight or volume) initially available as well as the amount remaining should be reported so selectivity coefficients can be determined for size groups comparable to those used in ecosystem models (commonly, large vs small organisms).

Since food selection is probably also a function of properties other than size, experiments offering diatoms, greens, blue-greens, unicellular and colonial algal groups, detritus, and bacteria should also be carried out (*e.g.*, Semenova, 1974). Also from a systems analysis standpoint, the constructs and coefficients of the selective feeding process are most meaningful when determined for functional groups of zooplankton.

Respiration

The constructs used to simulate zooplankton respiration have not changed substantially over the past several years. Riley (1947) related zooplankton respiration rate to temperature and, since that time, all models have included some temperature effect. DiToro *et al.,* (1971), Thomann *et al.* (1975) and Canale *et al.* (1976) all use linear relationships, whereas Chen and Orlob (1975) and DiToro *et al.* (1975) use exponential forms. Other models have included more complex relationships that account for exponential effects below an optimal temperature and a rate reduction above this optimum (Park *et al.,* 1974; MacCormick *et al.,* 1974; Scavia and Park, 1976; Scavia *et al.,* 1976a). The relationships between respiration rate and temperature have been demonstrated for several zooplankton species (*e.g., Diaptomus* spp.—Comita, 1968; *Mesocyclops edax* and *Cyclops vicinus*—Brandl and Fernando, 1975) and the exponential form seems to be more consistent with the effects of temperature on other physiological processes.

The relationship between respiration rate and animal size has been shown to be of the general form:

$$R = aW^b \qquad (18)$$

where   R = respiration rate per individual
          W = weight per individual

The value of a for zooplankton seems to vary with species and temperature; however, the coefficient b (usually between 0.7 and 1.0) has the same value as in respiration-weight relationships for animals in general (see Parsons and Takahashi, 1973; Zeuthen, 1947, 1953). If respiration is to be expressed as a weight-specific rate (imperative from a modeling perspective), one must divide both sides of the above equation by W, which results in values of b between 0.0 and -0.25. This indicates that the weight-specific respiration rate generally decreases as animal size (weight) increases; knowledge of this fact then assists in determining coefficients in models of several functional groups of zooplankton. (In fact, it appears that size itself may be an important criterion for separation of functional groups.) Conover (1960) has also indicated that carnivores have higher respiration rates than herbivores.

Conover (1959) suggested that respiration rates may be affected by the recent feeding history of an animal. Subsequent studies have, in fact, demonstrated this for two marine calanoids (Conover, 1961; Omori, 1970), for two of four *Diaptomus* species tested (Comita, 1968), and for *Daphnia hyalina* (Blazka, 1966). Marshall and Orr (1966) observed increased respiration rates for several small marine copepods following a spring diatom bloom and attributed this to an assumed increase in body size during the bloom, although some of the variability may have been caused by changes in food concentration, as Conover (1961) suggests.

The models of Steele (1974), Scavia and Park (1976) and Scavia *et al.,* (1976a) include the influence of feeding rate on respiration. Steele considered the impact on the model results of including functions that represent either a constant rate or a rate proportional to feeding. He concluded that the effect of feeding on respiration is important in describing zooplankton dynamics. The latter two models include two respiration coefficients, r_1 and r_2, which represent, respectively, a base respiration rate and a proportionality constant for feeding. These models calculate respiration (RESP) as follows:

$$RESP = f(T) (r_1 + r_2F)B \tag{19}$$

where $f(T)$ =   temperature function
         F =   a measure of ingestion or assimilation
         B =   zooplankton biomass

Unfortunately, little work has been done in recent years to verify this relationship or to determine the values of its coefficients (especially values for r_2). As the importance of zooplankton production in nutrient recycling becomes more evident, the need for more research relating respiration (and excretion) to other physiological processes becomes obvious.

Vertical Migration

The distribution of planktonic invertebrates throughout the water column, in fact the patchiness of all aquatic constituents, has been assumed to be unimportant in most models. The models have usually been based on average daily conditions and any movements (particularly diurnal) have been ignored. Recent works (*e.g.,* Steele and Mullin, 1977) have emphasized the importance of vertical migration and have suggested methods to cope with this process in models. It is perhaps time for modelers to begin investigating the effects of vertical migration and zooplankton patchiness on ecosystems. Bowers (Chapter 3) discusses the effects of vertical migration on zooplankton grazing and McNaught (Chapter 1) discusses considerations of space and time scales in general.

BACTERIA

Most ecosystem models implicitly include the effects of decomposers. That is, nonliving particulate and/or dissolved organic matter are assumed to decompose into inorganic nutrients through first-order decay kinetics:

$$- \frac{dC}{dt} = rC \tag{20}$$

where $\frac{dC}{dt}$ = the time rate of change of organic matter C due to decomposition into a particular nutrient

r = a transfer or decay rate coefficient

It has been assumed that r is a linear (Scavia *et al.*, 1976a; Thomann *et al.*, 1975) or an exponential (Chen and Orlob, 1975) function of temperature.

Saunders (1972) suggests that the reaction can be expressed as:

$$- \frac{dC}{dt} = (r_1 C)(r_2 B) \tag{21}$$

where B = concentration of bacteria

r_1, r_2 = functions (possibly nonlinear) of temperature, inorganic nutrients, biodegradability and many other physiological and environmental factors

However, only a few models have included bacterial biomass explicitly. Bloomfield *et al.*, (1973), Park *et al.*, (1974) and Scavia and Park (1976) have included part of the decomposer submodel developed by Bloomfield (1975) in ecosystem models. DiToro and Matystik (1976) assume (Equation 20) is a linear function of phytoplankton biomass in an attempt to account for both hydrolysis of nutrients by algae prior to uptake and fluctuations in bacterial biomass that are assumed to be proportional to phytoplankton biomass. DePinto (Chapter 2) reviews other models of bacterial dynamics.

The lack of quantified microbial kinetics is often cited as a reason for not modeling bacteria explicitly. Although there is a wealth of information on bacterial physiology, there appears to have been little work done to measure uptake and utilization rates of substrates by natural populations (Wetzel, 1975). Some work, however, has been done on uptake rates of dissolved organic material by planktonic bacteria (*e.g.*, Parsons and Strickland, 1962; Wright and Hobbie, 1966). This process is described generally by Michaelis-Menten kinetics:

$$v = V_m \frac{S}{K + S} \tag{22}$$

where v = uptake rate

V_m = maximum uptake rate

S = substrate concentration

K = half-saturation constant

Jannasch (1970) demonstrated a threshold carbon concentration, below which uptake did not occur (similar to the minimum food threshold for zooplankton grazing). Parsons and Takahashi (1973, p. 103) cite several other studies where meaningful relationships between uptake and substrate concentrations could not be obtained, due possibly to populations competing for substrates, as well as competitive inhibition by similar substrates. It appears that although the detritus-bacteria link may be the most important component of many ecosystems (Wetzel, 1975), it may also be the most complex and, therefore, the most difficult to investigate experimentally and to describe mathematically.

The microbial decomposition submodel of Bloomfield (1975) was structured after the results of Wetzel *et al.* (1972) and coupled to previously developed ecosystem models (Scavia, 1974; Park *et al.*, 1974) to better describe and simulate the decomposers of Lake George, NY. Since this submodel included several mechanistic process descriptions of microbial decomposition in a whole ecosystem context, some of its features will be outlined below. This will provide a framework against which new models can be compared and from which experimentally testable hypotheses can be generated. This is the most appropriate use for models proposed in fields not yet sufficiently quantified.

The submodel includes two decomposer populations, one attached to particles and one free in the water. Dissolved organic matter is assumed to exist in three classes: (1) sorbed to particulate organic matter, (2) dissolved labile forms, and (3) dissolved refractory forms. The major by-products of microbial metabolism are soluble inorganic carbon, nitrogen and phosphorus; oxygen is consumed as a consequence of bacterial metabolism.

Uptake of organics was modeled as a hyperbolic relationship, similar to that described in Equation 22. This equation was modified to include "effective substrate" weighting factors similar to the W_i factors described earlier for zooplankton selective feeding. This allowed for growth on, and selection among, several substrates. Remineralization (excretion of inorganic nutrients) was taken to be a function of both uptake, U and respiration, R:

$$\text{Remin} = \alpha U + RB \qquad (23)$$

where α = fraction of organic material taken up that is immediately remineralized
   B = bacteria biomass

The second term in Equation 23 represents excretion of metabolized cell material. Secretion or leakage of dissolved organic matter was assumed to be zero under aerobic conditions based on the work of Otuski and Hanya (1972). Loss to the sediments of decomposer biomass associated with detritus was modeled as a linear function of temperature for a nonstratified water column. Mortality losses other than by zooplankton predation were modeled as a simple first-order loss term. Zooplankton grazing was simulated in a fashion similar to that previously described in the zooplankton section. Respiration, uptake and predation were nonlinear functions of temperature.

The final processes simulated in this model were the sorption/desorption of dissolved organic matter and the active attachment of decomposers onto particles. The attachment process is controlled by placing an upper limit to the number of "active" sites for attachment by decomposers on particulate organic matter. Sorption-desorption of dissolved organics is modeled as a diffusive process controlled by both detrital-associated and water column-associated dissolved organic material.

Bloomfield (1975) based these equations on hypotheses generated from the aquatic microbial literature. The validity of his constructs will remain in question until further work is done to measure the processes for natural populations. In fact, the difficulty encountered in determining decomposer biomass in the field will limit further development severely. One should always attempt to include variables in models that can be verified with field observations. The added realism supplied by detailed mechanistic submodels is not useful unless they are coupled with experimental programs designed to measure the pertinent variables and processes. Bloomfield (1975) discusses this matter and indicates that direct measurement of microbial activity is one alternative to obtaining biomass estimates.

The functions described above, when assembled into a differential equation, require 26 coefficient values for one decomposer group feeding on one substrate and grazed by one predator. With only a cursory look at the functions described, it becomes obvious that most of the coefficients have not been determined for natural populations and in fact some may never be measured. This limits the applicability of such a model for predictive purposes; however, it in no way eliminates its usefulness for examining the framework and details of quantitative microbial decomposition. This is especially true for examination of the roles played by the decomposer group in nutrient cycles and phytoplankton dynamics (DePinto, Chapter 2). As mentioned above, this model can serve as a framework for describing decomposition and nutrient recycling. The constructs have been set forth and should now be substantiated, modified or replaced based on experimental and field studies on natural populations. Several studies are needed to measure bacterial dynamics in natural waters. Research should emphasize both the effects of bacterial activity (i.e., transformation of organic and inorganic compounds) and the controls of bacterial dynamics.

The model described above is for aerobic conditions. Very little model conceptualization has been done for anaerobic environments. It is important to develop models capable of handling anoxia because under those conditions quite different organisms and, therefore, different kinetics, become important. In fact, under anoxic conditions decomposers dominate community production because they are practically the only organisms present (Brock, 1966). Also, the mechanisms and dominant pathways of the nutrient cycles will be strongly affected when passing from an aerobic to an anaerobic environment. A model of anaerobic processes is important in describing not only the instances when the hypolimnion of a stratified lake becomes anaerobic but also the dynamics within the sediment, where oxygen is almost always absent.

FISH

Fisheries biology has become separated from limnology, perhaps because of the need to focus sufficient attention on these economically important organisms (Wetzel, 1975). In recent years, some limnologists have emphasized the importance of fish in the recycling of nutrients, fluxes of carbon, and predatory control of zooplankton and fish species succession (e.g., Brooks, 1969; Wells, 1970; McNaught, 1975; Wetzel, 1975; Shapiro, 1977).

Although a wealth of physiological information exists in the extensive fisheries literature, most fisheries models have been developed to describe population dynamics. There have been models developed more recently that use constructs for specific physiological processes (e.g., Ursin, 1967; Kerr, 1971a, b, c; Jørgensen, 1976; Kitchell et al., 1974); however, very few of these types of fish models address much of the rest of the ecosystem. Also, ecosystem models rarely include fish biomass dynamics at all.

Coupling fish submodels that simulate physiological processes with plankton-nutrient models could be an ideal framework for reuniting the fisheries and limnology fields. By including the top trophic levels in ecosystem models, one could begin to investigate entire system functions and properties (e.g., Robertson and Scavia, Chapter 11). The effects of nutrient loads on fish species succession could also be examined in multiple-group fish models. The effects on native fish and zooplankton composition (e.g., McNaught and Scavia, 1976), as well as second-order effects on phytoplankton biomass and composition, of introducing exotic fish species could be examined with such a model. Ecosystem models including fish dynamics will be able to force a common basis for investigation into the relative control of, and contribution to, eutrophication from atop the food web, as well as from its base.

The research needed in this field is presently a responsibility of both the modelers and the experimentalists. Fish biomass dynamics based on physiological process equations should be included in ecosystem models. By doing this, the most important aspects of fish physiology and ecology with respect to the whole system will be emphasized and research needs can be identified. Also, in order to adequately test fish submodels, estimates of the size of field populations will be necessary.

SEDIMENT

The processes operating in the surficial sediments of lakes can significantly influence the overlying water. Bannerman *et al.* (1975) have estimated sediment-derived phosphorus loads to Lake Ontario to be 10% of the stream loads. Burns and Ross (1972) estimated the load of phosphorus from Lake Erie sediments to be 137% of the external load during two critical summer months when the hypolimnion of the central basin was anoxic. Shagawa Lake, MN, receives an average of 2 mg-P m^{-2} day^{-1} from external sources (Malueg *et al.*, 1975) and has an average measured sediment release rate during two summer months of 7 mg-P m^{-2} day^{-1} (Sonzogni *et al.*, 1977).

Unfortunately, the actual properties of sediment nutrients and nutrient exchange processes are not yet well understood (Lee, 1970). In summarizing the control mechanisms and rate of transfer of nutrients (particularly phosphorus), the following parameters and processes appear to be most important:

1. The diffusion coefficients within the sediment, within the water column, and across the sediment-water interface are critical to any description of the exchange process. Consequently, the scale of the physical system model is quite important, as the diffusive process is scale-dependent.

2. The pH of the sediment and the hypolimnetic water controls various precipitation and sorption processes and can be an important control of the relative concentrations of certain chemical species.

3. Realistic descriptions of biological-chemical cycles of nutrients in both the water column and the sediment are critical for determining the concentrations, and therefore the gradients, of nutrients in the system. Diffusive processes are driven by such gradients.

4. The in-lake concentrations and dynamics of specific metal ions (*e.g.*, Fe (II), Fe (III), Mn) are important. The metals often participate in complexation and precipitation of specific nutrient ions (*e.g.*, PO$_4^{3-}$).

5. The oxygen concentrations, or redox potentials, of the water column and sediments are probably the most important criteria in determining the magnitude of the exchange of nutrients between the water column and the sediments.

Although few ecosystem models have included dynamic interactions between the water column and sediments, some have parameterized the process as an external load (Larsen *et al.,* 1973; Huff *et al.,* 1973; Hydroscience, 1976; Larsen and Mercier, 1975). There have also been conceptual frameworks (Stumm and Leckie, 1970; Mortimer, 1971; Williams and Mayer, 1972), submodels (Jørgensen *et al.,* 1975), and empirical correlations (Jacobsen and Jørgensen, 1975; Kamp-Nielsen, 1975) developed for the transfer of phosphorus and nitrogen under aerobic and anaerobic conditions. Some work has also been done to simulate phosphorus exchange in ecologically simplistic phosphorus models (Sonzogni, 1974; Lung *et al.,* 1976).

Figure 3 summarizes the major phosphorus forms and transformation pathways in the sediments. Similar systems may be conceptualized for nitrogen, silicon, carbon and oxygen. Research is needed to quantify the biological and chemical kinetics of the cycling of nutrients within the sediment.

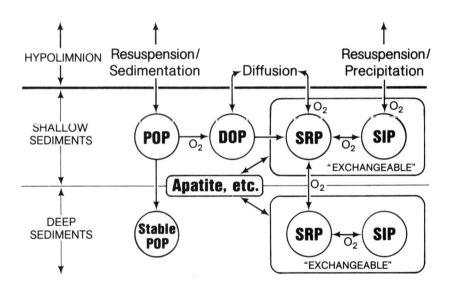

Figure 3. Suggested mechanism of phosphorus transformation and movement in the deep water/sediment zone. POP = particulate organic phosphorus, DOP = dissolved nonreactive phosphorus, SRP = soluble reactive phosphorus, SIP = sorbed inorganic phosphorus; arrows labeled with O_2 are affected by oxygen content.

The effects of benthic macroinvertebrates on the sediments and on sediment-water exchange may be significant. The role of worm populations in preventing the buildup of organic matter and inducing nutrient cycling is very important, for example, in polluted Toronto Harbor, Lake Ontario (Brinkhurst, 1972). Wood (1975), using a dye tracer, found that macroinvertebrates had a significant influence on the transport of solutes between the sediment and the water column through pumping mechanisms most likely related to respiration. Thomas (1975) suggests the most important role of these animals in water column-sediment exchange is in resuspension of materials from below the superficial sediment. To include these organisms in whole ecosystem models, we will need more quantitative information on the physiological processes of the benthic community as functions of various environmental controls.

HYDRODYNAMICS

The effects of water movements on the ecology of a lake are often unclear, possibly because the physical processes are traditionally studied on different time and space scales than are the major biological processes. This is particularly true for large lakes, where biological and chemical measurements, extensive in both space and time, are seldom made.

It is well known that physical properties such as the depth of the mixing zone and thermocline set-up and leakage have profound effects on production in the water column (Mortimer, 1975). Almost all recent ecosystem models include some parameterization of the hydrodynamics in the vertical direction. The general tendency has been to model two vertical layers of constant thickness, with the effect of stratification parameterized as a diffusion coefficient between the two layers. Large coefficient values or even complete mixing are usually assumed during the nonstratified period; small coefficient values or no mixing are assumed during the stratified period (Figure 4). Temperatures are usually specified from averaged data. A different scheme has been used by Scavia *et al.* (1976a), wherein the diffusion coefficient, layer thicknesses and temperature vary with time. The values (Figure 5) are determined by aggregating output from a one-dimensional diffusion model (Sundaram and Rehm, 1973) calibrated to measured temperatures.

Even this second method ignores the possibility of "episodal control," that is, short-term strong physical phenomena, such as storm episodes, that may actually set the stage for and control production during the following weeks (Mortimer, 1975). This can be visualized by imagining a strong storm of short duration (several hours). The physical effects of this episode would most likely be lost during the time averaging required in the the present models. In nature, such a storm could increase the current velocities in the epilimnion, setting up a large velocity gradient with depth. This could reduce the Richardson number (a measure of stability) below its critical value, resulting in shear instabilities at the thermocline. The result would be a short period of mixing of the waters near the

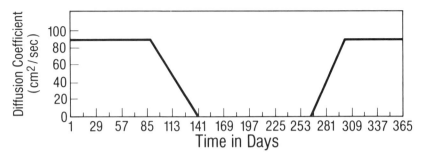

Figure 4. Simple assumed vertical diffusion coefficient between two layers (after Thomann *et al.*, 1975).

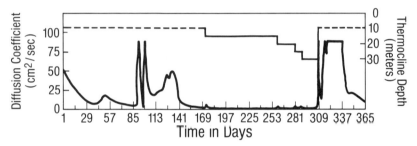

Figure 5. Simulated vertical diffusion coefficient and thermocline depth for two-layer model.

metalimnion and the transport of nutrient-rich hypolimnetic waters into the warm sunlit epilimnion. The stage would now be set for increased primary production if a previously limiting nutrient were resupplied.

The occurrence of such an episode is basically a wind-driven stochastic process and therefore predicting such an episode is impossible. Because of this, only longer-term average conditions (several days to weeks) can be predicted. One could, however, use a simulation model to assist in explaining a given biological and chemical data set and this analysis could be done for a shorter time frame. However, the impact of an episode on the biology and chemistry of a lake (especially a large lake) is difficult to observe since sampling cruises are generally not carried out often enough to discriminate the effects of an episode from the effects of other factors. To expand our understanding of physical and ecological process interactions, we must proceed with the two-pronged approach suggested by Mortimer (1975), that is, coupled experimental and modeling work. Ecological models should be expanded to include more refined hydrodynamics;

however, the verification of such models will be limited by the extent of field work done to measure the impact of episodal phenomena.

The ultimate step in coupling ecological and hydrodynamic models would be to implement the ecological models in three physical dimensions. It is in this framework that spatial and temporal variances can be examined, localized or regional impacts on a water body can be simulated, and modeling of truly mobile organisms makes sense.

In any model used to assess the impact of loads on a lake, one should be concerned with both the immediate and long-term fate of such loads. Eadie (1978), Herdendorf and Zapotosky (1978), and Wyeth and Ploscyea (1978) each found very little land-derived sediment remaining in the near-shore zone of an area in the Great Lakes. Such material is repeatedly re-suspended and transported lakeward, allowing sorption and remineralization to occur in the water column. Using a hydrodynamic model of Lake Erie, Lam and Jaquet (1976) demonstrated the importance of nearshore wave-induced sediment resuspension for phosphorus regeneration. The processes controlling nearshore resuspension and transport are thus especially important from a management perspective.

Recent attempts to implement and verify three-dimensional models that combine hydrodynamics and ecology have demonstrated that presently available modeling techniques are sufficient to begin to attack the problem (Simons, 1976; Thomann and Winfield, 1976; Chen and Smith, Chapter 10). Thomann and Winfield (1976) point out the importance of significant verification tests for their model and provide the first "standardized" test procedures. However, there exist very few three-dimensional ecological models that have been sufficiently compared to lake data to assess their validity. Again, usually the time and space resolution of the available data, as well as the verification techniques, limit development.

There are two other substantial reasons why few ecological models have included the effects of hydrodynamics in three dimensions. The computational requirements of such a model are quite large. Presently, computers capable of handling such large systems exist and some models have been implemented; however, techniques need to be developed for the efficient handling of mixed difference-differential equation models and for the use of different time steps for different components. These suggestions come from a report of one work group session (Parsons et al., 1975) held during the NATO Conference on Modelling of Marine Systems (Nihoul, 1975). The second reason cited by the work group as limiting the development of three-dimensional ecological models is the lack of reasonable estimates of hydrodynamic velocity fields. Hydrodynamic models with one and two physical dimensions have been reasonably successful in simulating water movements; however, this has generally not been the case for three-dimensional models (Stewart, 1975). The lack of verification data has caused the development of these models to proceed less rapidly than that of their meteorological counterparts. Until more detailed data become available, the refinement of the models and improvements in their predictive capability will be impeded (Stewart, 1975).

To proceed along productive lines in coupling three-dimensional hydrodynamics to ecological models, it is important that the modelers and experimentalists from both fields work together toward a common goal. Traditionally, the two types of models have been developed independently and under separate rules and constraints. This has resulted in models almost completely incompatible in space and time resolution. Hydrodynamic models have generally been tested for relatively short periods of time (a few days), whereas ecological models have been tested on seasonal cycles. Hydrodynamic models tend to become unstable over long simulation periods, while ecological models often have no credibility for less than weekly or biweekly averages.

No modeler, or group of modelers, from one discipline will be able to just plug in information, equations or data from models from another field. The modeling branch of the two-pronged approach to system study must include both ecological and hydrodynamic models *and* modelers.

RESEARCH NEEDS

The intent of the above presentation was to outline experimental results and related model constructs for several processes occurring within major components of the ecosystem so that weak points in our knowledge of the system could be identified and research into specific areas suggested. Because the development of our understanding and subsequent development of model constructs have not proceeded equally in all areas, the extent of construct resolution varies among components of the models. This is evident in the types of information necessary for the various components for further model development. For example, research needs in phytoplankton growth constructs mainly involve coefficient determination and discrimination among functional groups, whereas in parts of models involving bacterial effects, a clearer identification of the processes acting within the component are needed first.

The information gaps generally lie within three categories: (1) component descriptions, (2) process constructs and (3) coefficient values. These three categories form a hierarchy of information and construct development. Category 3 becomes important only after the process constructs (category 2) have been postulated, and these constructs are only relevant after the component descriptions (category 1) are laid out (Figure 6).

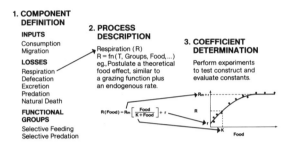

Figure 6. An example of the hierarchical structure of research needs for modeling zooplankton dynamics.

The research needs for the following components of the ecosystem fall into category 1. That is, these components have not yet been described quantitatively in terms of their interrelationships within the whole system.

Bacteria

The role of these organisms in the food chain and in nutrient cycles has been described above. Little work has been done to outline quantitatively their major sources and sinks, their functional groups and their interrelationships. The model discussed above (Bloomfield, 1975) provides an initial framework, but more conceptual information is necessary, especially for anaerobic environments. Fruitful studies would involve field and laboratory research coupled to a modeling framework. Much work is needed to determine the size of field populations of bacteria for model verification. It is also necessary to determine when explicit bacterial dynamics in models may be better than the commonly used implicit formulations. Further discussion of bacteria are presented by DePinto (Chapter 2) and Park *et al.* (Chapter 5).

Fish

These upper trophic levels can have a substantial impact on the composition of the lower levels in the food web. It is also possible that fish play an important role in nutrient cycling. Process information seems to exist for this component, and it appears to be the modelers' turn to push the state-of-the-art. Ecosystem models must begin to include the dynamics of fish biomass, to assess the importance of fish based on extant knowledge, and then to provide the experimentalists with the next round of hypotheses or questions to be addressed. The importance of obtaining reliable biomass estimates for major groups of fish (not just commercially important species) for verification tests is obvious. The implementation of equations for fish biomass dynamics in ecological models is discussed by Park *et al.* (Chapter 5) and Chen and Smith (Chapter 10).

Sediments

Lake sediments have received considerable attention as isolated chemical systems. Several models have been developed that describe processes of diffusion and chemical equilibrium; however, little work has been done to quantify relationships among basic biological processes within the surficial sediments. The actions of bacteria and benthic invertebrates have substantial impact on the production and movement of inorganic nutrients vital to the productivity of the overlying water. Modelers should begin coupling sediment and water colum compartments to help assess the importance of the former as a source or sink of nutrients. Concurrently, more research is needed to identify and to describe mathematically the important metabolic processes of the benthic micro- and macroinvertebrates and the influence of these organisms on transformation and transportation of sedimentary materials. The effects of anoxia on both the pathways and rates of nutrient transformations also require further study.

Hydrodynamics

The movement of water in lakes is understood theoretically; however, measurement and simulation of the movement on fine spatial and extended temporal scales have presented difficulties. To increase our understanding of system controls, as well as our ability to simulate and predict system dynamics, we must begin to couple the full dynamics of water movements with present ecological models.

Progress along these lines will be hampered at first by the difficulty of precisely simulating current velocities and by computational problems with regard to coupling processes with very different time frames. Substantial research is being carried out in both these areas and the outlook is encouraging.

The most significant recommendation for further progress in this area is that both ecological and hydrodynamical modelers should begin to structure their models to facilitate interfacing. The interactions between the two fields should begin at the model development stages, rather than after two distinct products have been produced.

Additional field work will also be necessary to provide verification data for these models. Data should be collected with fine space resolution and coarse time resolution (seasonal), as well as coarse space resolution and fine time resolution (episodes). In this way salient features of both models (ecological and hydrodynamical) can be tested.

The next set of research areas is from the second category, process constructs and interrelationships. These constructs represent processes, such as respiration and sinking, that operate within a system component and processes, such as grazing and excretion, that interrelate system components. They are important functions controlling production or nutrient cycles and require further definition. The process should be studied in terms of cause-and-effect relationship with other processes, and experiments should be designed to test equations based on theoretical hypotheses.

Use of Dissolved Organic Material

Dissolved organic matter can constitute a large fraction of the total organic material in an aquatic ecosystem. Often the dissolved organic form of a particular element is found to be in excess of the dissolved inorganic form. The use of this material by the primary producers should be examined. This should be done as preference studies designed to determine the extent of dissolved organic matter use in the presence of varying concentrations of inorganic nutrients. Of course, the results from such studies will become even more useful when the overall fluxes of the dissolved organic matter pool are determined.

More information is also needed regarding processes within nutrient cycles. For example, the rates of nutrient adsorption and desorption on particulate material should be studied in relation to the particulates' sinking rate. The kinetics of these processes should be studied to develop better understanding, and thus better models, of the aquatic ecosystem.

Nitrogen System

Experimental work should be done to determine the best formulation to describe the relative uses of ammonium, nitrate and molecular nitrogen under varying environmental conditions (*cf.* Walsh and Dugdale, 1972, concerning ammonium and nitrate). The measurement of total nitrogen uptake, as well as the fractions of ammonium, nitrate and molecular nitrogen assimilated, should be measured in environments consisting of different ratios of the nitrogen forms and for different functional groups of algae, such as various size ranges of diatoms, greens and blue-greens with and without heterocysts.

Additional studies are also needed to determine the effects of competition between phytoplankton and bacteria for ammonium. In these studies, ammonium uptake rates by phytoplankton and nitrification rates should be measured simultaneously.

Silicon Cycling

Detailed investigations into the mechanisms and kinetics of silicon cycling both in the water column and the surficial sediments are important. The uptake and incorporation of silicon into diatom frustules, with subsequent sinking and loss to the sediments, is thought to be the fate of dissolved silicon loaded to a lake. Recent studies of Lake Michigan by Conway *et al.* (1977) and Parker *et al.* (1977a,b) have indicated that a major portion of the silicon in sinking diatoms frustules may dissolve in the water column and thus not be lost to the sediments. Additional work should be done to determine the mechanisms and rates of this cycle (especially for incorporation into models looking toward long-term effects of external loads), because the dynamics of nutrients (in this case, silicon) will determine long-term trends in plankton succession.

Luxury Consumption of Nutrients

Field measurements should be used to determine the magnitude of luxury consumption (*i.e.,* nonsteady-state nutrient uptake and growth kinetics) in the natural environment. One estimate of this would be changes in the internal nutrient pools of natural assemblages of algae throughout a season. By determining the extent and duration of these nonsteady-state conditions and by comparing these results with rates of nutrient supplies through recycling, one could compare the relative influence of the two processes on natural populations.

Multiple Nutrient Limitation

Rhee (1978) has performed experiments to test the dual limitation effects of nitrogen and phosphorus on a pure culture of algae. He found that, for dual limitation, the effect on a pure culture was best described as the minimum of the two limitation terms, rather than their product or sum. One could visualize, however, that for a group of different organisms growing together the switching point from N-limited to P-limited may not be as clear as Rhee found. Extensions of Rhee's work are necessary to define the appropriate mathematical expression to simulate multiple nutrient limitation of an assemblage of species.

Sinking Rates

While the dynamics of sedimentation may not be very important on a short-term seasonal scale, long-term (multiyear) simulations are quite dependent on sinking rates.

Considerable work has been done to determine sinking rates for many species of algae. Also, considerable work has been done in the laboratory to define some of the physiological controls of buoyancy. Two pieces of the sedimentation puzzle are still missing: the effects of a turbulent environment on sinking and a determination of detrital sinking rates.

The interaction of sinking rates and dispersion mechanisms appears to be quite important in models. The processes can be described theoretically rather well; however, field observations necessary for model verification do not exist. Although this is a formidable task, the measurement of sinking rates that are subject to minimal artificial constraints in the field are needed.

Also, the nature of detrital material in the environment must be determined before adequate sinking mechanisms can be implemented. The most critical properties to be determined seasonally are size, chemical composition and physical structure.

Grazing

Much work has been done to determine the mechanisms involved in filter feeding by zooplankton. Several constructs have been proposed to describe this feeding process. It is necessary to begin testing various proposed equations with grazing data from many zooplankton species, as Mullin *et al.* (1975) have done with *Calanus pacificus.* Some specific aspects of the feeding process should receive particular attention. The existence of a threshold food concentration clearly is one concept needing further definition. This phenomenon, as well as most others, should be examined for various functional groups of zooplankton.

The nature of selective feeding also is in need of further study. The mechanisms involved in the selection process are unclear at this time and, therefore, validated constructs are not readily available. Selection studies should be designed to include the entire food spectrum available to specific groups of animals.

Control of the efficiency of the overall feeding process should also be investigated. Both ingestion efficiency and assimilation efficiency may be variables rather than constants. Studies should address this problem for both filter feeders and raptors.

Respiration and Excretion

The effects of production (both primary and secondary) on respiration should be examined. The relationship between photosynthesis and respiration has been studied, but quantitative relationships for the phytoplankton have not been developed. Respiration rates of starved and fed zooplankton vary by as much as 50 to 60%. There have been no studies designed to examine this effect as a continuous function of feeding rates or food concentration and, therefore, no equation (other than linear) can be hypothesized. Respiration, both as a significant loss term to the organisms and as a means of relating excretion rates to metabolism, demands a better understanding of its variability and control.

Field Observations of Process Rates

Although modelers have translated process information determined in the laboratory into mathematical expressions used in simulating field conditions, there has been little opportunity to verify the models for these processes. The available verification data usually consist of state-variable measurements (*i.e.*, standing crops of plankton and ambient nutrient concentrations). These state variables are the net result of several processes adding to, and subtracting from, their pool over a period of time. If a model is verified with state-variable estimates, it is assumed, although not assured, that the process formulations are correct. One cannot rule out compensating errors, however, and a serious problem arises when a model fails to simulate accurately the state variables. At this point, one is left with several potential sources of errors: (1) poor coefficients, (2) improper process equations, and/or (3) inadequate component (state-variable) definition. Without additional verification data, it is difficult to determine where the errors lie.

The information most needed to solve this problem is process information. If several process rates (*e.g.*, production, respiration, sinking, grazing) were measured at the same time as the state variables, it would be possible to verify parts of the model at the process level. For example, if the simulation of algae does not match field observations, but production and sinking rates do match observations; then one is reasonably sure that the failure of the state-variable verification was due to incorrect simulation of respiration or grazing rates. This reduces the scale of the problem considerably.

The methods of measuring process rates in the field generally are time consuming, costly and presently not well defined. However, in comprehensive field surveys, it would be very useful to concentrate less on state-variable measurements so that process measurements could be made.

The third category of research needs, coefficient evaluations, involves all of the compartments and processes of all models. It is at this level that modelers and experimentalists have become most separate. In the above two categories, the experimentalists are themselves modelers, postulating and describing various components and functions as process models (actually subcomponents of larger models). This third category is tied closely with the process formulation category because, if the construct coefficients cannot be determined to be constant for a given environment and organism, then the construct itself needs modification.

Functional Groups

Functional groups of organisms within a trophic level can best be defined in terms of their responses to changes in environmental conditions. Although all groups in a trophic level may exhibit the same functional relationships (*i.e.*, category 2), the coefficient values will probably vary from group to group. It is in this way that functional groups should be defined and coefficients determined for them. One should not rely solely on taxonomic groupings for definition of functional groups. For example, two algal species of the same genus may have very different half-saturation constants, yet the constants for each may be similar to those for other species with similar surface area-to-volume ratios.

The processes listed in Table II should be investigated for many species to determine trends or groups of species with similar coefficient values. Groups having similar traits may be parameterized as functional groups in multigroup models. Of course other process equations and their associated coefficients are important, but further analysis must await developments in categories 1 and 2. Coefficient values should always be determined on a weight specific basis. Rates or other properties determined on an individual bases (*e.g.*, per cell or per animal) are not very useful from a modeling standpoint.

In this chapter, I have outlined developments, both experimental and theoretical, in research areas pertinent to whole ecosystem models. In some areas ideas appear to be converging and resulting in firm mathematical constructs representing the real world. In other areas there is little convergence. In synthesizing material, as in this chapter, we begin to see clearly gaps in our understanding of the system. It is then possible to direct research toward those areas. It is in this framework that the interrelationships between modeling and experimental limnology can be exploited to its fullest.

ACKNOWLEDGMENTS

The author wishes to thank the following people for providing critical reviews of this chapter: Andrew Robertson, Robert Thomann, Stephen Tarapchak, Janice Boyd, Brian Eadie, Thomas Nalepa and Henry Vanderploeg. These reviewers provided helpful suggestions and comments, especially in their fields of expertise; however, any errors or omissions are the responsibility of the author.

Table II. Important Model Coefficients

Process	Control Parameters	Coefficients
Primary Production and/or Nutrient Uptake	Temperature	Optimum temperature
		Upper lethal temperature
		Q_{10} Value
	Nutrients	Half-saturation constant
		Maximum growth rate
		Maximum uptake rate
	Light	Saturation light intensity
	Functional groups	All coefficients
Respiration	Temperature	(as above)
	Primary production	Proportionality constant
	Zooplankton consumption	Proportionality constant
	Functional groups	Endogenous rate at optimum temperature and low food
Grazing	Temperature	(as above)
	Total food	Half-saturation or Ivlev coefficients
		Maximum rate
		Threshold values
	Food types	Selectivity coefficients
	Functional groups	All coefficients
Feeding and Assimilation Efficiencies	Temperature	(as above)
	Total food	Half-saturation constant?
	Functional groups	Maximum efficiency
	Functional groups Food types	Similar to selectivity coefficients?
Nitrogen Uptake	Forms of nitrogen available	Preference constants
Excretion (inorganics and organics)	Temperature	(as above)
	Respiration	Proportionality/stoichiometry constants
	Consumption	
Nutrient Transformation (water column and sediments)	Temperature	(as above)
	Organic components	First-order decay constants
Sinking	Physiological state	Proportionality constant?

REFERENCES

Ambler, J. W., and B. W. Frost. "The Feeding Behavior of a Predatory Planktonic Cypepod, *Tortanus discaudatus,*" *Limnol Oceanog.* 19:446-451 (1974).

Bannerman, R. T., D. E. Armstrong, R. F. Harris and G. C. Holden. "Phosphorus Uptake and Release by Lake Ontario Sediments," U.S. Environmental Protection Agency, Corvallis, Oregon, EPA-600/375006, (1975).

Banse, K. "Rates of Growth, Respiration and Photosynthesis of Unicellular Algae as Related to Cell Size—A Review," *J. Phycol.* 12:135-140 (1976).

Bates, S. S. "Effects of Light and Ammonium on Nitrate Uptake by Two Species of Estuarine Phytoplankton," *Limnol. Oceanog.* 21:212-218 (1976).

Bell, W. H., J. M. Lang and R. Mitchell. "Selective Stimulation of Marine Bacteria by Algal Extracellular Products," *Limnol. Oceanog.* 19:833-839 (1974).

Berman, M. S., and S. Richman. "The Feeding Behavior of *Daphnia pulex* from Lake Winnebago, Wisconsin," *Limnol. Oceanog.* 19:105-109 (1974).

Bierman, V. J. "Dynamic Mathematical Model of Algal Growth in Eutrophic Freshwater Lakes," Ph.D. Thesis, University of Notre Dame, Notre Dame, IN (1974).

Bierman, V. J. "Mathematical Model of the Selective Enrichment of Blue-Green Algae by Nutrient Enrichment," in *Modeling of Biochemical Processes in Aquatic Ecosystems,* R. P. Canale, Ed. (Ann Arbor, MI: Ann Arbor Science Publishers, Inc., 1976), p. 1.

Bierman, V. J., F. H. Verhoff, T. L. Paulson and M. W. Tenny. "Multi-Nutrient Dynamic Models of Algal Growth and Species Competition in Eutrophic Lakes," in *Modeling the Eutrophication Process,* E. J. Middlebrooks, D. H. Falkenborg and T. E. Maloney, Eds. (Ann Arbor, MI: Ann Arbor Sicence Publishers, Inc., 1973), p. 89.

Blazka, P. "Metabolism of Natural and Cultured Populations of *Daphnia* Related to Secondary Production," *Verh. Intl. Ver. Limnol.* 16:380-385 (1966).

Bloomfield, J. A. "A Mathematical Model for Decomposition and Nutrient Cycling in the Pelagic Zone of Lake George, New York," Ph.D. Thesis, Department of Environ. Eng., Rensselaer Polytechnic Institute, Troy, NY (1975).

Bloomfield, J. A., R. A. Park, D. Scavia and C. S. Zahorcak, "Aquatic Modeling in the EDFB, US-IBP," in *Modeling the Eutrophication Process,* E. J. Middlebrooks, D. H. Falkenborg and T. W. Maloney, Eds. (Ann Arbor, MI: Ann Arbor Science Publishers, Inc., 1973), p. 139.

Bogdan, K. G., and D. C. McNaught. "Selective Feeding by *Diaptomus* and *Daphnia,*" *Verh. Intl. Ver. Limnol.* 19:2935-2942 (1975).

Brandl, Z., and C. H. Fernando. "Food Consumption and Utilization in Two Freshwater Cyclopoid Copepods *(Mesocyclops edax* and *Cyclops vicinus),*" *Int. Rev. Ges. Hydrobiol.* 60:471-494 (1975).

Brezonik, P. L. "Nitrogen: Sources and Transformations in Natural Waters," in *Nutrients in Natural Waters,* H. E. Allen and J. R. Kramer, Eds. (New York: Wiley Interscience, 1972), p. 1.

Brinkhurst, R. O. "The Role of Sludge Worms in Eutrophication," U.S. Environmental Protection Agency, Corvallis, Oregon, EPA/R3-72-004, (1972).

Brock, T. D. *Principles of Microbial Ecology,* (Englewood Cliffs, NJ: Prentice-Hall, Inc., 1966), 306 p.

Brooks, J. L. "Eutrophication and Changes in the Composition of the Zooplankton," in *Eutrophication: Causes, Consequences, Correctives* (Washington, DC: National Academy of Science, 1969), p. 236.

Burns, C. W. "Particle Size and Sedimentation in the Feeding Behavior of Two Species of *Daphnia," Limnol. Oceanog.* 14:392-402 (1969).

Burns, C. W., and F. H. Rigler. "Comparison of Filtering Rates of *Daphnia Rosea* in Lake Water and in Suspensions of Yeast," *Limnol. Oceanog.* 12:492-502 (1967).

Burns, N. M., and A. E. Pashley. *"In situ* Measurement of the Settling Velocity Profile of Particulate Organic Carbon in Lake Ontario," *J. Fish. Res. Bd. Can.* 31:291-297 (1974).

Burns, N. M., and C. Ross. "Oxygen-Nutrient Relationships Within the Central Basin of Lake Erie," in *Nutrients in Natural Waters,* H. E. Allen and J. R. Kramer, Eds. (New York: Wiley-Interscience, 1972), p. 193.

Canale, R. P., L. M. DePalma and A. H. Vogel. "A Plankton-Based Food Web Model for Lake Michigan," in *Modeling Biochemical Processes in Aquatic Ecosystems,* R. P. Canale, Ed. (Ann Arbor, MI: Ann Arbor Science Publishers, Inc., 1976), p. 33.

Canale, R. P., D. F. Hinemann and S. Nachippan. "A Biological Production Model for Grand Traverse Bay," University of Michigan Sea Grant Program, *Technical Report No. 37,* (Ann Arbor, MI, University of Michigan Sea Grant Program, 1974).

Canale, R. P., and A. H. Vogel. "Effects of Temperature on Phytoplankton Growth," *J. Env. Eng. Div., ASCE* 100(EEI):231-241 (1974).

Caperon, J., and J. Mayer. "Nitrogen-Limited Growth of Marine Phytoplankton I. Changes in Population Characteristics with Steady State Growth Rate," *Deep-Sea Res.* 19:601-618 (1972a).

Caperon, J., and J. Meyer. "Nitrogen-Limited Growth of Marine Phytoplankton II. Uptake Kinetics and their Role in Nutrient Limited Growth of Phytoplankton," *Deep-Sea Res.* 19:619-632 (1972b).

Chen, C. W. "Concepts and Utilities of Ecologic Models," *J. San. Eng. Div., ASCE* 96(SAS):1,085-1,086 (1970).

Chen, C. W., and G. T. Orlob. "Ecological Simulation for Aquatic Environments," in *Systems Analysis and Simulation in Ecology* Vol. III, B. C. Patton, Ed. (New York: Academic Press, Inc., 1975), p. 475.

Chisholm, S. W., R. G. Stross and P. A. Nobbs. "Environmental and Intrinsic Control of Filtering and Feeding Rates in Arctic *Daphnia," J. Fish. Res. Bd. Can.* 32:219-226 (1975).

Comita, G. W. "Oxygen Consumption in *Diaptomus," Limnol. Oceanog.* 13:51-57 (1968).

Conover, R. J. "Regional and Seasonal Variations in the Respiratory Rate of Marine Copepods," *Limnol. Oceanog.* 4:259-268 (1959).

Conover, R. J. "The Feeding Behavior and Respiration of some Marine Planktonic Crustacea," *Biol. Bull.* 119:399-415 (1960).

Conover, R. J. "Metabolism and Growth in *Calanus hyperboreus* in Relation to its Life Cycle," *Symp. on Zooplankton Production,* ICES No. 11, (1961), p. 190.

Conover, R. J. "Factors Affecting the Assimilation of Organic Matter by Zooplankton and the Question of Superfluous Feeding," *Limnol. Oceanog.* 11:346-354 (1966).

Conway, H. L., J. I. Parker, E. M. Yaguchi and D. L. Mellinger. "Biological Utilization and Regeneration of Silicon in Lake Michigan," *J. Fish. Res. Bod. Can.* 34:537-544 (1977).

Cowen, W. F., and G. F. Lee. "Algal Nutrient Availability and Limitation in Lake Ontario during IFYGL Part 1. Available Phosphorus in Urban Runoff and Lake Ontario Tributary Waters," U.S. Environmental Protection Agency, Corvallis, Oregon, EPA-600/3-76-094a (1976).

Dagg, M. J. "Loss of Prey Body Contents during Feeding by an Aquatic Predator," *Ecology* 55:903-906 (1974).

Davidson, R. S., and A. B. Clymer. "The Desirability and Applicability of Simulating Ecosystems," *Ann. NY Acad. Sci.* 128:790-794 (1966).

DePinto, J. V., V. J. Bierman and F. H. Verhoff. "Seasonal Phytoplankton Succession as a Function of Species Competition for Phosphorus and Nitrogen," in *Modeling Biochemical Processes in Aquatic Ecosystems,* R. P. Canale, Ed. (Ann Arbor, MI: Ann Arbor Science Publishers, Inc., 1976), p. 141.

DiToro, D. M. "Combining Chemical Equilibrium and Phytoplankton Models—A General Methodology," in *Modeling Biochemical Processes in Aquatic Ecosystems,* R. P. Canale, Ed. (Ann Arbor, MI: Ann Arbor Science Publishers, Inc., 1976), p. 233.

DiToro, D. M., and W. Matystik. "Phytoplankton Biomass Model of Lake Huron and Saginaw Bay," in *Environmental Modeling and Simulation,* W. R. Ott, Ed. U.S. Environmental Protection Agency, Washington, D.C., EPA 600/9-76-016 (1976).

DiToro, D. M., D. J. O'Connor, R. V. Thomann and J. L. Mancini. "Phytoplankton—Zooplankton Nutrient Interaction Model for Western Lake Erie," in *Systems Analysis and Simulation in Ecology,* Vol. III B. C. Patton, Ed. (New York: Academic Press Inc., 1975), p. 423.

DiToro, D. M., R. V. Thomann and D. J. O'Connor. "A Dynamic Model of Phytoplankton Population in the Sacramento-San Joaquin Delta," in *Advances in Chemistry Series 106: Nonequilibrium Systems in Natural Water Chemistry,* R. F. Gould, Ed. (Washington, DC: American Chemical Society, 1971), p. 131.

Dodson, S. I. "Predation Rates of Zooplankton in Arctic Ponds," *Limnol. Oceanog.* 20:426-433 (1975).

Dolan, D., V. J. Bierman, M. H. Dipert and R. D. Geist. "Statistical Analysis of the Spacial and Temporal Variability of the Ratio Chlorophyll *a* to Phytoplankton Cell Volume in Saginaw Bay, Lake Huron," *J. Great Lakes Res.* (in press).

Droop, M. R. "Vitamin B12 and Marine Ecology, IV. The Kinetics of Uplake, Growth and Inhibition in *Monochrysis lutheri,*" *J. Mar. Biol. Assoc. (UK)* 48:689-733 (1968).

Dugdale, R. C. "Biological Modeling I," in *Modeling of Marine Systems,* Oceanography Series, Vol. 10, J. C. J. Nihoul, Ed. (New York: Elsevier Series Vol. 10, 1975), p. 187.

Dugdale, R. C. "Nutrient Modeling," in *The Sea, Vol. 6: Marine Modeling,* E. D. Goldberg, I. N. McCave, J. J. O'Brien and J. H. Steele, Eds. (New York: Wiley-Interscience, 1977), p. 789.

Dugdale, V. A., and R. C. Dugdale. "Tracer Studies of the Assimilation of Inorganic Nitrogen Sources," *Limnol. Oceanog.* 10:53-57 (1965).

Eadie, B. J. "The Effects of the Grand River Spring Runoff on Lake Michigan," International Joint Commission Report, Windsor, Ontario (1978).

Eppley, R. W., and W. H. Thomas. "Comparison of Half-Saturation Constants for Growth and Nitrate Uptake of Marine Phytoplankton," *J. Phycol.* 5:375-379 (1969).

Eppley, R. W., J. N. Rogers and J. J. McCarthy. "Half-Saturation Constants for Uptake of Nitrate and Ammonia by Marine Phytoplankton," *Limnol. Oceanog.* 14:912-920 (1969).

Ferrante, J. G. "The Role of Zooplankton in the Intrabiocenotic Phosphorus Cycle and Factors Affecting Phosphorus Excretion in a Lake," *Hydrobiol.* 49:203-214 (1976).

Ferrante, J. G., and J. I. Parker. "Transport of Diatom Frustules by Copepod Fecal Pellets to the Sediments of Lake Michigan," *Limnol. Oceanog.* 22:92-98 (1977).

Fleming, R. J. "The Control of Diatom Populations by Grazing," *J. Cons. Explor. Mer.* 14:210-227 (1939).

Foree, E. G., W. J. Jewell and P. L. McCarthy. "The Extent of Nitrogen and Phosphorus Regeneration from Decomposing Algae," in *Advances in Water Pollution Research,* Vol. 2, S. H. Jenkins, Ed. (New York: Pergamon Press, 1970), p. III-27/1-15.

Frost, B. W. "Effects of Size and Concentration of Food Particles on the Feeding Behavior of the Marine Planktonic Copopod *Calanus pacificus,*" *Limnol. Oceanog.* 17:805-815 (1972).

Frost, B. W. "A Threshold Feeding Behavior in *Calanus pacificus,*" *Limnol. Oceanog.* 20:263-266 (1975).

Frost, B. W. "Feeding Behavior of *Calanus pacificus* in Mixtures of Food Particles," *Limnol. Oceanog.* 22:472-491 (1977).

Fuhs, G. W. "Phosphorus Content and Rate of Growth in the Diatoms *Cyclotella nana* and *Thalassiosira fluviatilis,*" *J. Phycol.* 5:312-321 (1969).

Fuller, J. L. "Feeding Rate of *Calanus firmarchicus* in Relation to Environmental Conditions," *Biol. Bull.* 72:233-246 (1937).

Ganf, G. G., and P. Blazka. "Oxygen Uptake, Ammonia and Phosphate Excretion by Zooplankton in a Shallow Equatorial Lake (Lake George, Uganda)," *Limnol. Oceanog.* 19:313-325 (1974).

Gaudy, R. "Feeding Four Species of Pelagic Copepods under Experimental Conditions," *Mar. Biol.* 25:125-141 (1974).

Gavis, J. "Munk & Riley Revisited: Nutrient Diffusion Transport and Rates of Phytoplankton Growth," *J. Mar. Res.* 34:161-179 (1976).

Goldman, J. C., W. J. Oswald and D. Jenkins. "The Kinetics of Inorganic Carbon Limited Algal Growth," *J. Water Poll. Control Fed.* 46:554-573 (1974).

Gutel'makher, B. L. "Relative Fraction of Individual Algal Species in Primary Production of Plankton," *Hydrobiol. J.* 10(3):1-5 (1974) (Translation series).

Hallman, M., and M. Stiller. "Turnover and Uptake of Dissolved Phosphate in Freshwater, A Study in Lake Kinneret," *Limnol. Oceanog.* 19:774-783 (1974).

Hargrave, B. T., and G. H. Geen. "Effects of Copepod Grazing on Two Natural Phytoplankton Populations," *J. Fish Res. Bd. Can.* 27:1395-1403 (1970).

Harvey, H. W. "Notes on Selective Feeding by *Calanus*," *J. Mar. Biol. Assoc. (UK)* 22:97-100 (1937).

Harvey, H. S. *The Chemistry and Fertility of Seawater* (Cambridge, England: Cambridge University Press, 1955), 240 p.

Harvey, H. W., L. H. N. Cooper, M. V. Lebour and F. S. Russell. "Plankton Production and its Control," *J. Mar. Biol. Assoc.* 20:407-441 (1935).

Herbes, S. E. "Biological Utilization of Dissolved Organic Phosphorus in Natural Waters," Ph.D. Thesis, University of Michigan, Ann Arbor, MI (1974).

Herdendorf, C. E., and J. E. Zapotosky. "Effects of Tributary Loading to Western Lake Erie during Spring Runoff Events," International Joint Commission Report, Windsor, Ontario (1978).

Horne, A. J., and C. R. Goldman. "Nitrogen Fixation in Clear Lake, California I. Seasonal Variation and the Role of Heterocysts," *Limnol. Oceanog.* 17:678-692 (1972).

Huff, D. D., J. F. Koone, W. R. Ivarson, P. R. Weiler, E. H. Dettman and R. F. Harris. "Simulation of Urban Runoff, Nutrient Loading, and Biotic Response of a Shallow Eutrophic Lake," in *Modeling the Eutrophication Process*, E. J. Middlebrooks, D. H. Falkenborg and T. E. Maloney, Eds. (Ann Arbor, MI: Ann Arbor Science Publishers, Inc., 1973), p. 33.

Hutchinson, G. E. *A Treatise on Limnology. Vol. 2. Introduction to Lake Biology and the Limnoplankton* (New York: Wiley-Interscience, 1967), 1,115 p.

Hydroscience, Inc. "Assessment of the Effects of Nutrient Loadings on Lake Ontario using a Mathematical Model of the Phytoplankton," International Joint Commission, Great Lakes Water Quality Board Report, Windsor, Ontario (1976).

Infante, A. "Untersuchungen über die Ausnutzbarkeit verschiedener Algen durch das Zooplankton," *Arch. Hydrobiol. Suppl.* 42:340-405 (1973).

Ivlev, V. S. "The Biological Productivity of Waters," *J. Fish. Res. Bd., Can.* 23(11):1727-1759 (1966).

Jacobsen, O. S., and J. L. Jørgensen. "A Submodel for Nitrogen Release from Sediments," *Ecol. Model.* 1:147-152 (1975).

Jannasch, H. W. "Threshold Concentrations of Carbon Sources Limiting Bacterial Growth in Sea Water," in *Organic Matter in Natural Waters*, D. W. Hood, Ed. (Fairbanks: University of Alaska Press, 1970), p. 321.

Jassby, A. D., and T. Platt. "Mathematical Formulation of the Relationship between Phytosynthesis and Light for Phytoplankton," *Limnol. Oceanog.* 21:540-547 (1976a).

Jassby, A. D., and T. Platt. "The Relationship between Photosynthesis and Light for Natural Assemblages of Coastal Marine Phytoplankton," *J. Phycol.* 12:421-430 (1976b).

Jørgensen, S. E. "A Model of Fish Growth," *Ecol. Model.* 2:303-314 (1976).

Jørgensen, S. E., L. Kamp-Nielsen and O. S. Jacobson. "A Submodel for an Aerobic Mud-Water Exchange of Phosphate," *Ecol. Model.* 1:133-146 (1975).

Kamp-Nielsen, L. "A Kinetic Approach to the Aerobic Sediment—Water Exchange of Phosphorus in Lake Esrom," *Ecol. Model.* 1:153-160 (1975).

Kerr, P. C., S. E. Herbes and H. E. Allen. "The Carbon Cycle in Aquatic Ecosystems," in *Nutrients in Natural Waters*, H. E. Allen and J. R. Kramer, Eds. (New York: Wiley-Interscience, 1972), p. 101.

Kerr, S. R. "Analysis of Laboratory Experiments on Growth Efficiency of Fishes," *J. Fish. Res. Bd. Can.* 28:801-808 (1971a).

Kerr, S. R. "Prediction of Fish Growth Efficiency in Nature," *J. Fish. Res. Bd. Can.* 28:809-814 (1971b).

Kerr, S. R. "A Simulation Model of Lake Trout Growth," *J. Fish. Res. Bd. Can.* 28:815-819 (1971c).

Kersting, K., and W. Holterman. "The Feeding Behavior of *Daphnia magna* Studied with the Coulter Counter," *Verh. Int. Ver. Limnol.* 18:1434-1440 (1973).

Kersting, K., and W. van der Leeuw. "The Use of the Coulter Counter for Measuring the Feeding Rates of *Daphnia magna*," *Hydrobiol.* 49(3): 233-237 (1976).

Kersting, K., and C. van der Leeuw-Leegwater. "Effects of Food Concentration on the Respiration of *Daphnia magna*," *Hydrobiol.* 49(2):137-142 (1976).

Ketchum, B. H. "The Absorption of Phosphate and Nitrate by Illuminated Cultures of *Nitzschia closterium*," *Am. J. Bot.* 26:399-407 (1939).

Kilham, P. "A Hypothesis Concerning Silica and the Freshwater Planktonic Diatoms," *Limnol. Oceanog.* 16:10-18 (1971).

King, D. L. "Carbon Limitation in Sewage Lagoons," in *Nutrients and Eutrophication Special Symposia*, Vol. I (Am. Soc. Limnol. and Oceanog., 1972), pp. 98-112.

King, D. L., and J. T. Novak. "The Kinetics of Inorganic Carbon-Limited Algal Growth," *J. Water Poll. Control Fed.* 46:1812-1816 (1974).

Kitchell, J. F., J. F. Koonce, R. V. O'Neill, H. H. Shugart, J. J. Magnason and R. S. Booth. "Model of Fish Biomass Dynamics," *Trans. Am. Fish Soc.* 103:786-798 (1974).

Koonce, J. F. "Seasonal Succession of Phytoplankton and a Model of the Dynamics of Phytoplankton Growth and Nutrient Response," Ph.D. Thesis, University of Wisconsin, Department of Biology, Madison, WI (1972).

Lam, D. C. L., and J.-M. Jaquet. "Computations of Physical Transport and Regeneration of Phosphorus in Lake Erie, Fall 1970," *J. Fish. Res. Bd. Can.* 33:550-563 (1976).

Lam, R. K., and B. W. Frost. "Model of Copepod Filtering Response to Changes in Size and Concentration of Food," *Limnol. Oceanog.* 21: 490-500 (1976).

Larsen, D. P., and H. T. Mercier. "Shagawa Lake Recovery Characteristics as Depicted by Predictive Modeling," in *Water Quality Criteria Research of the U.S. Environmental Protection Agency*, U. S. Environmental Agency, Corvallis, Oregon, EPA-660/3-76-079 (1975).

Larsen, D. P., H. T. Mercier and K. W. Malueg. "Modeling Algal Growth Dynamics in Shagawa Lake, Minnesota, with Comments Concerning Projected Restoration of the Lake," in *Modeling the Eutrophication Process*, E. J. Middlebrooks, D. H. Falkenborg and T. E. Maloney, Eds. (Ann Arbor, MI: Ann Arbor Science Publishers, Inc., 1973), p. 15.

Lassiter, R. R., and D. K. Kearns. "Phytoplankton Population Changes and Nutrient Fluctuations in a Simple Aquatic Ecosystem Model," in *Modeling the Eutrophication Process*, E. J. Middlebrooks, D. H. Falkenborg and T. E. Maloney, Eds. (Ann Arbor, MI: Ann Arbor Science Publishers, Inc., 1973), p. 131.

Lean, D. R. S. "Phosphorus Dynamics in Lake Waters," *Science* 179: 678-680 (1973).

Lee, G. F. "Factors Affecting the Transfer of Materials between Water and Sediments," Eutrophication Information Program, Literature Review No. 1, University of Wisconsin, Water Resources Center, (1970).

Lehman, J. T. "The Filter-Feeder as an Optimal Forager, and the Predicted Shapes of Feeding Curves," *Limnol. Oceanog.* 21:501-516 (1976).

Lehman, T. D., D. B. Botkin and G. E. Likens. "The Assumptions and Rationales of a Computer Model of Phytoplankton Population Dynamics," *Limnol. Oceanog.* 20:343-364 (1975).

Lewis, W. M., Jr. "Surface Volume Ratio: Implications for Phytoplankton Morphology," *Science* 92(4242):885-887 (1976).

Lung, W. S., R. D. Canale and P. L. Freedman. "Phosphorus Models for Eutrophic Lakes," *Water Res.* 10:1,101-1,114 (1976).

MacCormick, A. S. A., O. L. Loucks, J. F. Koonce, J. F. Kitchell and P. R. Weiler. "An Ecosystem Model for the Pelagic Zone of Lake Wingra," Eastern Deciduous Forest Biome (IBP) Report EDFB-IBP-7-47 (1974).

Malueg, K. M., D. P. Larsen, D. W. Schultz and H. T. Mericer. "A Six-Year Water, Phosphorus, and Nitrogen Budget for Shagawa Lake, Minnesota," *J. Environ. Qual.* 4:236-242 (1975).

Maly, E. J. "A Laboratory Study of the Interaction between the Predatory Rotifer *Asplanchna* and *Paramecium*," *Ecology* 50:59-73 (1969).

Marshall, S. M., and A. P. Orr. "Experimental Feeding of the Copepod *Calanus finmarchicus* Gunner on Phytoplankton Cultures Labelled with Radioactive Carbon," *Paper Mar. Biol. Oceanog.* Suppl. to *Deep-Sea Res.* 3:110-114 (1956).

Marshall, S. M., and A. P. Orr. "Respiration and Feeding in some Small Copepods," *J. Mar. Biol. Assoc. (UK)* 46:513-530 (1966).

McAllister, D. C. "Zooplankton Rations, Phytoplankton Mortality, and Estimation of Marine Production," in *Marine Food Chains*, J. H. Steele, Ed. (Berkeley, CA: University of California Press, 1970), p. 419.

McMahon, J. W., and F. G. Rigler. "Mechanisms Regulating the Feeding Rate of *Daphnia magna* Straus," *Can. J. Ecol.* 41:321-332 (1963).

McMahon, J. W., and F. G. Rigler. "Feeding Rate of *Daphnia magna* Straus in Different Food Labeled with Radioactive Phosphorus," *Limnol. Oceanog.* 10:105-113 (1965).

McNaught, D. C. "A Hypothesis to Explain the Succession from Calanoids to Cladocerans during Eutrophication," *Verh. Int. Ver. Limnol.* 19:724-731 (1975).

McNaught, D. C., and D. Scavia. "Application of a Model of Zooplankton Composition to Problems of Fish Introduction to the Great Lakes," in *Modeling of Biochemical Processes in Aquatic Ecosystems,* R. P. Canale, Ed. (Ann Arbor, MI: Ann Arbor Science Publishers, 1976), pp. 281-304.

McQueen, D. J. "Grazing Rates and Food Selection in *Diaptomus oregonensis* (Copepoda) from Marion Lake, B. C.," *J. Fish. Res. Bd. Can.* 27:13-20 (1970).

Monakov, A. V. "Review of Studies on Feeding of Aquatic Invertebrates Conducted at the Institute of Biology of Inland Waters, Academy of Science, USSR," *J. Fish. Res. Bd. Can.* 29:363-383 (1972).

Mortimer, C. H. "The Exchange of Dissolved Substances between Mud and Water in Lakes I and II," *J. Ecol.* 29:280-329 (1941).

Mortimer, C. H. "The Exchange of Dissolved Substances between Mud and Water in Lakes III and IV. Summary and Refs.," *J. Ecol.* 30:147-201 (1942).

Mortimer, C. H. "Chemical Exchanges between Sediments and Water in the Great Lakes—Speculations on Probable Regulatory Mechanisms," *Limnol. Oceanog.* 16:387-404 (1971).

Mortimer, C. H. "Modeling of Lakes as Physico-Biochemical Systems—Present Limitations and Needs," in *Modeling of Marine Systems,* J. C. J. Nihoul, Ed. (New York: Elsevier, 1975), p. 217.

Mullin, M. M. "Some Factors Affecting the Feeding of Marine Copepods of the Genus *Calanus,*" *Limnol. Oceanog.* 8:239-250 (1963).

Mullin, M. M., E. F. Stewart and F. J. Foglister. "Ingestion by Planktonic Grazers as a Function of Concentration of Food," *Limnol. Oceanog.* 20:259-262 (1975).

Munk, W. H., and G. A. Riley. "Adsorption of Nutrients by Aquatic Plants," *J. Mar. Res.* 11:215-240 (1952).

Nalepa, T. F. Great Lakes Environmental Research Laboratory (NOAA), Ann Arbor, MI. Personal Communication (1977).

Nalewajko, C., and D. R. S. Lean. "Growth and Excretion in Planktonic Algae and Bacteria," *J. Phycol.* 8:361-366 (1972).

Nalewajko, C., and D. W. Schindler. "Primary Production, Extracellular Release, and Heterotrophy in Two Lakes in the ELA, Northwestern Ontario," *J. Fish. Res. Bd. Can.* 33:219-226 (1976).

Nelson, D. N., J. J. Goering, S. S. Kilham and R. R. L. Guillard. "Kinetics of Silicic Acid Uptake and Rates of Silica Absorption in the Marine Diatom *Thalassiosira pseudonana,*" *J. Phycol.* 12:246-252 (1976).

Nihei, T., T. Sasa, S. Miyachi, K. Suzuki and H. Tamiya. "Change of Photosynthetic Activity of *Chlorella* Cells During the Course of their Normal Life Cycle," *Arch. Mikrobiol.* 21:156-166 (1954).

Nihoul, J. C. J., Ed. *Modeling of Marine Systems,* Oceanography Series, Vol. 10 (New York: Elsevier, 1975).

O'Connor, D. J., D. M. DiToro and R. V. Thomann. "Phytoplankton Models and Eutrophication Problems," in *Ecological Modeling in a Resource Management Framework,* Clifford S. Russell, Ed., (Washington DC: Resources for the Future, Inc., 1975), pp. 149-210.

Omori, M. "Variations of Length, Weight, Respiratory Rate, and Chemical Composition of *Calanus cristatus* in Relation to its Food and Feeding," in *Marine Food Chains,* J. H. Steele, Ed. (Berkeley, CA: University of California Press, 1970), p. 113.

O'Neill, R. V. "Indirect Estimation of Energy Fluxes in Animal Food Webs," *J. Theor. Biol.* 22:284-290 (1969).

Orlob, G. T. "Present Problems and Future Prospects of Ecological Modeling," in *Ecological Modeling in a Resource Management Framework*, C. S. Russell, Ed., (Washington, DC: Resources for the Future, Inc., 1975), pp. 283-312.

Otuski, A., and T. Hanya. "Production of Dissolved Organic Matter from Dead Green Algal Cells I Aerobic Microbial Decomposition," *Limnol. Oceanog.* 17:248-257 (1972).

Park, R. A., R. V. O'Neill, J. A. Bloomfield, H. H. Shugart, R. S. Booth, R. A. Goldstein, J. B. Mankin, J. F. Koonce, D. Scavia, M. S. Adams, L. S. Clesceri, E. M. Colon, E. H. Dettmann, J. Hoopes, D. D. Huff, S. Katz, J. F. Kitchell, R. C. Kohberger, E. J. LaRow, D. C. McNaught, J. Peterson, J. Titus, P. R. Weiler, J. W. Wilkinson and C. S. Zahorcak. "A Generalized Model for Simulating Lake Ecosystems," *Simulation* 23(2):33-50 (1974).

Parker, J. I., H. L. Conway and E. M. Yaguchi. "Dissolution of Diatom Frustules and Recycling of Amorphous Silicon in Lake Michigan," *J. Fish. Res. Bd. Can.* 34:545-551 (1977a).

Parker, J. I., H. L. Conway and E. M. Yaguchi. "Seasonal Periodicies of Diatoms, and Silicon Limitation in Offshore Lake Michigan, 1975," *J. Fish. Res. Bd. Can.* 34:552-558 (1977b).

Parker, R. A. "Simulation of an Aquatic Ecosystem," *Biometrics* 24:803 821 (1968).

Parsons, T. R., R. J. LeBrasseur and J. D. Fulton. "Some Observations on the Dependence of Zooplankton Grazing on the Cell Size and Concentration of Phytoplankton Blooms," *J. Oceanog. Soc. (Japan)* 23:10-17 (1967).

Parsons, T. R., D. Menzel, G. Radach, N. R. Andersen, G. Pichot, J. P. Mommaerts, C. Walters, J. Steele, D. Rodrigues and R. C. Dugdale. "Productivity of Sea Waters" Work Group Report, in *Modeling of Marine Systems*, Oceanography Series, Vol. 10, J C. J. Nihoul, Ed. (New York: Elsevier, 1975), p. 247.

Parsons, T. R., and J. P. H. Strickland. "On the Production of Particulate Organic Carbon by Heterotrophic Processes in Sea Water," *Deep-Sea Res.* 8:211-222 (1962).

Parsons, T., and M. Takahashi. *Biological Oceanographic Processes* (New York: Pergamon Press, 1973), 186 p.

Patten, B. C., D. A. Egloff, T. H. Richardson, *et al.* "Total Ecosystem Model for a Cove in Lake Taxoma," in *Systems Analysis and Simulation in Ecology* Vol. III, B. C. Patten, Ed., (New York: Academic Press, Inc., 1975), p. 206.

Platt, T., K. L. Denman and A. D. Jassby. "Modeling the Productivity of Phytoplankton," in *The Sea Vol. 6: Marine Modeling*, E. D. Goldberg, I. N. McCave, J. J. O'Brien and J. H. Steele, Eds. (New York: John Wiley & Sons, Inc., 1977), p. 807.

Redfield, A. C., B. H. Ketchum and F. A. Richards. "The Influence of Organisms on the Composition of Sea-Water," in *The Sea*, Vol. 2, (New York: Wiley Interscience, 1963), p. 26.

Reeve, M. R. "The Filter-Feeding of *Artemia* I. In Pure Cultures of Plant Cells," *J. Exp. Biol.* 40:195-205 (1963).

Rhee, G. Y. "Effects of N:P Atomic Ratios and Nitrate Limitations on

Algal Growth, Cell Composition, and Nitrate Uptake," *Limnol. Oceanog.* 23:10-25 (1978).

Richman, S. "The Effect of Phytoplankton Concentrations on the Feeding Rate of *Diaptomus oregonensis,*" *Verh. Int. Ver. Limnol.* 16:392-398 (1966).

Rigler, F. H. "The Relation between Concentration of Food and Feeding Rate of *Daphnia magna* Straus," *Can. J. Ecol.* 39:857-868 (1961).

Rigler, F. H. "Laboratory Measurements of Processes Involved in Secondary Production," in *Secondary Productivity in Fresh Waters,* W. T. Edmonson and G. G. Winberg, Eds., IBP Handbook No. 17 (Oxford: Blackwell Scientific Publications Ltd., 1971) p. 228-256.

Rigler, F. H. "A Dynamic View of the Phosphorus Cycle in Lakes," in *Environmental Phosphorus Handbook,* E. J. Griffith, A. Beeton, J. M. Spencer and D. T. Mitchell, Eds. (New York: John Wiley & Sons, Inc., 1973), p. 539.

Riley, G. A. "Factors Controlling Phytoplankton Populations on Georges Bank," *J. Mar. Res.* 6:54-73 (1946).

Riley, G. A. "A Theoretical Analysis of the Zooplankton Population of Georges Bank," *J. Mar. Res.* 6:104-113 (1947).

Riley, G. A., and D. F. Bumpus. "Phytoplankton—Zooplankton Relationships on Georges Bank," *J. Mar. Res.* 6:33-47 (1946).

Riley, G. A., H. Stommel and D. F. Bumpus. "Quantitative Ecology of the Plankton of the Western North Atlantic," *Bull. Bingham Oceanog. Cool.* 12:1-169 (1949).

Ryther, J. H. "Photosynthesis in the Ocean as a Function of Light Intensity," *Limnol. Oceanog.* 1:61-70 (1956).

Saunders, G. W. "Summary of the General Conclusions of the Symposium," *Mem. Ist. Ital. Idrobiol.* 29(Suppl.):533-540 (1972).

Scavia, D. "Implementation of a Pelagic Ecosystem Model for Lakes," M.S. Thesis, Rensselaer Polytechnic Institute, Troy, NY, (1974).

Scavia, D. "An Ecological Model for Lake Ontario," Great Lakes Environ. Res. Lab., Ann Arbor, MI, [manuscript in review (1978a)].

Scavia, D. "Examination of Phosphorus Cycling and the Control of Phytoplankton Productivity in Lake Ontario During IFYGL," Great Lakes Environ. Res. Lab., Ann Arbor, MI, [manuscript in review (1978b)].

Scavia, D., J. A. Bloomfield, J. S. Fisher, J. Nagy and R. A. Park. "Documentation of CLEANX: A Generalized Model for Simulating the Open-Water Ecosystems of Lakes," *Simulation* 23(2):51-56 (1974).

Scavia, D., and S. C. Chapra. "Comparison of an Ecological Model of Lake Ontario and Phosphorus Loading Models," *J. Fish. Res. Bd. Can.* 34:286-290 (1977).

Scavia, D., B. J. Eadie and A. Robertson. "An Ecological Model for Lake Ontario Model Formulation, Calibration, and Preliminary Evaluation," NOAA, Technical Report ERL 371GLERL 12, (Ann Arbor, MI: National Oceanic and Atmospheric Administration, 1976a), 63 p.

Scavia, D., B. J. Eadie and A. Robertson. "An Ecological Model for the Great Lakes," in *Environmental Modeling and Simulation,* W. R. Ott, Ed. EPA 600/9-76-016 (Washington, DC: U.S. Environmental Protection Agency, 1976b), p. 629-633.

Scavia, D., and R. A. Park. "Documentation of Selected Constructs and Parameter Values in the Aquatic Model *CLEANER,*" *Ecol. Model.* 2:33-58 (1976).

Schelske, C. L. "Silica and Nitrate Depletion as Related to Rate of Eutrophication in Lakes Michigan, Huron, and Superior," in *Coupling of Land and Water Systems, Ecological Studies, Vol. 10,* A. D. Hasler, Ed. (New York: Springer-Verlag New York, Inc., 1975), p. 277.

Schelske, C. L., and E. F. Stoermer. "Phosphorus, Silica and Eutrophication of Lake Michigan," in *Nutrients and Eutrophication,* G. E. Likens, Ed. (Am. Soc. Limnol. Oceanog., Spec. Symp. I, 1972), p. 167-171.

Schindler, D. W. "Production of Phytoplankton and Zooplankton in Cahadian Shield Lakes," in *Productivity Problems of Freshwater,* Z. Kajak and A. Hillbricht-Ilkowska, Eds. (Warsaw: Polish Scientific Publishers, 1972), pp. 311-332.

Semenova, L. M. "The Feeding Habits of *Bosmina coregoni* Baird (Cladocera)," *Hydrobiol. J.* 10(3):28-34 (1974) (translation series).

Shapiro, J. "A Statement on Phosphorus," *J. Water Poll. Control. Fed.* 42:772-775 (1970).

Shapiro, J. "Blue-Green Algae: Why They Became Dominant," *Science* 179:382-384 (1973).

Shapiro, J. "Biomanipulation—A Neglected Approach?," presented at the 40th Annual Meeting of the American Society of Limnology and Oceanography, East Lansing, MI, June 1977.

Simons, T. J. "Analysis and Simulation of Spacial Variations of Physical and Biochemical Processes in Lake Ontario," *J. Great Lakes Res.* 2:215-233 (1976).

Smayda, T. J. "The Suspension and Sinking of Phytoplankton in the Sea," *Oceanog. Mar. Biol. Ann. Rev.* 8:353-414 (1970).

Smayda, T. J. "Some Experiments on the Sinking Characteristics of Two Freshwater Diatoms," *Limnol. Oceanog.* 19:628-635 (1974).

Smayda, T. J., and B. J. Boleyn. "Experimental Observations on the Floatation of Marine Diatoms, II. *Skeletonema costatum & Rhizosolenia setigera,*" *Limnol. Oceanog.* 11:18-34 (1966a).

Smayda, T. J., and B. J. Boleyn. "Experimental Observations on the Floatation of Marine Diatoms, III *Bacteriastrum hyalinum* and *Chaetoceros lauderi,*" *Limnol. Oceanog.* 11:35-43 (1966b).

Sonzogni, W. C. "Effect of Nutrient Input Reduction on the Eutrophication of the Madison Lakes," Ph.D. Thesis, University of Wisconsin, Madison, WI (1974).

Sonzogni, W. C., D. P. Larsen, K. W. Malueg and M. D. Schuldt. "Use of Large Submerged Chambers to Measure Sediment—Water Interactions," *Water Res.* 11:461-464 (1977).

Stadelmann, P., and A. Fraser. "Phosphorus and Nitrogen Cycle on a Transect in Lake Ontario during the International Field Year 1972-1973 (IFYGL)," in *Proc. 17th Conference Great Lakes Research* (1974), pp. 92-107.

Stadelman, P., J. E. Moore and E. Pickett. "Primary Productivity in Relations to Temperature Structure, Biomass Concentration, and Light Conditions at an Inshore and Offshore Station in Lake Ontario," *J. Fish. Res. Bd. Can.* 31:1215-1232 (1974).

Steele, J. H. "Environmental Control of Photosynthesis in the Sea," *Limnol. Oceanog.* 7:137-150 (1962).

Steele, J. H. "Notes on Some Theoretical Problems in Production Ecology," in *Primary Production in Aquatic Environments,* C. R. Goldman, Ed. (Berkeley, CA: University of California Press, 1965), p. 383.

Steele, J. H. *The Structure of Marine Ecosystems,* (Cambridge, MA; Harvard University Press, 1974), 128 p.

Steele, J. H., and M. M. Mullin. "Zooplankton Dynamics," in *The Sea Vol. 6: Marine Modeling,* E. D. Goldberg, I. N. McCave, J. J. O'Brien and J. H. Steele, Eds. (New York: Wiley-Interscience, 1977), p. 857.

Stewart, R. W. "Physical Modeling," in *Modeling of Marine Systems,* Oceanography Series, Vol. 10, J. C. J. Nihoul, Ed. (New York: Elsevier, 1975), p. 155.

Stoermer, E. F., M. M. Bowman, J. C. Kingston and A. L. Schaedel. "Phytoplankton Composition and Abundance in Lake Ontario during IFYGL," U.S. Environmental Protection Agency, Corvallis, Oregon, EPA-660/3-75-004 (1975), 373 p.

Strickland, J. D. H. "Measuring the Production of Marine Plankton," Bull. No. 122 *Fish. Res. Bd. Can.* (1960).

Strumm, W., and J. O. Leckie. "Phosphate Exchange with Sediments, its Role in the Productivity of Surface Waters," in *Advances in Water Pollution Research, Vol. 2,* S. H. Jenkens, Ed. (New York: Pergamon Press, 1970), pp. III-26/1-16.

Sundaram, Y. R., and R. G. Rehm. "The Seasonal Thermal Structure of Deep Temperate Lakes," *Tellus* 25:157-167 (1973).

Sushchenya, L. M. "Food Rations, Metabolism and Growth of Crusta-ceans," in *Marine Food Chains,* J. H. Steele, Ed. (Berkeley, CA: University of California Press, 1970).

Taguchi, S. "Relationship between Photosynthesis and Cell Size of Marine Diatoms," *J. Phycol.* 12:185-189 (1976).

Tarapchak, S. J., and E. F. Stoermer. "Environmental Status of the Lake Michigan Region," *Vol. 4 Phytoplankton of Lake Michigan,* Argonne National Laboratory, Environmental Central Technology and Earch Sciences, Argonne, IL, (UC-11) ANL/ES-40 (1976).

Thomann, R. V., D. M. DiToro and D. J. O'Connor. "Preliminary Model of Potomac Estuary Phytoplankton," *J. Env. Eng. Div.,* ASCE 100 (EE2):699-715 (1974).

Thomann, R. V., D. M. DiToro, R. P. Winfield and D. J. O'Connor. "Mathematical Modeling of Phytoplankton in Lake Ontario, Part 1. Model Development and Verification," U.S. Environmental Protection Agency, Corvallis, Oregon, EPA-660/3-75-005 (1975).

Thomann, R. V., and R. P. Winfield. "On the Verification of a Three Dimensional Phytoplankton Model of Lake Ontario," in *Environmental Modeling and Simulation,* W. T. Ott, Ed. EPA-600/9-76-016, (Washington DC: U.S. Environmental Protection Agency, 1976), pp. 568-572.

Thomas, N. A. "Accumulation and Transport of Energy-Related Pollutants by Benthos," in *Proc. of the Second Federal Conf. on the Great Lakes,* (Great Lakes Basin Commission, 1975), pp. 361-365.

Tilman, D., and S. S. Kilham. "Phosphate and Silicate Growth and Uptake Kinetics of the Diatoms *Asterionella formosa* and *Cyclotella meneghiniana* in Batch and Semicontinuous Culture," *J. Phycol.* 12:375-383 (1976).

Titman, D., and P. Kilham. "Sinking in Freshwater Phytoplankton: Some Ecological Implications of Cell Nutrient Status and Physical Mixing Processes," *Limnol. Oceanog.* 21:409-417 (1976).

Ursin, E. "A Mathematical Model of some Aspects of Fish Growth, Respiration, and Mortality," *J. Fish Res. Bd. Can.* 24:2355-2392 (1967).

Vanderhoef, L. N., C. Huang, R. Musil and J. Williams. "Nitrogen Fixation (Acetylene Reduction) by Phytoplankton in Green Bay, Lake Michigan, in Relation ot Nutrient Concentrations," *Limnol. Oceanog.* 19:119-125 (1974).

Vanderploeg, H. A., and D. Scavia. "Calculation and Use of Selectivity Coefficients of Feeding: Zooplankton Grazing," *Ecol. Model.* (in press).

Vollenweider, R. A. "Calculation Models of Photosynthesis—Depth Curves and some Implications Regarding Day Rate Estimates in Primary Production Measurements," in *Primary Productivity in Aquatic Environments*, C. R. Goldman, Ed. (Berkeley, CA: University of California Press, 1965), p. 425.

Walsh, J. J. "A Biological Sketchbook for an Eastern Boundary Current," in *The Sea Vol. 6: Marine Modeling*, E. D. Goldberg, I. N. McCave, J. J. O'Brien and J. H. Steele, Eds. (New York: Wiley-Interscience, 1977), p. 923.

Walsh, J. J., and R. C. Dugdale. "Nutrient Submodels and Simulation Models of Phytoplankton Production in the Sea," in *Nutrients in Natural Waters*, H. E. Allen and J. R. Kramer, Eds. (New York: Wiley-Interscience, 1972), p. 171.

Welch, E. B., D. A. Rock and J. D. Krull. "Long-Term Recovery Related to Available Phosphorus," in *Modeling the Eutrophication Process*, E. J. Middlebrooks, D. H. Falkenborg and T. W. Maloney, Eds. (Ann Arbor, MI; Ann Arbor Science Publishers, Inc., 1973), p. 5.

Wells, L. "Effects of Alewife Production on Zooplankton Populations in Lake Michigan," *Limnol. Oceanog.* 15:556-565 (1970).

Wetzel, R. G. *Limnology* (Philadelphia, PA: W. B. Saunders Company, 1975), 743 p.

Wetzel, R. G., P. H. Rich, M. C. Miller and H. L. Allen. "Metabolism of Dissolved and Particulate Detrital Carbon in a Temperate Hard-Water Lake," in *Detritus and its Role in Aquatic Ecosystems*, Nelchiorri-Santolini, and J. W. Hopton, Eds. (Memorie Dell'Instituto Italiano di Idrobiologia, 29 Suppl., 1972), pp. 185-244.

Williams, J. D. H., and T. Mayer. "Effects of Sediment Diagenesis and Regeneration of Phosphorus with Special Reference to Lakes Erie and Ontario," in *Nutrients in Natural Waters*, H. E. Allen and J. R. Kramer, Eds. (New York: John Wiley & Sons, Inc., 1972), p. 281.

Wilson, D. S. "Food Size Selection among Copepods," *Ecology* 54:909-914 (1973).

Wood, L. W. "Role of Oligochaetes in the Circulation of Water and Solutes across the Mud-Water Interface," *Verh. Int. Ver. Limnol.* 19:1530-1533 (1975).

Wright, R. T., and J. E. Hobbie. "Use of Glucose and Acetate by Bacteria and Algae in Aquatic Ecosystems," *Ecology* 47:447-464 (1966).

Wyeth, R. K., and J. Ploscyea. "Effects of Genesee River Discharge and Wind-Induced Resuspension on the Nearshore Area of Lake Ontario," International Joint Commission Report, Windsor, Ontario (1978).

Zeuthen, E. "Body Size and Metabolic Rate in the Animal Kingdom," *Comp. Rend. Lab. Carlsberg, Ser. Chim.* 26:17-161 (1947).

Zeuthen, E. "Oxygen Uptake as Related to Body Size in Organisms," *Quart. Rev. Biol.* 28:1-12 (1953).

SECTION THREE

MODELS IN MANAGEMENT

The use of mathematical models for environmental management purposes has become common; however, techniques for testing model adequacy for these purposes have not been sufficiently developed and standardized and, therefore, the validity of using a specific model for a specific purpose is often unknown. Standard procedures for testing model adequacy, especially for management purposes, are certainly needed; the first two chapters in this section address this need. The third chapter addresses an equally important management problem—surveillance and trend detection.

In Chapter 7, W. J. Snodgrass reviews several Great Lakes' models in regard to their adequacy for management purposes and demonstrates how some models achieve and others fall short of their stated goals. In this chapter, he outlines the general procedures for model development and testing that ensure a model that is acceptable for predictive use. In Chapter 8, K. H. Reckhow describes the use of first-order uncertainty analysis for establishing bounds on predictions made by mathematical models. He uses the analysis technique to compare prediction confidence among several models of total phosphorus dynamics. In the final chapter of this section, L. M. DePalma, R. P. Canale and W. F. Powers combine a stochastic model of total phosphorus with linear filter theory and optimization procedures to assist in designing a minimum-cost surveillance program for Lake Michigan. The combined procedures are used to determine the spatial and temporal frequency of sampling required to detect year-to-year trends in total phosphorus and chloride.

CHAPTER 7

PREDICTIVE WATER QUALITY MODELS FOR THE GREAT LAKES: SOME CAPABILITIES AND LIMITS

William J. Snodgrass

Departments of Civil and Chemical
   Engineering
McMaster University
Hamilton, Ontario
Canada L8S 4L7

INTRODUCTION

The eutrophication of the Great Lakes is a major concern of water quality managers. While universal agreement about its definition, causes and socioeconomic consequences does not exist, the detrimental effects of eutrophication are generally undeniable. Eutrophication is characterized by excess growth of algae, detritus accumulation, partial or total depletion of oxygen in the hypolimnion, fish kills and other associated nuisances. It can destroy the utility of lakes as a water resource.

Cultural eutrophication arises from increased inputs of nutrients due to man's activities. Control can also be accomplished by man. Many procedures for controlling eutrophication are presently available; it is plausible that new methods will be developed. Existing methods for controlling nutrient inputs include: (1) treatment of municipal and industrial wastes for phosphorus and/or nitrogen removal, (2) removal of phosphorus from detergents and (3) modification of agricultural practices, such as regulating the procedures, timing and extent of fertilizer application and prohibiting fertilizer use within a narrow border zone along all water courses. Existing methods for direct management of lakes include: (1) mixing the water with air or thermal discharge to achieve destratification

171

(*e.g.,* Hamilton Harbor, Ontario Ministry of Environment, 1976), (2) dredging the sediments, and (3) withdrawing nutrient-rich bottom waters from reservoirs for downstream discharge.

In any given situation, the costs of these management procedures can be determined with relative ease and accuracy. However the results (benefits) of such procedures can only be predicted with assurance if quantitative models, capable of predicting the response of a lake to a given management stretegy, are available. The purposes of this chapter are to present a framework for assessing the capabilities and limitations of various predictive water quality models for the Great Lakes and to apply this framework to various models presently found in the literature.

A model is an abstraction of reality that describes, in either qualitative or quantitative terms, a certain set of the complex interrelationships of the system being studied. A quantitative model is described herein as being either empirical or fundamental. An empirical quantitative model involves a statistical relationship between two or more variables. A fundamental quantitative model is based upon the law of conservation of mass.

All models reviewed below are fundamental quantitative models describing part or all of the St. Lawrence Great Lakes System. These models have the following objectives: (1) to provide a useful predictive basis for making decisions concerning water quality management alternatives and/or (2) to describe mathematically the most important interactions between various nutrient cycles and biological species. The utility of such models for assessing the effects of water quality management alternatives on eutrophication is determined by the degree to which they describe either useful indicators of eutrophication or the dynamics of eutrophication.

THE MODELING PROCESS

The modeling process used in deriving a fundamental quantitative model can by described by several steps. Various investigators have their own set of steps (*e.g.,* Orlob, 1975); however, this author finds it most useful to use the following six steps (Snodgrass, 1974):

1. selection of model objectives
2. system discretization
3. model construction
4. model calibration
5. model verification
6. model prediction

A model is constructed to fulfill certain objectives. Different objectives necessitate different spatial and temporal scales. For example, one's objective may be to develop a model capable of predicting the average annual concentration of total phosphorus in a lake for a given rate of nutrient supply from land-based sources. Alternatively, one's objective may be to develop a model capable of predicting the timing and extent of an algal bloom in a lake. The former objective involves a temporal scale of the order of six months to one year, whereas the latter objective involves a temporal scale of days to weeks. Also, one may wish to predict average total phosphorus concentrations in either the nearshore zone or the whole lake. The former case involves a spatial scale of the order of the nearshore zone, whereas the latter case involves a spatial scale of the order of the whole lake.

System discretization entails the physical, biological and chemical representation of the water body. The system boundaries are selected based on surveys of the physical, biological and chemical characteristics of the system and surveys of the flows of mass or energy into and out of the system. Selection of the system boundaries is made to be consistent with the model objectives. Herein, a box (*e.g.,* epilimnion) is considered as one of a spatially distributed set of entities each of which has the capacity to store mass or energy; the total number of boxes describes the physical structure of the entire system. A box is chosen such that flows of mass or energy occur between interconnected boxes, but such that there is no spatial variation of mass or energy within the box. Thus, the lake is divided into volumes (boxes) which are each sufficiently homogeneous such that physical, chemical or biological detail necessary for the realization of a model's objectives is not lost. A box is subdivided into compartments, each describing a different biological or chemical entity deemed to be important to fulfill the model's objective(s). Thus, each compartment describes a different form of mass or energy found within a box (*e.g.,* orthophosphate, detrital phosphorus and all biological forms of phosphorus); compartments within a box describe all of the forms of mass or energy stored within that box. Physical, biological and chemical flows of mass or energy may occur between interconnected compartments.

As Okubo (1971) notes, "the box model treats mixing 'averaged' over each box and attempts to see changes only as between boxes. The mixing processes at the interfaces of the boxes are parametrically disguised as exchange - or transfer-rate constants with the dimensions of (t^{-1})." Okubo also notes that other investigators have found the well-mixed assumption unnecessary for successful box model application, and have related the box transfer-rate coefficients to advective and eddy diffusivity concepts of models for plug flow with dispersion. Treatment of a lake as one box is similar to the concept of a continuously stirred tank reactor (CSTR) used by chemical engineers (*e.g.,* see Levenspiel 1962).

Construction of a model involves five basic steps:

1. identifying the physical, biological and chemical processes and the corresponding physical laws, and biological and chemical kinetics governing the rates of mass flow between compartments and/or boxes (*e.g.,* photosynthesis, advective transport);
2. listing the assumptions made, including simplifications of physical laws;
3. constructing systems of mathematical equations which describe the behavior of the system. The system of equations is constructed by writing a statement of conservation of momentum, energy or mass for each compartment in each box. For example, the rate of change of mass equals the difference between the rate of mass input and the rate of mass output;
4. evaluating the boundary conditions; and
5. solving the mathematical equations.

Model calibration consists of selecting a set of coefficient values from field measurements and/or literature values such that the model output duplicates a set of *in situ* measurements of some known system. The coefficients are constants describing empirical relations where fundamental laws are unavailable. Their uniqueness is partially a function of the spatial and temporal structure of the model. Selection of values for coefficients for which field measurements have not been made is usually done by varying the coefficient values over the range of reported literature values until satisfactory agreement is reached between model predictions and environmental observations. Coefficient values obtained from field measurements may also be varied if the modeler decides that a measurement does not describe conditions throughout a compartment. Alternatively coefficient values may be determined using an optimization method which minimizes the difference between model predictions and observations.

Model verification consists of using the calibrated model to predict one or more sets of system conditions independent of the first set. If the verification is acceptable, then the model may be used for the predictive purposes described in the model objectives. If the verification is not acceptable, it may be necessary to modify the model using one of the first three steps listed above, particularly model construction. Model predictions may be used with confidence for those objectives for which model is verified.

For verification, several tests are possible. The first (least severe and insufficient) test is to compare the model structure—compartments and intercompartment mass flows—with the real world to determine that functional responses are reasonable and that all major factors are included. A second test involves comparing model predictions and environmental observations for one lake for one time period (*e.g.,* for a month or year) equal to or longer than the time scale of the model's objectives. The third (more severe) test involves two possibilities: comparisons between model predictions and environmental observations for several lakes and/or comparisons between model predictions and observations from one lake for a time period equal to several time scales. A fourth test involves comparison between model predictions and observations for both several time scales on one lake and one time scale on several lakes. The standard used to determine the adequacy of agreement between model predictions and observations may be either qualitative (*e.g.*, judgment, by eye) or quantitative (*e.g.*, use of appropriate statistics and lack of fit test); to the present time, it has generally been qualitative. (See Reckhow, Chapter 8, for a discussion of some quantitative techniques.)

CRITIQUE OF SOME PREDICTIVE WATER QUALITY MODELS FOR THE GREAT LAKES

The literature contains many excellent water quality models for the Great Lakes, which have objectives of varying degrees of complexity. These models represent a spectrum of completeness. This critique is not designed to be comprehensive. Rather, selected quantitative box models are critiqued to illustrate various points concerning the modeling process.

Models Describing Chemical Species

O'Connor and Mueller (1970) were among the first investigators to calibrate and verify successfully the box model approach for predictive water quality purposes in lakes. Their objective was to predict the temporal and lake-by-lake distribution of a conservative material (chloride) for the Great Lakes system. Their system selection is shown in Figure 1a. Each lake was treated as one box, within which the rate of accumulation of mass was determined by the rates of mass inflow and mass outflow. The term conservative defines the condition of no internal addition or removal of the material within the lake (*e.g.*, due to sedimentation or chemical or biological reaction). Each box was treated as a subsystem of the Great Lakes system, with mass from the Lake Superior and Lake Michigan boxes flowing into the Lake Huron box and then through the Lake St. Clair box into the Lake Erie box and into the Lake Ontario box and, finally, out of this box via the St. Lawrence River. Conditions in Lake St. Clair were not considered, presumably since it has a relatively short detention time compared to the other lakes. The lakes were treated as two CSTRs in parallel connected to three CSTRs in series. The model was calibrated using data for 1960. The general applicability of this box or CSTR approach was verified by using water quality data and chloride inputs for over a 60-year period. Model predictions over time (1900 to

1960) were made for the five lakes; model predictions for Lake Ontario are shown in Figure 1b. Agreement was good and hence the model was considered to be verified for predictive purposes. This verification permitted predictions of future chloride concentrations that would result if alternative management plans to control various kinds of chloride discharges in the different lake basins were initiated.

The assumption that a substance supplied to a lake does not enter into a chemical or biological reaction is only rarely met in a natural system. Even chloride in lakes may be nonconservative, due to slight losses to the sediments. Phosphorus, of course, must be modelled as a nonconservative substance.

Snodgrass (1974) and Snodgrass and O'Melia (1975) developed and verified a model that predicts the average concentrations of total phosphorus in a lake when the rate of nutrient supply, the mean depth and the hydraulic loading are considered. Phosphorus compartments utilized for the summer period are shown in Figure 2a. For summer stratification, the lake was divided into two boxes—the epilimnion overlying the hypolimnion. Winter circulation was approximated by combining the two layers into one box. Each box was composed of only two compartments— orthophosphate [OP] and particulate phosphorus [PP]. Each arrow in Figure 2a represents a mass flow of phosphorus between compartments. The hydraulic input and discharge occur respectively to and from the epilimnion during summer, but to and from the whole lake during winter. The summer equations are joined to the winter equations by appropriate boundary conditions. The model was calibrated using data for Lake Ontario, and model verification is presented in Figure 2b. The agreement between observed and predicted phosphorus concentrations was considered excellent. This verification permitted the authors to conclude that, for lakes whose hypolimna remain oxic, their model is verified to make predictions concerning two questions: (1) What will be the response of total phosphorus concentrations to a change in phosphorus inputs? (2) For a desired concentration of phosphorus in the lake, what reduction in inputs is necessary? Lack of sufficient field data prevented verification of the model's predictions of temporal response trends.

The U.S. Army Corps of Engineers (1975) applied a model to predict the long-term total phosphorus concentration in Lake Erie. The model was based upon that of Lorenzen et al. (1976). The temporal scale of the model was annual, the spatial scale was the whole lake. The lake was spatially divided into a box for the water (in contrast to Snodgrass and O'Melia, 1975; Thomann et al., 1975, etc., who divided the lake water into two boxes) and a box for the sediments. The model included hydraulic inflow and outflow and mass exchange of total phosphorus

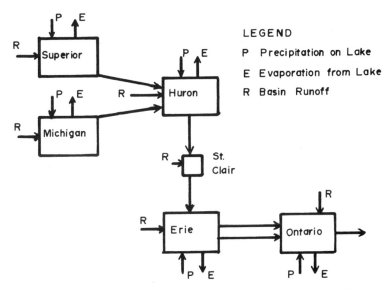

Figure 1a. Boxes of Great Lakes (after O'Connor and Mueller, 1970).

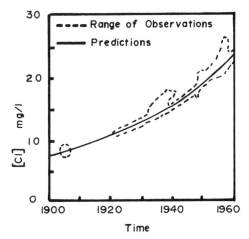

Figure 1b. Comparison of observed and predicted chloride concentrations (after O'Connor and Mueller, 1970) for Lake Ontario.

SUMMER STRATIFICATION (A)

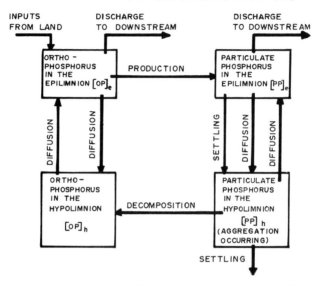

Figure 2a. Compartment model for P during summer stratification (after Snodgrass and O'Melia, 1975).

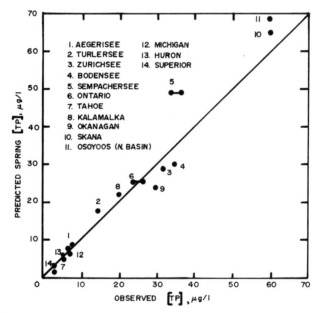

Figure 2b. Comparison of model predictions and observations (after Snodgrass and O'Melia, 1975).

between the water and the sediments and between the shallow sediments and the deeper nonexchanging sediments. The model was calibrated by Lorenzen *et al.* for the period 1941 to 1950 for Lake Washington. The model was verified by comparing observations to predictions for the period 1950 to 1970. During this period phosphorus concentrations increased (due to population expansion) and then decreased (due to diversion to Lake Sammamish) by threefold. The model was applied to Lake Erie by considering the lake as three CSTRs (the western basin, the central basin and the eastern basin) in series with advective flow in only one direction. The model was recalibrated for Lake Erie in 1973 and estimates made of expected phosphorus concentrations for different management strategies for the period 1975 to 1990.

Chapra (1977) developed a model to predict total phosphorus concentration in each of the Great Lakes. Its objective has a spatial scale of a whole lake or basin and a temporal scale of a year (variability within a year is ignored). A hydraulic flow model similar to O'Connor and Mueller's (1970) was employed, except Lake Erie was considered as three basins in series (west, central and east) rather than as one lake. Estimates of phosphorus inputs were made for 1950 to 1975 using estimates of inputs from diffuse, point and atmospheric sources. The only sink (apparent setting) was calibrated using observations of others. Lack of adequate historical data for total phosphorus from 1950-1965 prevented verification of the time-response predictions of the model. Comparison of model predictions for total phosphorus with observations during 1965-1975 showed excellent agreement (correlation coefficient $r = 0.94$) for the seven water bodies. Chapra then used the model to predict the magnitude (extent) and rate of recovery of the lakes in response to phosphorus removal programs currently in progress (reduction of total phosphorus in effluents of all point sources to 1 mg-P l^{-1}).

Richardson (1976) and Canale and Squire (1976) developed models to predict spatial distribution of a substance. They independently developed models to simulate chloride transport in Saginaw Bay, Lake Huron. Both investigators: (1) spatially divided the bay into interconnected boxes among which dispersive and advective transport occurred, (2) considered vertical stratification insignificant, (3) used steady-state conditions, and (4) assigned land-based inputs to the appropriate boxes. Richardson (1976) calibrated the transport model by determining the set of advective and dispersive rates which caused model predictions of chloride to correspond to 1965 observations. Subsequent comparison with 1974 data showed poor agreement. For July to November 1974, the model showed good steady-state agreement, but for March to June 1974, poor agreement. When solving the model for other than steady-state conditions, the effects of thermal bar development caused the need for a new calibration. Accordingly, the transport model is considered to be calibrated for the spring period and verified for the summer-fall period.

Canale and Squire (1976) calibrated the transport and dispersive terms of their model with a numerical circulation model of the bay developed by others which uses stage height, current meter and drogue data. Model predictions of spatial chloride distribution showed reasonable agreement with observations for September 1974 and November 1974. This permitted the conclusion that the transport model is verified. Treating total phosphorus as unreactive and considering land-based inputs, poor agreement was obtained between predictions and observations—predictions were too low. Inclusion of atmospheric sources had an insignificant effect. The authors concluded that scour and resuspension sources for total phosphorus 2-3 times the input rate from land-based sources are needed. With this input, predictions of spatial distribution agreed with observations and hence the loading for total phosphorus was recalibrated.

Models Describing Biological-Chemical Species

Thomann et al. (1975) developed a model to predict the dynamics of phytoplankton in Lake Ontario. The lake was spatially divided into two horizontally well-mixed layers, the epilimnion and the hypolimnion. Each layer was divided into ten biological or chemical compartments including four trophic levels above the phytoplankton, chlorophyll a as a measure of the phytoplankton biomass, two phosphorus compartments and three nitrogen compartments (see Figure 3a). Preliminary and detailed calibrations were carried out by extensively varying various model coefficients until a set of parameter values were obtained which satisfactorily explained the observed data for 1967-1970. Model verification was attempted by comparing model predictions with observations from 1967-1970 for six epilimnetic variables and four hypolimnetic variables (see Figure 3b for epilimnetic phytoplankton verification). Agreement between observations and model output for the timing and extent of the spring peak of the subsequent midsummer die-off and of the fall bloom of epilimnetic phytoplankton was considered quite good. The comparison for the other five variables was considered satisfactory. This agreement allowed Thomann et al. to conclude that their model is verified for the objective of making predictions concerning the timing and extent of phytoplankton blooms in a lake.

Simons (1976) used the model of Thomann et al. (1975) as a diagnostic tool for Lake Ontario. A hydrodynamic model, previously verified, was used to estimate horizontal and vertical transport among twenty-one

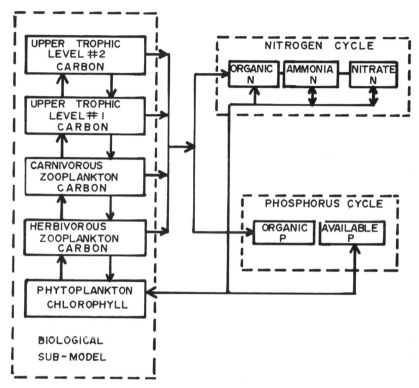

Figure 3a.   System diagram for biological/chemical model (after Thomann *et. al.*, 1975).

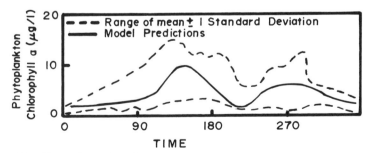

Figure 3b.   Comparison of model predictions and observations for lakewide mean of chlorophyll (observations are for period 1967-1970; after Thomann *et. al.*, 1975).

epilimnetic boxes and the corresponding boxes for three lower layers.
The biological and chemical model of Thomann *et al.* (1975), calibrated
to 1967-1970 data, was applied to each box. Comparison of the timing
of the peak of chlorophyll *a* in a nearshore epilimnetic zone to that in
a deep-water epilimnetic zone indicated that transport processes explain
the difference in timing (the nearshore peak of chlorophyll *a* occurs before that in the offshore zone). The essential features of average lakewide conditions derived from the segmented model were reproduced by
a horizontally mixed model. However, while the model predicted two
chlorophyll *a* peaks (spring and fall peaks were observed in the calibration
data set), the observations suggested that only one peak occurred in 1972.
This discrepancy led Simons to suggest model modifications which might
simulate the 1972 observations; however, no attempt was made to modify
or recalibrate the model.

Thomann *et al.* (1977) applied the model of Thomann *et al.* (1975)
to predict the long-term (20-yr) behavior of peak phytoplankton chlorophyll in Lake Ontario under four alternative loading schemes—continuation
of present loading, an estimated loading in the distant past and two proposed schemes for reducing inputs. Of interest here are their predictions
for present loadings; that is, those used for calibration. Thomann *et al.*
(1977) indicate that the model calibration of Thomann *et al.* (1975),
based on one year of simulation, provided an estimate only for loss of
phytoplankton to the sediments by settling. For this calibration, the
peak chlorophyll increased by approximately 50% over 20 years (*i.e.*, the
model was initial-value constrained—the calibrated model was not at steady
state with observed conditions). Only with assumptions that both organic
and inorganic forms of dissolved phosphorus and dissolved nitrogen decay
(analogous to radioactive decay and presumably due to some precipitation
mechanism) was the lake at steady state with present loadings.

Scavia *et al.* (1976) developed a model with objectives to examine
basic ecological properties of Lake Ontario and to identify deficient areas
of knowledge. For this purpose, a three-box model (epilimnion, thermocline and hypolimnion) was employed. For each box, phytoplankton (four
types), zooplankton (six types), detritus, nitrogen (organic, ammonia and
nitrate), phosphorus (available), benthic invertebrates and the carbonate
system were considered. The model was calibrated by assuming literature
values for the model coefficients and partially verified by comparing observations and predictions for various processes. Predictions for primary
production, epilimnetic turnover rate and carbon dioxide exchange rate
with the atmosphere showed reasonable agreement with observations;
predictions of settling rates were within the range of observations. Comparison of predictions with observations for various state variables (the

same variables used by Thomann *et al.*, 1975) showed reasonable agreement. These comparisons permitted the authors to have some confidence in using their model to examine various ecological theories and to make suggestions concerning the relative importance of various processes.

Canale *et al.* (1976) constructed a food-web model to describe the ecosystem of Lake Michigan. The long-term objective of the model was to use it as a management tool in assessing strategies designed to ameliorate eutrophication and the deterioration of Michigan's fishery. Their model, based upon forage-fish predation, described the lake as two boxes (the epilimnion and hypolimnion) and considered compartments for three elements (silicon, nitrogen and phosphorus), four phytoplankton types (small diatoms, large diatoms, blue-greens and greens), seven zooplankton types (including herbivores, omnivores and predators) and the alewife. While they considered that models including compartments for phytoplankton and herbivorous and predatory zooplankton (*e.g.*, Thomann *et al.*, 1975) were adequate to describe most of the critical interactions occurring during eutrophication, compartments for forage fish and omnivorous zooplankton were needed to describe the ecosystem of Lake Michigan. The size of the lake and lack of sufficient data precluded application to Lake Michigan. Thus data from Grand Traverse Bay were used for model calibration and to examine the chemical-biological interactions of the model.

For calibration, two approaches were used by Canale *et al.* (1976): (1) where coefficient values reported from the literature had a narrow range, the range was *not* violated, (2) where the reported range was so broad as to indicate scientific uncertainty, the coefficient was varied widely to achieve agreement between the model output and observations. Canale *et al.* noted that, particularly for zooplankton predation rates, the coefficient values require further experimental validation. The calibrated model simulated the seasonal trends of concentrations of zooplankton types, of phytoplankton types, of dissolved silica and of nitrate + nitrite for data from 1971 to 1973, composited to represent one year of data, but did not duplicate the seasonal trend of total dissolved phosphorus due to large scatter in the observed data. Canale *et al.* examined the effects of higher initial concentrations of phosphorus and forage fish on concentrations of plankton and nutrients. Such simulations, while instructive in a relative sense, must be viewed by water quality managers with great caution, as the model is not verified. Use of the model as a research guide, in which the most important model mechanisms are determined and then compared to weaknesses in scientific knowledge, is its most powerful use at present.

Chen *et al.* (1975) had as their objective the development of a comprehensive water quality ecological model for Lake Ontario. Physically, the lake was divided into seven layers with a sufficient number of of horizontal elements in each layer (*e.g.,* 715 elements in the surface layer) to permit predictions of mass transport (diffusion and advection) among boxes. These mass transport rates were agglomerated into a coarser series of boxes (41 for the surface layer), each of which contained biological and chemical compartments. The major compartments were nitrogen, phosphorus, silicon, algae, zooplankton, detritus, fish and organic sediment. Vertical exchange was permitted among the atmosphere, water column and sediments. The fish compartment was subdivided into three age groups and four major types. At present, the major effort has been directed towards conceptualization, model construction and some model simulation. Such simulations tested program logic and assured that the model structure simulated processes such as thermal stratification and the general spatial patterns of algal concentrations.

DISCUSSION

To make decisions among different management alternatives concerning the water quality of lakes, the decision maker can use mathematical models for two main purposes: (1) to predict the ultimate extent of lake change for a given control strategy of changing the inputs to the lake and (2) to predict the rate of response of the lake to the control strategy. In assessing different models for such uses, the decision maker should examine: (1) the model's objectives, (2) whether these objectives relate to the decision maker's purposes, (3) whether the model is verified to make predictions concerning these objectives, and (4) the temporal and spatial scales utilized in the model. These four points will be used to examine the models summarized above with respect to their capabilities and limitations for use by a decision maker.

O'Connor and Mueller (1970) had the objective of predicting the temporal response of the Great Lakes, on a whole-lake basis, to loadings of chlorides. Utilizing a time scale of 1-2 years and a spatial scale of the whole lake, they successfully verified the model's predictions not only for concentration but also for time-response characteristics. Hence their model is verified to predict on a yearly basis the extent and rate of recovery of a whole lake if control of discharges of chloride or similar conservative substances to the lake is initiated.

Developing models capable of **being** verified for nonconservative substances presents more difficulties. Problems with verification are caused in part by the nature of the substance being modeled. For example, phosphorus, a biologically reactive nonconservative substance, has cycling pathways and input sources which are less well characterized than those for chloride. Nitrogen is more difficult than phosphorus due to its potential exchange with the atmosphere. Models which attempt to simulate ecological behavior are even more complex and require more data for verification. Often the limiting factor for verification is lack of sufficient appropriate field data.

The model of Snodgrass and O'Melia (1975) provides a contrast to the verification achieved by O'Connor and Mueller. It is verified to make predictions concerning the extent of change of annual concentrations of total phosphorus in a whole lake upon initiation of a control strategy for reducing the amount of phosphorus entering the lake; however, the model is not verified to make predictions concerning the rate of recovery. Other predictions are also possible. To illustrate some of these, consider the following statements that can be made based on model results.

1. Deep lakes are able to assimilate higher nutrient loadings than shallow ones, a morphometric effect observed by several other limnologists.
2. Model predictions of total phosphorus concentrations are sensitive, in decreasing order, to the land-based phosphorus loading to the lake, the settling rate in the hypolimnion, the hydraulic outflow rate and the decomposition rate of particulate phosphorus in the hypolimnion. They are insensitive to the rate of production of particulate phosphorus in the epilimnion.
3. Vertical transport of soluble phosphorus from the hypolimnion to the epilimnion considerably exceeds the land-based loading of phosphorus to the epilimnion of Lake Ontario during the stratification period.
4. For Lake Ontario, it is predicted that 2.3 years will be required after initiation of phosphorus removal from wastewater before 50% of the expected water quality improvement is attained; 10.1 years will be required to achieve 95% of the expected improvement.

All of the predictions indicated above are derivable from the model; however, the model is verified to make some of these statements but not others. The model is verified to make predictions concerning total phosphorus concentrations; hence predictions concerning morphometric effects (Statement 1) and the sensitivity* of the model predictions to different coefficients (Statement 2) may be treated as valid. Due to lack of sufficient data, the model is not verified to make statements concerning fluxes. Hence any attempt to compare fluxes due to vertical transport to those from land-based sources (Statement 3) must be viewed as tentative. Similarly, lack of sufficient field data precludes

---

*Predictions concerning the sensitivity of model predictions allow statements concerning priorities in lake research, *i.e.,* to which parameters is the model most sensitive and hence which parameters are most important for characterizing factors affecting total phosphorus concentrations in lakes.

verification of model predictions for temporal trends[*], hence predictions concerning the response rate of Lake Ontario to water quality management (Statement 4) must also be viewed as tentative. These examples point out the need to use model predictions with much care. Inferences may be drawn from model predictions which are not verified, but such predictions can be considered as indicative of only model behavior until the predictions are further tested and verified. Such unverified predictions should not be used by decision makers for management purposes without extreme caution.

The U.S. Army Corps of Engineers (1975) applied a model to Lake Erie which was calibrated and verified for a smaller lake (Lake Washington, Lorenzen *et al.,* 1975). Such verification provides some confidence in its use for Lake Erie but the model was recalibrated for Lake Erie. The same essential ability for predicting the timing and extent of recovery of Lake Washington can also be obtained using the model of Snodgrass and O'Melia (1975) and Snodgrass (1974). Since the model of Snodgrass and O'Melia does not consider resolubilization of phosphorus from the sediments, the sediment-water interactions probably have a small effect in the model of Lorenzen *et al.* (1976), and hence the sediment-water part needs further testing. Accordingly, the absolute predictions of the U.S. Corp's model should be viewed with much caution, but, when these are used in a relative sense to test alternate management strategies, this author believes that some confidence is permitted.

Chapra (1977) calibrated and verified a model describing present total phosphorus concentrations in the Great Lakes; verification of the rate of response of the lakes to changes in phosphorus inputs was not achieved due to lack of appropriate field data. Chapra's model is similar to the work of Snodgrass and O'Melia (1975). Hence, predictions of Chapra's model concerning rate of recovery of the Great Lakes may be used in the same way as those from the U.S. Army Corps of Engineers model (1975), that is, to test the relative effect of alternative management strategies. The verification of the magnitude of phosphorus concentrations over a range may allow one more confidence in the use of Chapra's model than that of the U.S. Army Corps of Engineers.

Richardson's (1976) work, which predicts spatial variation of substances in bays, illustrates that even verification of a chloride model is not assured after its calibration. The more extensive the testing of a model, the more confident one may be of its predictions. The verification

---

[*]This lack of sufficient field data has been a general problem that has been partially corrected in the past few years. Several lakes have had wastewater inputs diverted, but lack of measurements of inputs of phosphorus to the lake prior to diversion prevent complete confidence in using these lakes as sources for verifying the model's predictions of temporal trends.

achieved by Canale and Squire (1976) under two different conditions permits confidence in their chloride model. Canale and Squire assert that their phosphorus model is verified. Since the phosphorus model was recalibrated to obtain estimates of inputs due to processes such as scour, it needs to be verified with another set of conditions before its predictions can be used with confidence. Since estimates of inputs due to scour vary and they are not explained by the model, further model development may be in order.

Thomann *et al.* (1975) successfully calibrated their model for predicting the timing and extent of phytoplankton peaks in Lake Ontario using data for different years during 1967-1970. Using composite data for the same period, they assert that their model verified on the basis of good agreement between model predictions and observed values of the trends and concentrations of several compartments. This assertion may be criticized, however. In model building, one uses one set of data for calibrating the model and an *independent* set of data for verifying the model. Until a model is verified, it is only a calibrated model which should not be used for predictive purposes with complete confidence. Since Thomann *et al.* used data from the same set for both calibration and verification, their "verification procedure" is not model verification in the strict sense. Their model is, at present, only a well-calibrated model and its predictions should be used with some caution by the decision maker.

The studies of Simons (1976) and Thomann *et al.* (1977) show methods other than using an independent data set for model testing. Simon's work was mainly diagnostic, particularly comparing the effects of transport in a segmented model to those in a horizontally well-mixed model. The work of Thomann *et al.* permits a partial verification without long-term data if a theory exists which describes whether or not Lake Ontario is at steady state with its current input rates of nutrients.

Two other points concerning the model verification of Thomann *et al.* (1975) are relevant. The same basic model structure was used to predict the response of phytoplankton in one area of the San Joaquin estuary (DiToro *et al.*, 1971). Successful calibration of the model of DiToro *et al.* for Lake Ontario and its portability to this lake system provide some verification for the generality of the model structure. The second point concerns the length of time between data sets needed to assure independence for verification purposes. If the detention time of a lake is a number of years (*e.g.*, eight years for Lake Ontario) and if calibration is made using data from one year, use of data from the next year will not constitute sufficient verification for conservative materials (*e.g.*, chloride) or slowly reacting materials as the model predictions for the second year will be dominated by the initial conditions of the model for the first year. The appropriate time between data sets to assure independence depends upon the material. One possible time scale is the hydraulic detention time of the lake for materials whose main sources are from land-based inputs (*e.g.,* phosphorus). A less conservative time scale is the detention time of the substance itself (detention time of substance = mass in water column/rate of input of mass from land-based sources; *e.g.,* for Lake Ontario, the detention time of phosphorus is approximately three years). For substances which exchange with the atmosphere, (*e.g.,* nitrogen, oxygen) the appropriate time scale may be the substance's detention time defined as mass divided by rate of input from land-based and atmospheric sources. For substances which are formed *in situ* (*e.g.,* phytoplankton, chlorophyll), the appropriate time scale can range from the substance's turnover time to the time between spring oscillations. For oscillatory phenomena, several criteria (*e.g.,* for different mixing conditions, wind conditions and rates of input of phosphorus) should be used for model verification to assure confidence in its use by the decision maker.

The models of Scavia *et al.* (1976), Canale *et al.* (1976) and Chen *et al.* (1975) have been calibrated to varying degrees. That of Scavia *et al.* appears to be well-calibrated and partially verified, but various predictions of ecological behavior need further testing. In fact, the model structure is, in part, similar to that of Thomann *et al.* (1975). Use of literature values for model coefficients for calibration plus the degree of agreement achieved between model predictions and observations suggest that that there may be a consistent set of values for several coefficients which is valid for ecological models of varying degrees of complexity. This writer suggests that such consistency (or generality) may result because predictions of the different models are sufficiently insensitive to these coefficients. The model of Canale *et al.* (1976) is calibrated but some of the values of coefficients, particularly for zooplankton, require further experimental testing. The model's greatest utility is as a research guide in which the most important model mechanisms are determined and compared to ecological knowledge to detect weaknesses. The model of Chen *et al.* (1975) requires extensive testing and calibration. Due to the small time steps required for numerical calculation, simulation of one month on the computer is feasible, but it is very expensive to simulate the model for five full years. Adequate field data for verification of Chen *et al.'s* model may not be obtainable unless another very large field study similar to the 1972 IFYGL (International Field Year for the Great Lakes) study is conducted. Since, at this time, the objectives of all three models center around understanding ecological cycles, their value is as a research tool rather than as a tool for making water quality predictions for the immediate use of decision makers.

From the point of view of contrasting model objectives, consider the models of Snodgrass and O'Melia (1975), Chapra (1977) and Thomann *et al.* (1975). The objective of Snodgrass and O'Melia and of Chapra was to predict total phosphorus concentrations, not to construct an eutrophication model. Their models are valuable, hence, to the extent that phosphorus is an adequate parameter for measuring the trophic state of a lake. In contrast, the model of Thomann *et al.* attempts to be more representative of eutrophication by predicting not only total phosphorus concentrations but also algal dynamics; both are indicators of eutrophication. The greater difficulties of Thomann *et al.* in attempting to verify their model relative to those of Snodgrass and O'Melia and of Chapra point out the difficulties of successfully building a model whose objectives include predicting several different parameters for short time scales.

Two contrasting strategies for model construction are apparent. Both result from the model objective. For an objective of constructing a phytoplankton biomass model, DiToro (personal communication) deliberately avoids using more compartments than are necessary, choosing instead to use broad general classes (*e.g.,* phytoplankton, rather than diatoms, greens and blue greens) and to test the capabilities of a particular set of compartments to simulate the magnitude and temporal scale of concern. If the detail is insufficient, then a compartment may be subdivided. In addition, attempts are made to choose only compartments for which field data are either available for verification or could conceivably become available. In contrast, the models of Scavia *et al.* (1976), Canale *et al.* (1976) and Chen *et al.* (1975) attempt to describe many more aspects of an ecological system, implying that, by this method, one can be assured that the model will satisfy a simpler objective such as predicting the extent and timing of a phytoplankton bloom. For this simpler objective, the model can be tested to determine which parts have no effect on its predictions and these parts may then be discarded. However, lack of sufficient field data usually precludes complete testing of the more complex model and confidence in the resultant simpler model needs examination. Both the "building-up" procedure of DiToro and the "simplification" procedure from the very complex models should result in the same model if both have been adequately verified.

SUMMARY

A framework for assessing some capabilities and limitations of predictive water quality models for the Great Lakes is presented. The framework describes the modeling process as consisting of six steps: definition of model objectives, system discretization, model construction, model calibration, model verification and model prediction. Using this framework, several models for water quality are critiqued.

From this analysis, the following conclusions are drawn:

1. The box approach for modeling lakes is a useful and verified approach for considering spatial scales of the whole lake and temporal scales on the order of a year.
2. The smaller the time or spatial scale and the more nonconservative the material of concern, the more difficult it is to verify the model.
3. The decision maker is often concerned with the rate and extent of a lake's recovery upon the initiation of a water quality management scheme. Within this context, models exist which are verified to predict the extent of lake recovery for control of chloride and phosphorus concentrations and to predict the rate of recovery of chloride concentration. Due to the lack of appropriate field data, current models are not yet verified to predict the rate of recovery of a lake in response to control of phosphorus or to predict the response of phytoplankton to water quality management schemes.
4. Fundamental models derived from the law of conservation of mass should only be used with confidence by decision makers for those objectives for which model predictions are verified. Confidence in their use in other circumstances (*e.g.,* for other objectives or for different assumptions) cannot be assured.

REFERENCES

Canale, R. P., L. M. DePalma and A. H. Vogel. "A Plankton-Based Food Web Model for Lake Michigan," in *Modeling Biochemical Processes in Aquatic Ecosystems,* R. P. Canale, Ed. (Ann Arbor, MI: Ann Arbor Science Publishers, Inc. 1976), pp. 33-74.

Canale, R. P., and J. Squire. "A Model for Total Phosphorus in Saginaw Bay," *J. Great Lakes Res.* 2(2):364-373 (1976).

Chapra, S. C. "Total Phosphorus Model for the Great Lakes," *Envir. Eng. Div., ASCE* 103 (EE2):147-161 (1977).

Chen, C. W., M. Lorenzen and D. J. Smith. "A Comprehensive Water Quality-Ecological Model for Lake Ontario," Final report to Great Lakes Environmental Research Laboratory, NOAA, Ann Arbor, MI (1975), 202 p.

DiToro, D. M., D. J. O'Connor and R. V. Thomann. "A Dynamic Model of the Phytoplankton Population in the Sacramento-San Joaquin Delta." in *Advances in Chemistry Series 106: Nonequilibrium Systems in Natural Waters,* R. F. Gould, Ed. (Washington, DC: American Chemical Society, 1971), pp. 131-180.

Levenspiel, O. *Chemical Reactor Engineering.* (Toronto: John Wiley & Sons Canada, Ltd., 1962).

Lorenzen, M. W., D. J. Smith and L. V. Kimmel. "A Long-Term Phosphorus Model for Lakes: Application to Lake Washington," in *Modeling Biochemical Processes in Aquatic Ecosystems,* R. P. Canale, Ed. (Ann Arbor, MI: Ann Arbor Science Publishers, Inc., 1976), pp. 75-92.

O'Connor, D. J., and J. A. Mueller. "A Water Quality Model of Chlorides in Great Lakes," *Proc. Amer. Soc, Civ. Eng., J. San. Eng. Div.* 96(1): 955-975 (1970).

Okubo, A. "Horizontal and Vertical Mixing in the Sea." in *Impingement of Man on the Oceans,* D. W. Hood, Ed. (Toronto: John Wiley & Sons Canada, Ltd., 1971), p. 89.

Ontario Ministry of Environment. *Hamilton Harbor Studies, 1975* (Toronto: Ontario Ministry of Environment, 1977).

Orlob, G. T. "Present Problems and Future Prospects of Ecological Modeling." in *Ecological Modeling,* C. S. Russel, Ed. (Washington, DC: Resources for the Future, Inc. 1975), pp. 283-313.

Richardson, W. L. "An Evaluation of the Transport Characteristics of Saginaw Bay Using a Mathematical Model of Chloride." in *Modeling Biochemical Processes in Aquatic Ecosystems,* R. P. Canale, Ed. (Ann Arbor, MI: Ann Arbor Science Publishers, Inc., 1976), pp. 113-140.

Scavia, D., B. J. Eadie and A. Robertson. "An Ecological Model for Lake Ontario: Model Formulation, Calibration, and Preliminary Evaluation," *NOAA Technical Report ERL 371-GLERL 12,* (Washington, DC: U.S. Government Printing Office, 1976), 63 p.

Simons, J. J. "Analysis and Simulation of Spatial Variations of Physical and Biochemical Processes in Lake Ontario," *J. Great Lakes Res.* 2: 215-233 (1976).

Snodgrass, W. J. "A Predictive Phosphorus Model for Lakes—Development and Testing," Ph.D. Dissertation, University of North Carolina at Chapel Hill (1974).

Snodgrass, W. J., and C. R. O'Melia. "Predictive Model for Phosphorus in Lakes," *Environ. Sci. Technol.* 9:937-944 (1975).

Thomann, R. V., D. M. DiToro, R. P. Winfield and D. J. O'Connor. "Mathematical Modeling of Phytoplankton in Lake Ontario, 1. Model Development and Verification," U.S. Environmental Protection Agency 600/3-75-005, Corvallis, Oregon (1975), 177 p.

Thomann, R. V., R. P. Winfield and D. S. Szumski. "Estimated Responses of Lake Ontario Phytoplankton Biomass to Varying Nutrient Levels," *J. Great Lakes Res.* 3(1-2):123-131 (1977).

United States Army Corps of Engineers "Lake Erie Wastewater Management Study," (Buffalo, NY: U.S. Army Corps of Engineers, 1975), 172 p.

CHAPTER 8

EMPIRICAL LAKE MODELS FOR PHOSPHORUS: DEVELOPMENT, APPLICATIONS, LIMITATIONS AND UNCERTAINTY*

Kenneth Howland Reckhow

Department of Resource Development
Michigan State University
East Lansing, Michigan 48824

INTRODUCTION

There has been a proliferation of mathematical models in water quality literature in recent years. This has corresponded with the development of strong feelings both pro and con as to the usefulness of the models. It must be said that, at times, the skepticism directed toward the value of a mathematical model exists due to confusion as to the proper use of the model and its output. This, in turn, is somewhat the fault of the modeler who may not clearly document and discuss the use and misuse of a proposed model.

A model is a simplified representation of reality; thus it does not reproduce natural behavior exactly. Models range widely in complexity, but many current models cluster at two extreme points on the complexity scale. In lake water quality analysis, ecosystem simulation models are designed to describe the interactions among the components of the aquatic ecosystem. They are complex, theoretically based models that require considerable data for parameter estimation and generally examine seasonal or short-term changes on a single lake. Ecosystem simulation models are

---

*Some of the material in this chapter has been modified from "Quantitative Techniques for the Assessment of Lake Quality," written by the same author for the Michigan Department of Natural Resources, 1978.

developed both to increase understanding of the aquatic ecosystem and to predict the impact of changes in various control variables. Black box or empirical models, on the other hand, are highly aggregated and are designed to examine the concentration, or annual changes in concentration, of a single component (generally phosphorus or chlorophyll a). They are developed from multi-lake, cross-sectional analyses (an examination of data on multiple cases, or lakes, covering one point in time), yet are often applied for single-lake, longitudinal prediction (forecasting over time) purposes.

Each of these two basic types of lake models is useful in the understanding and management of lake ecosystems. Therefore, it is somewhat unfair to contrast the two approaches, yet it is important to understand the uses and limitations of each model type. With that as a goal, this chapter will discuss the black box models for phosphorus in lakes.

However, before a critical review of this class of lake models is undertaken, a few cautionary words, as alluded to earlier, are in order. It is the modeler's responsibility to clearly document the appropriate use of his or her model. This documentation should include a discussion of the limitations (geographic, trophic, geomorphologic) of the model and possible biases within the range of application. In addition, it would clearly be valuable for the modeler to provide a format that may be used to assess the value of the information supplied by the model. This can be done using first-order uncertainty analysis which, through explicit consideration of error, may be used to estimate confidence limits for the model prediction. The inverse of the variance associated with the prediction is then a measure of information content. This approach is discussed later in this chapter.

It would be wise for the prospective model user to check carefully any documented model "verification" presented by the modeler. Verification, or validation, is defined as the establishment of truth, accuracy or reality. In practice, this has resulted in quite different approaches for different authors. One fundamental problem has been noted by Popper (1963) who observed that, in fact, it is impossible to verify a theory through induction (the process used to define general laws through the observation of specific points). A researcher may, at best, *in*validate a theory using induction. Caswell (1973) proposes, therefore, that the validation process apply to predictive models and *corroboration* refer to the evaluative process for theoretical models. This still does not aid in determining a process definition for verification, but it does underscore the problem.

In the absence of a statistical criterion for successful verification of predictive models, it is suggested here that the modeler use an *independent* test for the model. Specifically, verification for a cross-sectional model should, at minimum, consist of evaluation of the model on an independent data set (different from the data set used to establish the model parameters) with no significant reduction in predictive success (vs that for the original set). For a longitudinal model, the verification process should specify that a model reproduce events that occur under conditions different from those used to establish the model parameters (*e.g.,* changes in hydrology, nutrient loading). It must be noted that no precise statistical definition was offered for either "no significant reduction" or "reproduce." At this point, these terms must remain undefined (statistically) because of their subjective nature. Clearly, though, the prospective model user must establish and carry out a satisfactory verification process for the proposed application of the model. Some of the problems that may occur in the absence of this verification step are discussed later in this chapter.

One final caveat on the development and application of mathematical models concerns the tendency to make causal inferences from the "successful" application of a model. At best, based on Popper's observations, hypothesized causal relationships can be corroborated with a successful theoretical model. However, empirical or predictive models are highly simplified representations of the real world from which causal conclusions may be drawn only when coupled with a deep conceptual understanding of the system being modeled. As Blalock (1964) notes, "One admits that causal thinking belongs completely on the theoretical level and that causal laws can never be demonstrated empirically." It may be possible to devise and empirically test simple causal models, but only if the simplifying assumptions behind the models are well thought out and justifiable. Unless this conceptual documentation is included, however, causal conclusions should probably not be made from empirically derived models.

With those words of introduction, the remainder of this chapter contains a discussion of the black box or empirical models of phosphorus in lakes. It begins with the historical development of this approach from the mass balance for phosphorus and then leads into a comparison of several recently proposed empirical models. This comparison is stated in terms of predictive success, limitations, biases and sensitivities. Then, after general conclusions are made on the application of the black box models, a discussion of the importance and use of uncertainty analysis coupled with the models is presented.

LAKE MODELS

The Phosphorus Mass Balance Model: Historical Development

Biffi (1963) was among the first to propose a simple mass balance model for chemical substances in lakes. Since the focus of this chapter is on an examination of phosphorus models, all models will be expressed in terms of phosphorus. Biffi's equation may be written in the form:

$$\frac{d(PV)}{dt} = M - (Q/V)(PV) \tag{1}$$

where
- P = lake phosphorus concentration (mgl^{-1})
- M = annual mass rate of phosphorus inflow $(10^3 \text{ kg yr}^{-1})$
- Q = annual volume rate of water inflow $(10^6 m^3 \text{ yr}^{-1})$
- V = lake volume $(10^6 m^3)$

The time-dependent solution is:

$$P = P_{in} [1 - (1 - P^o/P_{in}) e^{-t/\tau}] \tag{2}$$

- P_{in} = average influent phosphorus concentration (mgl^{-1})
- P^o = initial lake phosphorus concentration (mgl^{-1})
- τ = hydraulic detention time (yr)

and the steady-state solution is:

$$P = \frac{M}{Q} = P_{in}. \tag{3}$$

Biffi's model considers the lake to be a continuously stirred tank reactor (CSTR) with complete and instantaneous mixing and reaction. Stratification and biological and chemical reactions are explicitly ignored. This is the basic form of the black box phosphorus model. Biffi, however, did not include a term in his model for sedimentation of the substance. As a result, at steady state, the substance is modeled as a conservative material; the lake concentration, which equals the inflow concentration, equals the outflow concentration. Unfortunately, phosphorus cannot be modeled successfully in this manner since, in many lakes, a large fraction of the input is deposited in the sediments. Therefore, this approach is valid only for truly conservative substances such as chloride.

Piontelli and Tonolli (1964) proposed the first black box model that included sedimentation. They suggested that a fixed fraction (f_s) of the influent phosphorus mass would be deposited in the sediments:

$$\frac{dm_s}{dt} = f_s M \tag{4}$$

where m_s = mass of phosphorus in the sediments (kg)

Thus, the rate of change of the phosphorus mass in the lake could be expressed as:

$$\frac{d(PV)}{dt} = (1 - f_s) M - QP \tag{5}$$

with the time-dependent solution:

$$P = (1 - f_s) \, P_{in} \, (1 - e^{-t/\mathcal{T}}) + P^{\circ} e^{-t/\mathcal{T}} \qquad (6)$$

and steady-state solution:

$$P = (1 - f_s) \, P_{in}. \qquad (7)$$

Piontelli and Tonolli's consideration of sedimentation of phosphorus provided the basis for Vollenweider's (1964) early work on this type of model. However, unlike Piontelli and Tonolli, who modeled sedimentation as a function of input, Vollenweider proposed that sedimentation be made a function of lake concentration. Thus:

$$V \frac{dP}{dt} = (1 - f_o) \, M - \sigma P V \qquad (8)$$

where f_o = fraction of influent phosphorus lost through the outflow
σ = sedimentary loss coefficient (yr^{-1})

This has the time-dependent solution:

$$P = \frac{(1 - f_o) \, M/V}{\sigma} \left[1 - \left(1 - \frac{\sigma P^{\circ}}{(1 - f_o) \, M/V} \right) e^{-\sigma \mathcal{T}} \right] \qquad (9)$$

and the steady-state solution:

$$P = \frac{(1 - f_o) \, M}{A \sigma z} = \frac{L'}{\sigma z} \qquad (10)$$

where A = lake surface area (km^2)
z = lake mean depth (m)
$L' = \frac{1}{A} (1 - f_o) \, M$ = net "specific loading" $(g \cdot m^{-2} \, yr^{-1})$

Vollenweider calculated L'/P for five Swiss lakes, and he found this term to be relatively constant despite the fact that the lakes covered a wide range of trophic states. He took this result to be confirmation of his approach (which means, although it was not explicitly stated, that σ is inversely proportional to lake mean depth).

Vollenweider (1969) revised his model and for the first time focused on total phosphorus alone. Again, he hypothesized that the rate of deposition of phosphorus to the sediments is a function of the mass of phosphorus in the lake. However, in his revised model he assumed that the lake and outflow phosphorus concentration are identical. This results in:

$$V \frac{dP}{dt} = M - \sigma P V - Q P \qquad (11)$$

with the time-dependent solution:

$$P = \frac{L}{\sigma z + z/\tau}\left[1 - e^{-\left(\frac{1}{\tau} + \sigma\right)t}\right] + P^{\circ}\, e^{-\left(\frac{1}{\tau} + \sigma\right)t} \tag{12}$$

and the steady-state solution:

$$P = \frac{L}{z\left(\frac{1}{\tau} + \sigma\right)} \tag{13}$$

where $L = M/A$ = annual areal phosphorus loading $(g{\cdot}m^{-2}\ yr^{-1})$

Alternatively, the phosphorus mass balance may be written:

$$V\frac{dP}{dt} = M - v_s\, PA - QP. \tag{14}$$

with the steady-state solution:

$$P = \frac{L}{v_s + z/\tau} \tag{15}$$

where v_s = apparent settling velocity $(m\ yr^{-1})$

The difference between Equations 13 and 15 (as expressed here) is simply that Equation 13 assumes a depth-dependent settling velocity (σz) while Equation 15 assumes a constant settling velocity. This difference results from the original differential Equations 11 and 14. The steady-state model in Equation 13 was developed from the assumption that phosphorus deposition to the sediments is a function of the total mass of phosphorus in the lake or of the total lake volume. The model in Equation 15 results from the assumption that the sediments are an areal sink, and thus the rate of deposition of phosphorus should be a function of the surface (bottom) area (Chapra, 1975).

A third steady-state phosphorus model form results from the definition of R, the fraction of influent phosphorus retained in the lake (Dillon and Rigler, 1974):

$$R = \frac{M - QP_0}{M} \tag{16}$$

where P_0 = average outflow phosphorus concentration (mgl^{-1})

Since $M = LA$ and $Q/A = z/\tau$:

$$R = \frac{L - (z/\tau)P_0}{L} = \frac{\dfrac{L\tau}{z} - P_0}{\dfrac{L\tau}{z}}$$

where $\dfrac{L\tau}{z}$ = average influent phosphorus concentration (mgl^{-1})

If it is assumed that $P_O = P$ (lake and outflow concentrations are equal), then the third model becomes:

$$P_O = P = \frac{L\tau}{z}(1 - R) \qquad (17)$$

The three basic model forms, from which the empirical models are estimated, are Equations 13, 15 and 17. By setting Equation 13 equal to Equation 17, it may be shown that:

$$R = \frac{\sigma}{1/\tau + \sigma} = \frac{1}{1 + \dfrac{1}{\tau\sigma}} \qquad (18)$$

In a similar manner, the relationship between R and v_S (Equations 15 and 17) is:

$$R = \frac{v_S}{v_S + z/\tau} = \frac{v_S}{v_S + q_S} \qquad (19)$$

where $q_S = z/\tau$

Since the form of Equation 17 may easily be converted to one of the other two models using either Equation 18 or 19, the major difference among the three models (aside from the empirical extensions discussed in the next section) focuses on the depth-dependent (or independent) nature of the settling velocity. Based on the introductory comments on verification and causal modeling, the selection of the "best" model form should likely be termed an unresolvable issue on theoretical grounds. Considering the simplicity of the models and the complexity of the lake system, it is possible that either form of the black box model may predict well when fitted from a representative set of data. For the black box models, it seems considerably more important that the model form be selected on the basis of successful prediction as opposed to precise theoretical validity.

Empirical Models

The first attempt to develop an empirical fit to one of the mass balance models presented in the previous section was by Vollenweider (1975) for Equation 13. From an analysis of available data (not specified further), Vollenweider found that σ could be estimated by:

$$\sigma = \ln 5.5 - 0.85 \ln z \qquad (20a)$$
$$(r = 0.79)$$

or

$$\sigma \cong \frac{10}{z} \tag{20b}$$

Thus, based on Equation 20b, Vollenweider implicitly advocated a constant settling velocity ($v_S = 10$ m yr^{-1}), empirical model.

Dillon and Rigler (1974) examined Equation 17 in a paper that was presented as a test of a model predicting phosphorus concentration. They did not test an empirical formulation for R, but instead examined, through Equation 17, the frequently invoked assumption that lake and outflow concentrations are identical. It has been shown (Reckhow, 1977a) that this assumption is strongly supported by available data.

Equations 18 and 19 suggest that R may be empirically estimated as a function of z/τ or τ. Kirchner and Dillon (1975) confirmed this in a model for R, developed from a data set of 14 Canadian Shield lakes:

$$R_{KD} = 0.426 \exp (-0.271 \ z/\tau) + 0.574 \exp (-0.00949 \ z/\tau) \tag{21}$$
$$(r = 0.94)$$

Dillon and Rigler (1975) proposed that Equations 17 and 21 be combined and used to assess the impact of land use on lake quality.

Chapra (1975), in a comment on Equation 21, noted the relationships between R and σ (Equation 18) and R and v_S (Equation 19). Using the Canadian Shield lake data, Chapra found that $v_S = 16$ m yr^{-1} provided the best fit to Equation 19. This model has recently been applied by Chapra (1977) to evaluate and project the phosphorus concentrations in the Great Lakes. A comparison of results from Chapra's model and from Kirchner-Dillon's model are presented in Figure 1. Dillon and Kirchner (1975) responded to Chapra with a statement equating the v_S and σ models for P. Actually, the implicit assumption behind the differential Equations 11 and 14 is that σ (Equations 11-13) and v_S (Equations 14 and 15) are *constants*, not functions of limnological variables. Thus, despite Dillon and Kirchner's contention that the models presented in Equations 13 and 15 are equivalent, they are equivalent *only* if v_S is re-expressed in Equation 14 as a linear function of z. Dillon and Kirchner concluded their reply with a revised best-fit for v_S of 13.2 m yr^{-1}. This change from Chapra's estimate of 16 m yr^{-1} occurred because Dillon and Kirchner rejected data from two "outlier" lakes in Figure 1, based on suspected errors in data collection. It should be noted that this conclusion on data errors resulted from comparisons of observations with model predictions and not from known field or laboratory problems, so it is not immediately clear which of the two figures, 13.2 or 16, has more merit.

In a concurrent study, Larsen and Mercier (1975) examined data collected through the EPA National Eutrophication Survey (EPA-NES)

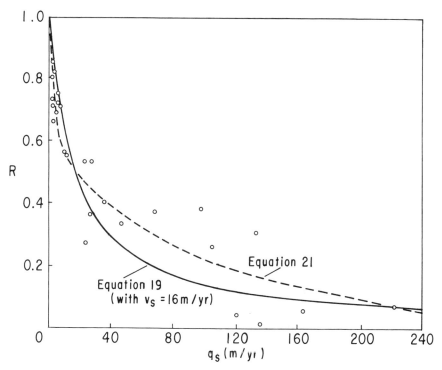

Figure 1. A comparison of the Kirchner and Dillon model for R with Chapra's constant apparent settling velocity approach.

for northeast and northcentral U.S. lakes. Using a data set from 20 lakes, they found the best fit for phosphorus retention to be:

$$R_{LM} = \frac{1}{1 + 1.12 \left(\frac{1}{\tau}\right)^{0.49}} \qquad (22)$$

or more simply:

$$R_{LM} = \frac{1}{1 + \sqrt{1/\tau}} \qquad (r = 0.94) \qquad (23)$$

Note that Equation 23 is identical to Equation 18 when σ becomes variable and is set equal to $1/\tau$.

Finally, in a recent paper, Jones and Bachmann (1976) examined data from a large number of north temperate lakes and proposed Equation 13 as the best model with $\sigma = 0.65$ yr^{-1} and a multiplicative correction factor of 0.84.

Many of the previous modeling efforts form the basis for work by Reckhow (1977a) and Walker (1977) who independently examined data from north temperate lakes and conducted comprehensive studies on the empirical phosphorus lake models. The remainder of this chapter will rely heavily on the work by Reckhow, although many of the same observations and conclusions were reached by Walker.

The basic mass balance on phosphorus can be re-expressed in the following general form:

$$V\frac{dP}{dt} = M_I - M_O - \phi\,(M_I, M_O, P, Ca, Fe, Al, pH, O_2, \tau, z, z/\tau, V, \ldots) \quad (24)$$

where M_I = annual mass influx of phosphorus

M_O = annual mass efflux of phosphorus

ϕ = annual net flux of phosphorus to the sediments

The removal of phosphorus from a lake occurs through two pathways, the outlet (represented by M_O) and the sediments (represented by ϕ). ϕ is a multidimensional variable; it is dependent upon the influent and effluent phosphorus mass, lake geomorphology and hydrology, the dissolved oxygen concentration and the pH at the sediment-water interface, the concentration of major cations (Ca, Fe and Al) that combine with phosphorus and transport it to or hold it in the sediments, and probably additional mechanisms that are important on a limited scale. Thus, the expression of the sediment removal term as a phosphorus retention coefficient, an apparent settling velocity, or a sedimentary loss coefficient is actually a simplification of the true form. The black box models presented in this chapter all employ a geomorphology-dependent or hydrology-dependent sediment removal term for phosphorus (in part because information and/or data are incomplete for characterization of the other mechanisms listed above). Statistics presented by Reckhow (1977a) show that z, τ and z/τ (the geomorphologic and hydrologic terms used in the black box models to represent sediment removal of phosphorus) are correlated with at least some of the other "dimensions" of ϕ (e.g., dissolved oxygen concentration). This multicolinearity of the attributes of ϕ suggests that ϕ can be represented by a subset of the variables presented in Equation 24, and this may be one explanation for the success of the one-box models, despite their simple form. In addition, Equation 24 implies that the earlier discussion of the "correct" form of the sedimentation term (σ or v_s) must be of an empirical, and not a theoretical, nature since σ and v_s are incomplete representations of ϕ.

Like the one-box models described previously, the models proposed on the following pages originate from a simplified version of Equation 24. The version of Equation 24 employed herein is presented below:

$$V\frac{dP}{dt} = M - v_s(z, \tau, z/\tau, L)PA - P_O Q \quad (25)$$

The models were developed from a data set consisting of measurements from 95 lakes from the northeast and northcentral U.S. (EPA-NES data), the Canadian Shield (from Dillon and Rigler, 1974), and from other north temperate regions in North America and Europe (Snodgrass, 1974). Some statistics for the entire data set are presented in Table I. A more detailed analysis of these data is available elsewhere (Reckhow, 1977a).

One significant aspect of the Table I statistics is the multicolinearity among the log-transformed values of z, τ and z/τ, and R. Rawson (1955) noted the relationship between mean depth and lake trophic state; Vollenweider (1968), Snodgrass (1974) and others have thought that this mean-depth/lake-quality relationship is a fundamental phenomenon that should be considered in the modeling of lakes. Vollenweider (1975) later revised his model to relate quality to surface overflow rate (q_s = z/τ = Q/A), while Kirchner and Dillon (1975) found that z/τ could be used to predict lake phosphorus retention, which in turn could be used to estimate lake phosphorus concentration. Larsen and Mercier (1975), on the other hand, estimated phosphorus retention as a function of τ. Thus, in a cross-sectional analysis of lakes, any one (or more) of z, τ and z/τ may be the best geomorphologic/hydrologic indicator(s) of phosphorus retention *and* resultant lake quality. Because of their multicolinearity, the choice cannot be made on the basis of theoretical considerations for these simple models. Rather, the variable(s) used and the model form selected should be based on predictive success.

The scatter of points in Figure 1, which led Dillon and Kirchner to remove some of the points due to hypothesized measurement error, may be explained in another manner. Aside from measurement error, faulty sampling design, and random fluctuations in natural phenomena, the data scatter may be due to an incorrect (or inadequate) model. Kirchner and Dillon and Chapra estimate R as a function of z/τ; these relations must be considered inadequate in light of the multidimensionality of ϕ. Extensive study of sedimentation basins by sanitary engineers has led to the conclusion that retention of material in these basins is, in part, a function of depth, detention time, surface overflow rate, influent loading and violations of ideal flow conditions (ideal with respect to the conceptual model of flow). Figures 2a and 2b indicate how sanitary engineers have attempted to model the multidimensionality of the response. Figure 2a (from Tebbutt and Christoulas, 1975) presents a family of curves for different influent concentrations on a plot of percent removal vs detention time. Figure 2b (from Fair *et al.*, 1968) shows another family of curves for flow variations (represented by n) on a plot of retention vs a length, depth or time factor. It therefore seems likely that the scatter of points in Figure 1 should be expected, and that it may be modeled as a family of curves

Table I. Statistics for the Data Set Used to Develop the General Model Proposed by Reckhow (1977a)

Variable	Minimum	Geometric Mean	Maximum
z (m)	0.78	7.8	100.0
τ (yr)	0.003	0.47	57.6
z/τ (m yr^{-1})	0.53	16.5	575.0
$L\tau/z$ (mg l^{-1})	0.0075	0.066	1.179
P (mg l^{-1})	0.004	0.033	0.958
R	-0.08	0.38[a]	0.95

Variable	log τ	log (z/τ)	log $(L\tau/z)$	log P	R
			Correlation Coefficients		
log z	0.751	-0.379	-0.151	-0.408	0.508
log τ		-0.896	0.119	-0.249	0.700
log z/τ			-0.260	0.074	-0.639
log $L\tau/z$				0.851	0.049
log P					-0.405
R					

[a]Arithmetic mean.

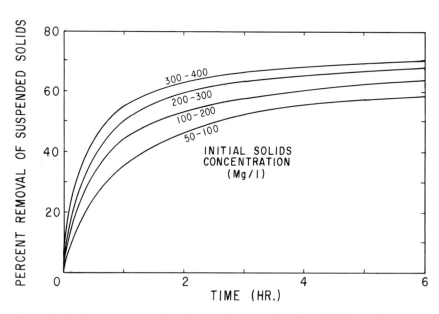

Figure 2a. Performance curves for sedimentation, reproduced from Tebbutt and Christoulas (1975).

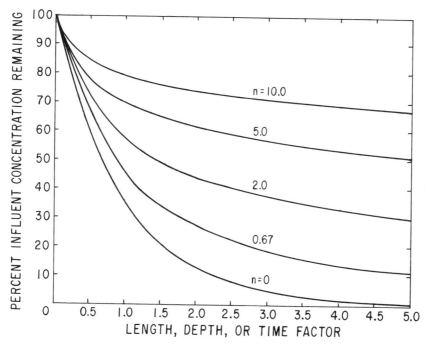

Figure 2b. Generalized time rates of purification (n represents flow variations), reproduced from Fair, Geyer and Okun (1968).

incorporating additional dimensions, as the sanitary engineers have done. However, given the present quality of data and understanding of limnological processes affecting R, the discrimination between data noise (inadequate sampling design, measurement error and natural random fluctuations) and other functional dependencies on measured limnological variables need to be improved. This might be a fruitful direction for additional refinement to the black box model.

One additional comment must be made on the relationships among z, τ and z/τ. The cross-sectional data analysis presented in Table I documents the multicolinearities among these variables. However, one can imagine a set of lakes where any one of the three variables is changed while either of the others (but not both) remains constant. Thus, the fact that z/τ is a function of z and τ does not mean that model with a z/τ term contains (or does not contain, for that matter) a depth-dependency. It is important that variable intercorrelations be examined to determine the likely changes in the dependent variable due to changes in another variable not explicitly included in the model. Thus, even though Equation 15 does not explicitly contain a z or τ term, based on the statistics in Table I, it is likely that a change in z/τ will be accompanied by changes in z and τ. Correspondingly, any one of the three variables may, in part, serve as a surrogate variable for the other two.

The basic differential equation from which the models were developed, Equation 25, distinguishes between lake concentration (P) and outflow concentration (P_O). Most empirical models have equated P_O and P, but an attempt has been made here to discover a systematic relationship between these two variables. Correlation coefficients of 0.3 to 0.4 were found between the ratio P_O/P and the logarithms of A, z, V and τ, suggesting a slight positive relationship between lake size and P_O/P. This may be due to data noise (as defined above), outflow scour (large lakes may tend to have large volumetric flowrates), cultural development at the outflow biasing P_O in large lakes (large lakes may be more attractive to development), or nonsteady-state conditions. More importantly from a predictive perspective, however, the magnitude of this relationship is small in comparison to the sediment deposition effect. Thus, it was found in developing the models that the sediment removal term completely obscures any attempt to quantify the relation between P_O and P. While there is obvious value in obtaining an understanding of any systematic reasons for the magnitude of the P_O/P ratio, for the purposes of this chapter P_O and P will be considered equal.

With $P_O = P$ and the assumption of steady state, the model becomes:

$$P = \frac{L}{v_s\,(z,\ \tau,\ z/\tau,\ L) + z/\tau} \tag{26}$$

If $v_s \neq f(L)$, then for any particular lake:

$$P \cong kL, \tag{27}$$

unless τ is significantly affected by annual precipitation fluctuations. When Equation 27 holds in a longitudinal sense, it becomes apparent that most empirical models for R, v_s and σ are actually designed to estimate a single lake constant, k, which equals the P/L ratio. Further, if time series input-output data are available for a lake, it may be possible to use these data to obtain a better estimate of k (than from cross-sectional regressions), that might be functionally dependent upon the "minor" attributes of ϕ (P, O_2, Fe, etc.) in Equation 24. Reckhow (1977b) has suggested a Bayesian approach which combines a regional regression estimate of R with that obtainable from time series data for single lake applications.

Lake Classification

Examination of statistics from the 95-lake data set indicated that there were nonlinearities in the response (phosphorus retention or settling) relative to the major predictor variables (z, τ and z/τ), and that other "minor" variables (such as dissolved oxygen) may be important (Reckhow, 1977a). One way these factors may be implicitly considered is through the classification of lakes according to those characteristics that may be difficult to incorporate explicitly into a model. This effectively controls those mechanisms that are significant for a particular class of lakes only.

It was thus decided that the following three classes of lakes would be separately studied and modeled:

1. oxic, $z/\tau < 50$ m yr^{-1} (33 lakes)
2. $z/\tau > 50$ m yr^{-1} (28 lakes)
3. anoxic (25 lakes)

While the division between oxic and anoxic lakes may be obvious, the dichotomy based on the magnitude of z/τ may be less apparent. Two reasons were behind this additional classification criterion. First, many shallow (high z/τ) lakes were inadequately sampled for dissolved oxygen (surface samples only), so the oxic-anoxic classification was uncertain. Second, lakes with high overflow rates may actually represent a distinct class, in that they are frequently just widened sections of rivers. For instance, two types of lakes were identified in one statistic by Reckhow (1977a); those with 90% or more of their inflow from one tributary, and those formed as the confluence of two or more tributaries, none of which provide more than 90% of the total inflow. It was found that all of the lakes with $z/\tau > 50$ m yr^{-1} are of the former type, while 56% of the oxic lakes with $z/\tau < 50$ m yr^{-1} are of the latter type. From a modeling standpoint, this division point at $z/\tau = 50$ m yr^{-1} occurs in the region where lake phosphorus retention changes from functional dependence on z/τ to near independence of z/τ (see Figure 1). Chapra (1975) also noted this change in functional relationship between R and z/τ, stating that "different mechanisms (may) govern phosphorus budgets in the two types of lakes." Some of the original 95 lakes could not confidently be placed in *any* class (often because the dissolved oxygen was thought to be inconclusive), and some lakes fell in two classes (Classes 2 and 3, anoxia did not appear to affect phosphorus removal in lakes with high overflow rates). A lake was classified as anoxic from the EPA-NES data set if there was one measurement of zero dissolved oxygen during the summer stratification.

Tables I and II present statistics for the data set used to develop a general model and for the data set associated with each of the lake classes, respectively. Since the models constructed from these data are empirical, the minimum and maximum for each of the variables represent bounds beyond which the empirical models may not apply. These

Table II. Statistics for the Three Model Classes

Variable	Minimum	Geometric Mean	Maximum
Oxic Lakes, $z/\tau < 50$ m yr^{-1} (33 lakes)			
z (m)	0.73	18.2	100.0
τ (yr)	0.065	2.57	57.6
z/τ (m yr^{-1})	0.96	7.1	48.6
$L\tau/z$ (mg l^{-1})	0.008	0.040	0.298
P (mg l^{-1})	0.004	0.013	0.060
R	0.11	0.56[a]	0.95
Lakes with $z/\tau > 50$ m yr^{-1} (28 lakes)			
z (m)	0.85	4.05	13.4
τ (yr)	0.003	0.028	0.252
z/τ (m yr^{-1})	51.8	147.0	575.0
$L\tau/z$ (mg l^{-1})	0.008	0.047	0.178
P (mg l^{-1})	0.007	0.034	0.135
R	-0.08	0.18[a]	0.37
Anoxic Lakes (21 lakes)			
z (m)	1.40	6.9	34.0
τ (yr)	0.008	0.52	4.5
z/τ (m yr^{-1})	3.3	13.3	425.0
$L\tau/z$ (mg l^{-1})	0.024	0.119	0.621
P (mg l^{-1})	0.017	0.088	0.610
R	-0.03	0.23[a]	0.56

[a]Arithmetic mean

constraints exist on the empirical models because, without a strong theoretical basis, there is no reason to assume that similar behavior (to that predicted by each model) exists outside these bounds. For instance, classification of lakes was used to control for the nonlinearity in z/τ vs R. Clearly, use of one of these class models beyond the z/τ bound may result in biased predictions due to this nonlinearity. A modeler should always present these and all other constraints on the use of a proposed model.

Comparison of Models

Models developed by Reckhow (1977a) for the general case and for the three lake classes are presented in Table III. The models were constructed using nonlinear regression for optimal parameter estimation after exploratory data analysis and other curve-fitting exercises suggested the terms and form of the models. Three of the models were fit under a logarithmic transformation. The fourth model, developed for lakes with $z/\tau > 50$ m yr^{-1},

Table III. Models Developed by Reckhow (1977a)

Class	Model	R^2	Standard Error of the Estimate
General	$$\dfrac{L}{\left(\dfrac{z}{\tau}\right)e^{0.0025z/\tau} + z\left[0.35 + 0.11(z^2/\tau)e^{-0.05(z^2/\tau)} + e^{-100L\tau/z}\right]}$$	0.885^a	0.184^a
Oxic Lakes, $z/\tau <$ 50 m yr^{-1}	$$\dfrac{L}{\dfrac{18z}{10+z} + 1.05\left(\dfrac{z}{\tau}\right)e^{0.012z/\tau}}$$	0.876^a	0.123^a
Lakes with $z/\tau >$ 50 m yr^{-1}	$$\dfrac{L}{2.77z + 1.05\left(\dfrac{z}{\tau}\right)e^{0.0011z/\tau}}$$	0.949	0.0088
Anoxic Lakes	$$\dfrac{L}{0.17z + 1.13z/\tau}$$	0.948^a	0.105^a

[a]Based on a logarithmic transformation of terms.

was estimated in the untransformed state. The decision on the use of a log transformation was based on the distribution of the model residuals, since model parameters should be estimated for the form of the model (transformed or untransformed) that results in residuals that best approximate a normal distribution.

Table IV presents statistics for a general model proposed by Walker (1977). The geographic, geomorphic, hydrologic and trophic constraints for this model are similar to those for the Reckhow general model. Parameter statistics and error estimates in Table IV are necessary for the next section on error analysis.

Table V presents a comparison of the empirical models discussed in this section. The evaluative criteria describe the appropriateness of each model for a given application and the value of the information provided by the model. Recall that empirical models should not be applied outside the bounds of the data set used to construct the model (without prior verification). These bounds are described, in part, under the Data Base and Known Constraints columns. The Known Biases and Comments column entries are taken from the study by Reckhow (1977a). The Uncertainty column indicates whether or not the modeler provided model standard error and parameter statistics so that error can be explicitly evaluated. Entries such as unspecified, unknown, not evaluated, and not considered must be taken as indicators of the need for cautious use of the model (to which they apply), or of that model's restricted and indeterminate value in application.

In addition to or in elaboration of the criteria evaluated in Table V, the following comments may be made on the empirical models:

Table IV. Model Proposed by Walker (1977)

Model	R^2 [a]	Standard Error of the Estimate
$P = \dfrac{L\tau}{z}\left[\dfrac{1}{1 + 0.824\,\tau^{0.454}}\right]$ or $P = P_{in}\left[\dfrac{1}{1 + 0.824\,\tau^{0.454}}\right]$	0.906	0.171

where P_{in} = average influent phosphorus concentration mg l^{-1}

Parameter Statistics:

Parameter	Standard Deviation	Correlation Coefficient
0.824	0.067	-0.0064
0.546 (= 1 -0.454)	0.046	

Contribution to Total Error Estimate Due to Parameter Error

$$S_p^2 = 0.001\ Y^4\ \tau^{0.908}\ [4.49 + 1.44\ (1n\ \tau)^2 + 0.032\ 1n\tau]$$

where $Y = 1/(1 + 0.824\,\tau^{0.454})$

1n = base e logarithm

[a]Based on a logarithmic transformation of terms. These statistics compare the predicted concentration with the observed *outlet* concentration.

1. Given the current state-of-the-art, and the quality of existing data, most *carefully constructed* empirical phosphorus lake models (developed from a good, representative data base) will achieve roughly comparable predictive success for a large *heterogeneous* sample of lakes. In a study of three proposed models (Kirchner-Dillon's, Larsen, Mercier's, and Reckhow's general), none of the three is *clearly* best for *all* lakes, based on a comparison of average prediction errors.

2. Many proposed models are biased for certain classes of lakes, particularly when the lakes are outside the constraints of the model. For instance, the models developed by Kirchner and Dillon and by Larsen and Mercier were constructed from data sets composed primarily of oligotrophic lakes. As a result, these models tend to underestimate the phosphorus concentration in enriched lakes.

3. Classification of lakes has proven to be effective in reducing the average percent error, probably because it produces a more homogeneous group of lakes (and thus controls for certain source/sink processes). Specifically, the model for oxic lakes ($z/\tau < 50$ m yr^{-1}) was found to be superior to the general models for this class (based on average percent error). This may be the most important of the three defined classes to model, since most lakes where water quality

Table V. A Comparison of Empirical Models

Model	Data Base	Known Constraints[a]	Known Biases[a]	Consideration of Uncertainty	Comments
Constant v_s					
1. Vollenweider $v_s = 10$ m yr[-1]	unspecified	unknown	not evaluated	not considered	Based on the range of apparent settling velocities observed in natural lakes (Reckhow, 1977a), these models will probably overestimate P in lakes with high z/τ and underestimate P in highly enriched lakes.
2. Chapra $v_s = 16$ m yr[-1]	14 Canadian Shield lakes	$P < 0.015$ mg l[-1] $L\tau/z < 0.050$ mg l[-1]	not evaluated	not considered	
3. Dillon & Kirchner $v_s = 13.2$ m yr[-1]	12 Canadian Shield lakes	$P < 0.015$ mg l[-1] $L\tau/z < 0.050$ mg l[-1]	not evaluated	not considered	
Constant σ					
Jones & Bachmann $\sigma = 0.65$ yr[-1]	51 natural lakes, primarily north temperate	unknown	not evaluated	not considered	This model will probably overestimate P in shallow lakes with high values of z/τ
R-regression					
1. Kirchner & Dillon R_{KD}	14 Canadian Shield lakes	$P < 0.015$ mg l[-1] $L\tau/z < 0.050$ mg l[-1]	significant underestimation of P in lakes with $L\tau/z > 0.300$ mg l[-1]	not considered	R_{LM} was found (Reckhow, 1977a) to be less biased than R_{KD} for phosphorus-enriched lakes.
2. Larsen & Mercier R_{LM}	20 north temperate lakes, primarily oligotrophic	$P < 0.012$ mg l[-1] $L\tau/z < 0.025$ mg l[-1]	significant underestimation of P in lakes with $L\tau/z > 0.300$ mg l[-1]	not considered	

Table V. (Continued)

Model	Data Base	Known Constraints[a]	Known Biases[a]	Consideration of Uncertainty	Comments
Reckhow Models					
1. General Model	95 north temperate lakes	$P < 0.9$ mg l^{-1} $L\tau/z < 1.0$ mg l^{-1}	none significant	Parameter errors and model error are presented so that the total uncertainty may be calculated. This permits the assignment	No major shortcomings, somewhat cumbersome for error calculations.
2. Oxic $z/\tau < 50$ m yr^{-1} Lakes	33 north temperate lakes	$P < 0.060$ mg l^{-1} $L\tau/z < 0.298$ mg l^{-1}	none significant		Smaller errors than R_{KD} and R_{LM} on independent data set (Reckhow, 1977a). Verification not conclusive (Reckhow, 1977a).
3. $z/\tau > 50$ m yr^{-1} Lakes	28 north temperate lakes	$P < 0.135$ mg l^{-1} $L\tau/z < 0.178$ mg l^{-1} $z < 13$m $\tau < 0.25$ yr	none significant		
4. Anoxic Lakes	21 north temperate lakes	0.017 mg l^{-1} $< P$ < 0.610 mg l^{-1} 0.024 mg l^{-1} $< L\tau/z$ < 0.621 mg l^{-1}	none significant	of confidence limits to a prediction.	Needs to be verified.
Walker Model	105 north temperate lakes	similar to those for Reckhow general model	not evaluated	Prediction confidence limits can be estimated.	No major shortcomings known.

[a]Constraints or biases that may significantly affect the value of the model for prediction purposes.

management is an important concern are oxic, or managed to produce oxic conditions. The other two class models should prove useful but remain to be conclusively verified on an independent data set.

ERROR ANALYSIS

It has been stated earlier in this chapter that quantification of all uncertainty associated with the application of a model can provide the decision maker with information that may be used to weigh the value of model output. This section discusses the use of first-order uncertainty analysis (Cornell, 1972) to estimate the confidence limits associated with the prediction of one of the empirical models. Error analysis, like model verification, should become a standard or routine step in comprehensive water quality management.

The technique used to estimate the prediction confidence limits derives from the Taylor series expansion, which provides an approximation for a function, $f(x)$. For a bivariate relationship, the Taylor series may be written as:

$$f(x) = f(x_0) + \frac{\partial f(x)}{\partial x}(x - x_0) + \frac{1}{2}\frac{\partial^2 f(x)}{\partial x^2}(x - x_0)^2 + \ldots \tag{28}$$

If it may be assumed that the error $(x - x_0)$ is small and the higher derivatives of $f(x)$ are not large (the relationship is not highly nonlinear in the region of interest), then Equation 28 reduces to:

$$f(x) - f(x_0) \cong \frac{\partial f(x)}{\partial x}(x - x_0) \tag{29}$$

For a multivariate relationship, this becomes:

$$f(x_i, x_j, x_k) - f(x_{io}, x_{jo}, x_{ko}) \cong \frac{\partial f(x_i, x_j, x_k)}{\partial x_i}(x_i - x_{io}) +$$

$$\frac{\partial f(x_i, x_j, x_k)}{\partial x_j}(x_j - x_{jo}) + \frac{\partial f(x_i, x_j, x_k)}{\partial x_k}(x_k - x_{ko}) \tag{30}$$

Squared and expressed in standard error form, Equation 30 may be converted to the error propagation equation given below:

$$s(P) \cong \sum_{i=1}^{n}\left[\frac{\partial P^2}{\partial x_i}s^2(x_i) + \sum_{j=i+1}^{n}2\frac{\partial P}{\partial x_i}\frac{\partial P}{\partial x_j}s(x_i)s(x_j)\rho(x_i, x_j)\right]^{1/2}\ldots \tag{31}$$

$s(x_i)$ is the estimate of the standard deviation of x_i, and $\rho(x_i, x_j)$ is the correlation between x_i and x_j. Berthouex (1975) and Cornell (1972) discuss this error analysis approach for sanitary engineering and hydrologic applications, respectively.

Equation 31 may be used to estimate the effect of model parameter error and input variable error. These terms are then combined with the standard error of the model for an estimate of the prediction confidence limits. The empirical models presented in the last section are linear (or nearly linear) in L and approximately linear in z/τ for most lakes. Since the variance in z is usually quite small, it is ignored. Therefore, Equation 31 may be used with an empirical model (providing model and parameter error estimates) for all lakes within the model's constraints, except for those lakes that produce highly nonlinear changes in P due to changes in z/τ over the estimated variance in z/τ for that lake. This nonlinearity should be checked before application of Equation 31; otherwise Equation 31 *may* significantly overestimate the total error.

As an example, Equation 31 and the oxic lake model ($z/\tau < 50$ m yr^{-1}) were used to predict the phosphorus concentration in Lake Ontario and assign confidence limits to this prediction. Specifically, this approach was used to examine the likelihood of violation of a hypothetical phosphorus standard for Lake Ontario. While few standards are presently stated in a probabilistic form, the models and the phosphorus loading estimates are well suited for probabilistic statements. The expression of water quality standards in probabilistic terms is more realistic for both mathematical models and conceptual models, which must also admit to an uncertainty about the behavior of a lake system. Given uncertainty in the applied model and in the independent variables, the probability of meeting a lake quality standard (for phosphorus) can be calculated. This was done in Figure 3 for Lake Ontario ($z = 89$ m, $z/\tau = 11.3$ m yr^{-1}) assuming a total phosphorus standard of 0.045 mg l^{-1} (clearly a hypothetical standard, selected at this level to best illustrate the effect of a high degree of uncertainty in the phosphorus loading) and assuming the independent variable error exists in the phosphorus loading estimate only. (The large volume should dampen out hydrologic budget fluctuations.) Curves are presented in Figure 3 covering a range of uncertainties in the phosphorus loading.

The effect of the parameter error on the model prediction may be estimated by applying Equation 31 to the model expressed in logarithmic form, since the parameters were estimated for the log-transformed model. Using the statistics presented in Table VII (statistics for error estimates for the other Reckhow models and the Walker model are presented in Tables III, IV, VI and VII), the model parameter error contribution to the total error may be approximated by (after removing negligible terms):

$$s_{m.p.}^2 = [(\tfrac{1}{A}) \ (z/\tau) \ (e^{0.012z/\tau}) \ (0.0927)]^2$$

$$+ \ [(\tfrac{1}{A}) \ (1.05(z/\tau)^2 e^{0.012z/\tau}) \ (0.00545)]^2$$

$$- \ 2(\tfrac{1}{A})^2 (z/\tau)^3 (e^{0.012z/\tau})^2 (1.05) \ (0.351) \ (0.0927) \ (0.00545)$$

$$(32)$$

where $\qquad A = \dfrac{18z}{10 + z} + 1.05(z/\tau)e^{0.012z/\tau}$

When the values for z and z/τ for Lake Ontario are substituted into Equation 32, $s_{m.p.}$ becomes 4.02×10^{-2}. This is combined with the model error (standard error) of 0.123 (Table III) and the input variable error contribution to the total error (which, in the untransformed state, is proportional to the phosphorus concentration, since the input variable error is proportional to the phosphorus loading and the model is linear in L) to yield the total error estimate for the model prediction expressed in logarithmic form:

$$s_T^2 = s_{m.p.}^2 + s_m^2 + s_{i.v.}^2 \qquad (33a)$$

Converted to untransformed variables, the error becomes:

$$10^{s_T^2} = s_T'^2 = \left[P - 10^{\log P \pm [(s_{m.p.})^2 + (s_m)^2 + (s_{i.v.})^2]^{1/2}} \right]^2 \qquad (33b)$$

$$s_T'^2 = \left[P - 10^{\log P \pm [(4.02 \times 10^{-2})^2 + (0.123)^2]^{1/2}} \right]^2 + (kP)^2 \qquad (33c)$$

$$s_T'^2 = (0.347P)^2 + (kP)^2 \ \text{for} \ P < 0.045 \ \text{mg} \ l^{-1} \qquad (33d)$$

$$s_T'^2 = (0.258P)^2 + (kP)^2 \ \text{for} \ P > 0.045 \ \text{mg} \ l^{-1} \qquad (33e)$$

where $\quad s_T$ = total error (logarithmic form)

$\qquad s_T$ = total error (untransformed), in mg l^{-1}

$\qquad s_m$ = standard error of the model (logarithmic form)

$\qquad s_{i.v.}$ = input variable error (logarithmic form) contribution to total error

$\qquad k$ = a constant which expresses the fractional loading error (k = 0, 0.25, 0.50, 1.0; $\sigma(L) = kL$ in Figure 3)

s_T thus provides an estimate of the standard error of the prediction.

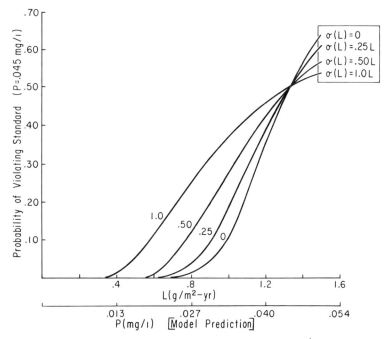

Figure 3. Lake Ontario: probability of standard (P = 0.045 mg l^{-1}) violation as a function of model error, parameter error and loading error, $\sigma(L)$.

Table VI. Parameter Statistics for General Model

Parameter	Standard Deviation	Correlations				
		0.0025	0.35	0.11	-0.05	-100.
0.0025	9.20x10^{-4}		0.184	-0.068	0.101	-0.052
0.35	2.87x10^{-2}			-0.087	-0.066	-0.407
0.11	4.81x10^{-2}				-0.888	0.092
-0.05	1.51x10^{-3}					-0.166
-100.	50.4					

Equations 33d and 33e are used to estimate the total error in mg l^{-1}. Then, assuming normally distributed errors, a phosphorus standard of 0.045 mg l^{-1} and an expected phosphorus concentration (from the model) of P, the probability of a standard violation may be estimated from the value of Z_n (the standard normal deviate) in Equation 34 and a cumulative normal frequency distribution table:

$$Z_n = \frac{0.045 - P}{s_T}. \tag{34}$$

Table VII. Parameter Statistics for the Three Class Models

Parameter	Standard Deviation	Correlations			
Oxic, $z/\tau < 50$ m yr^{-1} Lakes		18.0	10.0	1.05	0.12
18.0	2.41×10^{-3}		0.538	-0.274	0.133
10.0	6.56×10^{-3}			0.028	0.735
1.05	9.27×10^{-2}				-0.351
0.12	5.45×10^{-3}				
Lakes with $z/\tau > 50$ m yr^{-1}		2.77	1.05	0.0011	
2.77	1.24		-0.650	0.415	
1.05	9.27×10^{-2}			0.889	
0.0011	4.37×10^{-4}				
Anoxic Lakes		1.13			
0.17	2.61×10^{-2}	-0.656			
1.13	3.95×10^{-2}				

Figure 3 illustrates how the probability of meeting (or violating) the standard changes as the loading changes and/or as the loading uncertainty changes. If the standard itself is also expressed as a probability statement (say, a 0.30 chance of violating the phosphorus standard of 0.045 mg l^{-1}), the water quality modeler's recommendation to a decision maker would be less ambiguous (than with a phosphorus standard of 0.045 mg l^{-1} alone). Viewed in another way, the procedure quantifies the engineering safety factor required to achieve a specified probability of standard compliance. The approach presented in Figure 3 has recently been incorporated in the evaluation of lake management plans for a 208 Program in New Hampshire (Reckhow and Rice, 1976, 1977).

The procedure outlined above is most appropriate when the error in the variables in the data set used to develop the model contributes a small or negligible amount to the total prediction error. However, estimates of phosphorus loading, in particular, can be quite uncertain, as can the estimates of average phosphorus concentration and hydraulic detention time for selected lakes. Thus, it is likely that a nonnegligible fraction of the model standard error results from the errors in these variables (in the model development data set). In that case, two recommendations are made, given the existing models and model statistics.

1. If a model is used to assess current conditions in a lake for which the phosphorus input and hydraulic detention time are *direct measurements*, the level of uncertainty in these variables probably is approximately the same as the average uncertainty in these terms in the model development data set (for the models considered here). In that case, the uncertainty in these variables is incorporated in the model standard error, and the prediction confidence limits may be estimated using Equation 33b with $s_{i.v.} = 0$ ($k = 0$ in Equation 33c and on Figure 3).

2. If a model is applied to a lake for which the independent variables, particularly phosphorus loading, are *indirect estimates* (*e.g.,* literature export coefficients for phosphorus loading, or direct measurements for *future* predictions), then it is likely that the uncertainty in these estimates is quite large. The component of the model standard error contributed by the errors in the variables is probably small in comparison. In this case, the best estimate of the total prediction uncertainty would be provided by Equation 33b, unmodified.

In conclusion, among the issues for future work in this general topic of uncertainty analysis applied to lake modeling, is the impact, on a model and its use, of the error in the variables in the model development data set. The variable error, when significant, can affect the model standard error and bias the model parameters. Therefore, one approach that may be fruitful would be to: (1) estimate the "average" uncertainty in the model development data set variables; (2) "remove" these average uncertainty contributions from the model standard error; and (3) correct the model parameters for bias, if needed. This "revised" model standard error can then be inserted into Equation 33b, as s_m, for estimation of the total prediction error. Another alternative might be to develop models for various levels of variable uncertainty. Two obvious groups for this analysis are those lakes for which the independent variables are measured and those lakes for which the independent variables are estimated indirectly. This, in effect, "controls" for variable error and eliminates the need for an independent variable uncertainty term in Equation 33b (*if* each model is applied only to lakes within its group). Further refinements may become apparent as better data are gathered and more work is undertaken in uncertainty analysis.

SUMMARY

An attempt has been made, in this chapter, to trace the development of simple black box or empirical phosphorus lake models. A number of approaches have been examined, criteria have been suggested for model evaluation and discrimination, and the limitations of each model have been outlined. Finally, in the last section, first-order error analysis was suggested as a useful measure of model prediction information.

It was stated in the introduction that different types of lake models are developed for different purposes. The empirical models discussed herein may prove most useful in (1) suggesting general trends in lake quality (Chapra, 1977); (2) providing a quick assessment of quality and a first cut at phosphorus "carrying capacity" for a number of lakes (Reckhow, 1978); and (3) introducing useful quantitative techniques to planners and decision makers in lake management who, for various reasons (limited financial budget, lack of mathematical sophistication) have not previously applied mathematical models.

Since these models can be used by those unfamiliar with the theory and assumptions behind the techniques, it is critically important that limitations on their use be clearly documented, as has been done in this chapter. It must be remembered that the model output can only be as good as the input data and as applicable as the model assumptions. For instance, the models are aggregated in time and space and thus cannot be used to evaluate nearshore or seasonal effects. Yet with accurate input data, these models can provide a good indication of average (in time and space) lake quality conditions. When error analysis is used to assess the value of the model prediction, the information from an empirical model can be properly weighed and considered as one input to the lake management process.

ACKNOWLEDGMENTS

Much of the work discussed in the chapter was conducted while the author was a graduate student at Harvard University; it was supported by NSF Grant No. GI-35117. Appreciation is extended to Dr. Harold A. Thomas, Jr. of Harvard, Dr. William Walker, Jr. of Meta Systems, Inc., and Dr. Harbert Rice of Estimation Research Associates, Inc., for their critical review and comments on the research presented. In addition, the author wishes to express his thanks to Mr. Marvin Allum of the U.S. Environmental Protection Agency (Corvallis) for providing data from the National Eutrophication Survey. Helpful suggestions for the manuscript were provided by an anonymous reviewer and Coeditor, Donald Scavia.

REFERENCES

Berthouex, P. M. "Modeling Concepts Considering Process Performance, Variability, and Uncertainty" in *Mathematical Modeling for Water Pollution Control Processes,* T. M. Keinath and M. P. Wanielista, Eds. (Ann Arbor, MI: Ann Arbor Science Publishers, Inc., 1975), pp. 405-440.

Biffi, F. "Determining the Time Factor as a Characteristic Trait in the Self-Purifying Power of Lago d'Orta in Relation to a Continual Pollution," *Atti Ist. Ven. Sci. Lettl Arti.* 121:131-136 (1963).

Blalock, H. M., Jr. *Causal Inferences in Nonexperimental Research* (New York: W. W. Norton & Co., Inc., 1964), p. 6.

Caswell, H. "The Validation Problem," paper presented at the Symposium on Systems Ecology: The Modeling and Analysis of Ecosystems, University of Georgia, Athens, GA (1973).

Chapra, S. C. "Comment on 'An Empirical Method of Estimating the Retention of Phosphorus in Lakes,' " by W. B. Kirchner and P. J. Dillon." *Water Resources Res.* 2(6):1033-1034 (1975).

Chapra, S. C. "Total Phosphorus Model for the Great Lakes," *J. Environ. Eng. Div., ASCE* 103(EE2):147-161 (1977).

Cornell, C. A. "First-Order Analysis of Model and Parameter Uncertainty," in *Proceedings of the International Symposium on Uncertainties in Hydrologic and Water Resource Systems,* Vol. III. (Tucson, AZ: University of Arizona, 1972), pp. 1245-1274.

Dillon, P. J., and W. B. Kirchner. "Reply (to Chapra's Comment)," *Water Resources Res.* 2(6):1035-1036 (1975).

Dillon, P. J., and F. G. Rigler. "A Test of a Simple Nutrient Budget Model Predicting the Phosphorus Concentration in Lake Water," *J. Fish. Res. Bd. Can.* 31:1771-1778 (1974).

Dillon, P. J., and F. H. Rigler. "A Simple Method for Predicting the Capacity of a Lake for Development based on Lake Trophic Status," *J. Fish. Res. Bd. Can.* 32:1519-1531 (1975).

Fair, G. M., J. C. Geyer and D. A. Okun. *Water and Wastewater Engineering* (New York: John Wiley & Sons, Inc., 1968), pp. 22-27.

Jones, J. R., and R. W. Bachmann. "Prediction of Phosphorus and Chlorophyll Levels in Lakes," *J. Water Poll. Control Fed.* 48(9):2176-2182 (1976).

Kirchner, W. B., and P. J. Dillon. "An Empirical Method of Estimating the Retention of Phosphorus in Lakes," *Water Resources Res.* 2(1):182, 183 (1975).

Larsen, D. P., and H. T. Mercier. "Lake Phosphorus Loading Graphs: An Alternative," National Eutrophication Survey, Working Paper No. 174, U.S. Environ. Protection Agency, Corvallis, Oregon (1975).

Piontelli, R., and V. Tonolli. "The Time of Retention of Lucustrine Waters in Relation to the Phenomena of Enrichment in Introduced Substances, with Particular Reference to the Lago Maggiore," *Mem. Ist. Ital. Idrobiol.* 17:247-266 (1964).

Popper, K. *Conjectures and Refutations: The Growth of Scientific Knowledge* (New York: Harper & Row, Publishers, Inc., 1963).

Rawson, D. S. "Morphometry as a Dominant Factor in the Productivity of Large Lakes," *Verh. Int. Ver. Limnol.* 12:164-175 (1955).

Reckhow, K. "Phosphorus Models for Lake Management," Ph.D. dissertation, Harvard University (1977a).

Reckhow, K. Unpublished notes on Bayesian Approach to the Estimation of Lake Phosphorus Retention (1977b).

Reckhow, K. *Quantitative Techniques for the Assessment of Lake Quality* (Lansing, MI: Michigan Department of Natural Resources, 1978).

Reckhow, K., and H. Rice. *Nutrient Budget and Error Analysis for Lake Winnipesaukee and Allied Lakes.* report prepared for the New Hampshire Lakes Region Planning Commission, Meredith, NH (1977).

Snodgrass, W. J. "A Predictive Phosphorus Model for Lakes: Development and Testing," Ph.D. dissertation, University of North Carolina (1974).

Tebbutt, T. H. Y., and D. G. Christoulas. "Performance Relationships for Primary Sedimentation," *Water Res.* 9(3):347-356 (1975).

Vollenweider, R. A. "The Correlation Between Area of Inflow and Lake Budget," paper presented at the German Limnological Conference, Lunz (1964).

Vollenweider, R. A. "The Scientific Basis of Lake and Stream Eutrophication, with Particular Reference to Phosphorus and Nitrogen as Eutrophication Factors," Technical Report DAS/DSI/68.27, Organization for Economic Cooperation and Development, Paris, France (1968).

Vollenweider, R. A. "Possibilities and Limits of Elementary Models Concerning the Budget of Substances in Lakes," *Arch. Hydrobiol.* 66(1):1-36 (1969).

Vollenweider, R. A. "Input-Output Models with Special Reference to the Phosphorus Loading Concept in Limnology," *Sch. Zeit. Hydrologic* 37:53-84 (1975).

Walker, W. W., Jr. "Some Analytical Methods Applied to Lake Water Quality Problems," Ph.D. dissertation, Harvard University (1977).

CHAPTER 9

A MINIMUM-COST SURVEILLANCE PLAN
FOR WATER QUALITY TREND DETECTION
IN LAKE MICHIGAN

Leon M. DePalma

The Analytic Sciences Corporation
Reading, Massachusetts 01867

Raymond P. Canale

Department of Civil Engineering
University of Michigan
Ann Arbor, Michigan 48109

William F. Powers

Department of Aerospace Engineering
University of Michigan
Ann Arbor, Michigan 48109

INTRODUCTION

One objective of the sampling programs designed by the Great Lakes Water Quality Board (GLWQB) (1976) of the International Joint Commission is to detect trends in Lake Michigan water quality over future decades. In this paper we examine the cost-effectiveness of such programs and demonstrate how mass balance models and measurements of lake concentrations may be employed together in trend detection.

The uncertain parameters in a simple model for lakewide average concentrations of chloride and total phosphorus have been characterized statistically. The model is interpreted as nonlinear stochastic difference equations which can be used, along with lake measurements, in a linearized Kalman filter. If the filter is used to estimate the accuracy of lakewide average concentrations over the next 30 years and a sampling program is prespecified, then the standard deviations associated with the accuracy of the estimates can be precomputed. The estimated errors must be small enough for trend detection; thus the standard deviations must satisfy upper bound constraints. The objective of this study was to minimize the cost of the sampling program subject to these constraints. After an optimal sampling program was computed using algorithms developed by DePalma (1977), the sensitivity of this program to the accuracy of the model parameters was investigated to determine where future research should be directed to lower the cost of optimal sampling programs.

DERIVATION OF THE STOCHASTIC MODEL

Mass balance models have been used by O'Connor and Mueller (1970) and Chapra (1977) to predict the consequences of pollution control measures on Great Lakes chloride and total phosphorus concentrations. In this section, the accuracy of a stochastic model for chloride and total phosphorus in Lake Michigan is determined.

The Indicator of Trend

Chloride is a conservative pollutant that occurs primarily in dissolved form and is little affected by the annual biological cycle. Total phosphorus, however, includes phosphorus which is utilized in the biochemical cycles of phytoplankton and other organisms. As phytoplankton cells mature and die, they accumulate at the sediment-water interface resulting in removal of phosphorus from the water column. A measure of the phosphorus available for biological production during a particular year is provided by the lakewide average concentration of total phosphorus following spring overturn. Because ice at this time may prevent a sampling program from being implemented, it is convenient to assign May as the month when chloride and total phosphorus concentrations are relevant to trend detection.

The lakewide average concentrations of chloride, $x_1(n)$, and total phosphorus, $x_2(n)$, for May of year n are defined, respectively, by:

$$x_1(n) = m_1(n)/V(n) \text{ and } x_2(n) = m_2(n)/V(n) \tag{1}$$

where $m_1(n)$ and $m_2(n)$ are the pollutant masses in the lake, including the nearshore zone but excluding the sediments, and where $V(n)$ is the lake volume. Lakewide average concentrations may only be crude indicators of trend if severe spatial concentration gradients exist. However, if the lake is segmented spatially, then mass balance models must be derived for each segment average, and the resultant model must include uncertain parameters for advective and dispersive mass transfer among the segments.

In order to derive difference equations for the lakewide average concentrations, equations are first determined for the masses $m_1(n)$ and $m_2(n)$. During year n (May of year n to April of year n + 1), mass exits the lake through the Straits of Mackinac (SM) and the Chicago Sanitary and Ship Canal (CSSC) and enters from tributary, atmospheric and direct loading. Furthermore, a fraction of the total phosphorus that reaches the sediments during a year is incorporated into refractory deposits, resulting in a net sedimentation loss (Hutchinson, 1957). These sources and sinks are approximated on a net annual basis rather than on a seasonal basis, and their statistical characteristics are evaluated with information presently available.

Straits of Mackinac

The net annual flow through the SM is toward Lake Huron; however, at certain times of the year the flow is toward Lake Michigan. The net exchange of pollutant mass between the two lakes during one year is the difference between the flux towards Lake Michigan and the flux towards Lake Huron. Each flux is the product of an annual flow and an effective concentration. The effective concentration lies in the range of instantaneous concentrations during the year. If q_{in} and q_{out} are the components of annual flow toward and away from Lake Michigan, and if x_{in} and x_{out} are the effective concentrations on the Lake Huron and Lake Michigan sides of the SM, then the net loss from Lake Michigan is:

$$x_{out} \, q_{out} - x_{in} \, q_{in} \qquad (2)$$

If $x_{in} = x_{out}$ then the net loss is proportional to $q_{out} - q_{in}$ which is the net annual flow. However, chloride and total phosphorus concentrations in Lake Huron are less than those in Lake Michigan because of dilution by Lake Superior water. A conservative upper bound on the net loss is obtained by setting $x_{in} = 0$, in which case the loss is proportional to q_{out}. Therefore, the actual loss must lie between $x_{out} \, (q_{out} - q_{in})$ and $x_{out} \, q_{out}$, and is denoted by $x_{out} \, q_s$, where q_s is an uncertain effective outflow. Quinn (1977) has computed net annual flows, $q_{out} - q_{in}$, of 224 x 10^{11} liter to 522 x 10^{11} liter. The validity of his technique has been upheld by current meter data from Saylor and Sloss (1976). Quinn also suggests that the current meter data can be used to obtain the following relationship between gross flow and net flow:

$$q_{out} \approx 1.67 \, (q_{out} - q_{in}) \qquad (3)$$

If Equation 3 is used to estimate the gross outflow for the period 1950-1966 from Quinn's net flows, then a range of 374 x 10^{11} liter to 871 x 10^{11} liter results.

The effective flow for each year n, $q_s(n)$, is interpreted as a random variable, uniformly distributed between the minimum net flow and maximum gross flow for 1950-1966. The mean and standard deviation of this random variable are, respectively, 547 x 10^{11} liter and 186 x 10^{11} liter. The effective concentration on the Lake Michigan side of the SM for year n, x_{out} (n), must lie in the range of concentrations in the outflowing water during the year. It is assumed that this range is narrower than the range of concentrations over the entire lake for May of year n. The effective concentration is represented as the sum of the lakewide average concentration in May, x(n), and a deviation, ϵ_s (n). That is:

$$x_{out} \text{ (n)} = x \text{ (n)} + \epsilon_s \text{ (n)} \qquad (4)$$

The deviation is a zero mean random variable which is representative of the spatial heterogeneity of the lake. Its standard deviation can be estimated from unpublished data collected in the northern half of the lake by the Great Lakes Research Division (GLRD) of the University of Michigan for April and June 1976. A statistical summary of the chloride data from 37 stations sampled during the June cruise and similar results for total phosphorus for 33 stations sampled during the April cruise are provided in Table I. Only samples from station depths greater than 20 m have been included, to avoid bias due to nearshore pollution. For each station, the mean (unweighted for depth) and standard deviation of the data have been computed. The standard deviation of the station means is assumed to be representative of the variation in the data because the data suggest that the horizontal variation exceeds the vertical variation. These standard deviations are 0.32 mg-Cl l^{-1} and 1.6 μg-P l^{-1} . However, the total observed variation is not only due to spatial heterogeneity, but also to laboratory analysis error. Davis (1977) has suggested that standard deviations for laboratory errors are 0.10 mg-Cl l^{-1} and 1.0 μg-P l^{-1}. Therefore, the standard deviations of station concentrations for the lakewide average concentrations are 0.30 mg-Cl l^{-1} and 1.3 μg-P l^{-1}. These are used for the standard deviations of ϵ_{s_1} (n) and ϵ_{s_2} (n), respectively, where the subscripts have been introduced to distinguish chloride (subscript 1) from total phosphorus (subscript 2).

The net losses of chloride and phosphorus mass through the SM are, respectively:

$$[x_1(n) + \epsilon_{s_1}(n)] \; q_s(n) \text{ and } [x_2(n) + \epsilon_{s_2}(n)] \; q_s(n) \qquad (5)$$

Table I. Summary of GLRD 1976 Data for Northern Lake Michigan

	Chloride			Total Phosphorus		
Station No.	Number of Samples	Mean (mg l^{-1})	Standard Deviation (mg l^{-1})	Number of Samples	Mean (μg l^{-1})	Standard Deviation (μg l^{-1})
2	4	7.73	0.02	4	7.3	0.6
4	6	7.68	0.03	6	7.9	0.6
6	13	7.69	0.03	12	8.1	1.2
8	6	7.70	0.05	6	9.1	1.0
10	4	7.80	0.19	4	14.8	2.4
13	4	7.70	0.01	4	8.6	1.3
15	6	7.72	0.03	6	7.4	0.4
17	17	7.73	0.06	17	6.7	1.3
19	5	7.86	0.02	6	7.0	0.6
20	7	7.89	0.02	6	7.1	0.5
23	5	7.73	0.02	5	6.9	0.4
25	14	7.73	0.02	14	7.4	0.7
27	6	7.68	0.07	6	6.9	1.1
29	4	7.63	0.03	4	6.8	0.8
30	4	7.79	0.10	4	8.2	1.1
32	4	7.72	0.05	4	5.9	0.6
34	5	7.74	0.04	5	6.5	0.6
36	9	7.73	0.05	9	7.9	1.2
39	4	7.58	0.11	4	6.3	0.6
42	3	7.70	0.04	3	6.5	0.7
43	3	7.68	0.03	2	6.5	0.1
44	3	7.68	0.08	4	7.8	2.3
46	3	7.34	0.22	3	6.6	0.4
47	4	7.58	0.14	4	6.1	0.2
48	3	7.55	0.01	3	6.4	0.4
49	3	7.45	0.14	3	6.2	0.4
53	2	7.01	0.01			
54	3	7.10	0.03	3	8.4	2.0
55	4	7.32	0.19	4	5.9	0.6
58	3	6.58	1.13	3	5.8	0.7
59	5	7.07	0.06	3	6.1	0.6
60	3	7.06	0.12	2	6.0	0.1
63	5	7.00	0.26	3	6.4	0.6
64	3	6.89	0.10	3	5.2	0.4
67	6	7.84	0.06			
68	5	7.83	0.05			
69	3	7.90	0.06			

Chicago Sanitary and Ship Canal

The annual diversion of water from Lake Michigan to the Illinois River through the CSSC has been limited to 28.7×10^{11} since 1970 by U.S. Supreme Court decree. The outflow, denoted by $\bar{q}_c$, is interpreted as a known constant at this maximum value. The effective concentration in the outflow is interpreted in the same manner as for the SM with resulting losses:

$$[x_1(n) + \epsilon_{c_1}(n)] \; \bar{q}_c \text{ and } [x_2(n) + \epsilon_{c_2}(n)] \; \bar{q}_c \qquad (6)$$

The zero mean random variables, $\epsilon_{c_1}(n)$ and $\epsilon_{c_2}(n)$, are different random variables than $\epsilon_{s_1}(n)$ and $\epsilon_{s_2}(n)$ but have the same standard deviations, that is, 0.30 mg-Cl l^{-1} and 1.3 μg-P l^{-1}.

Refractory Sediment Deposits

For a lake in which the lakewide average phosphorus concentration in the spring is at steady state with the loads, the phosphorus retention coefficient (R_p) is:

$$R_p = (m_{in} - m_{out})/m_{in} \qquad (7)$$

where m_{in} and m_{out} are the mass loading input and mass outflow during a year. The phosphorus apparent settling velocity, S (m yr^{-1}), is that velocity such that the net sediment loss for a year is $Sm/\bar{h}$, where m is the lakewide mass of phosphorus in the spring and $\bar{h}$ is the mean depth. According to Chapra (1975), the apparent settling velocity can be related to the retention coefficient by Equation 8:

$$S = R_p \; (q/A)/(1 - R_p) \qquad (8)$$

where q is the total outflow and A is the lake surface area. Although R_p is difficult to measure for Lake Michigan, data have been compiled for smaller lakes and apparent settling velocities computed with Equation 8 are summarized in Table II. These data suggest a range of 0.7 to 37.9 m yr^{-1}.

The mechanisms which govern the formation of refractory sedimentary deposits are complex. The exchange between the sediments and the overlying waters of a lake depends on current velocities and degree of stratification, on phytoplankton species composition, and on biological and chemical cycles at the interface. Harrison et al. (1972), for example, discuss the effect of oxygen and bacteria on the exchange. It would be extremely difficult to obtain an estimate of S by relating exchange

Table II. Apparent Settling Velocity for Small Lakes

Lake	$S(m\ yr^{-1})^a$	Notes
Four Mile	11.0, 11.0	Kirchner and Dillon
Raven	11.3, 13.3	(1975)
Talbot	9.3	
Bob	16.2, 16.3	
Twelve Mile + Boshkung	22.9, 21.4	
Halls	25.5, 29.4	
Beech	8.6, 16.8	
Pine	5.0, 1.4	
Eagle + Moose	8.6, 14.6	
Oblong + Haliburton	14.5	
Rawson	5.6, 4.9, 6.0	
Clear	4.3	
Brewer	3.7	Dillon and Kirchner
Clarke	17.0	(1975)
Costello	8.7	
Kearney	6.5	
Aegerisse	12.1	
Bodensee	37.9	
Turlersee	26.0	
Zurichsee	11.3	
Hallwilersee	4.1	
Griefensee	15.3	
Pfaffikersee	23.1	
Baldeggersee	11.8	
Tahoe	5.7	
Found	7.2	
Little McCauley	8.8	
Mendota	11.1	Sonzogni (1974)
Joseph	8.8	Michalski et al. (1973)
Rousseau	13.6	
Conesus	35.8	Stewart and Markella
Canadice	30.9	(1974)
Honeoye	23.3	
Washington	27.0	Lorenzen et al. (1976)
Leman	3.2	Larsen and Mercier (1976)
Bay of Naples	14.1	
Canadaligua	5.1	
Carlos	4.5	
Carry Falls	20.2	
Cass	15.1	
Charlevoix	8.9	
Higgins	2.1	
Houghton	1.6	
Long (Aroostook Co.)	5.3	

Table II (Continued)

Lake	$S(m\ yr^{-1})^a$	Notes
Long (Cumberland Co.)	8.6	
Mattawamkegg	14.4	
Moosehead	4.6	
Pelican	0.7	
Rangeley	4.9	
Sebago	7.8	
Winnipesaukee	11.0	

[a]Multiple values represent different data sets.

mechanisms in Lake Michigan with those of the lakes in Table II. Therefore S will be interpreted as a random variable, uniformly distributed between 0.7 and 37.9 m yr^{-1}. The mean and standard deviation are, respectively, 19.3 and 10.7 m yr^{-1}. If $\bar{h}(n)$ is the mean depth of Lake Michigan, then the net loss of phosphorus to the sediments during year n is:

$$S(n)\ m_2(n)/\bar{h}(n) \qquad (9)$$

Tributary Loads

The GLWQB has identified 27 major tributaries in the Lake Michigan drainage basin. The mass of pollutant discharged into the lake during a year by a tributary is the integral of the instantaneous flux over the year. The Mean Value Theorem permits the integral to be replaced by a sum:

$$\text{Load} = \sum_{d=1}^{365} x_d\ q_d \qquad (10)$$

where q_d is the tributary discharge at the mouth during day d and x_d is an effective concentration of pollutant in the discharge at the mouth on day d.

Although daily discharges are available for the tributaries from the U.S. Geological Survey Water Supply Papers, concentration measurements are normally made only at monthly intervals. An estimate of the annual load for a tributary, based on a complete and accurate record of discharge but an incomplete and inaccurate record of concentration, may be obtained in the following manner. Assume that a concentration measurement is available for each day d in a set of D days, and that this measurement serves as an estimate for x_d. Furthermore, assume that a concentration measurement is not made on each of the other 365-D days in set $\overline{D}$, and that x_d in this case is estimated from past data on the tributary (for example, from a regression relationship between concentration and discharge).

The concentration estimate, x_d, for a day in D may include errors due to laboratory analysis, variation over the day, variation over the river cross section, and distance of the sampling station from the river mouth. The error is interpreted as a random variable with zero mean and standard deviation σ_1 = 5.0 mg l^{-1} for chloride and σ_2 = 50.0 μg l^{-1} for total phosphorus. The standard deviations were chosen as 10% of typical concentrations of 50.0 mg-Cl l^{-1} and 500.0 μg-P l^{-1}. The concentration estimation error for a day in $\overline{D}$ depends upon the availability of past data and upon the ability to use these data and the known discharge to estimate the concentration x_d. The error associated with this procedure is interpreted as a zero mean random variable whose standard deviation is equal to that of data available for the Maumee River (U.S. Army Corps of Engineers, 1975). The standard deviations of these data are 15.6 mg-Cl l^{-1} and 271.0 μg-P l^{-1}. These deviations are used because they are probably conservative upper bounds on the standard deviations of errors associated with estimates of Lake Michigan tributary concentrations. Standard deviations are not tributary specific, thus, for days in $\overline{D}$, $\overline{\sigma}_1$ = 15.6 mg-Cl l^{-1} and $\overline{\sigma}_2$ = 271 μg-P l^{-1} for each tributary.

The error variances for the annual tributary loads for each tributary are therefore given by:

$$\sigma_1^2 \sum_D q_d^2 + \overline{\sigma}_1^2 \sum_{\overline{D}} q_d^2 \quad \text{and}$$

$$\sigma_2^2 \sum_D q_d^2 + \overline{\sigma}_2^2 \sum_{\overline{D}} q_d^2, \tag{11}$$

for chloride and total phosphorus. The calculations in Equation 11 require measured tributary discharges q_d. However, in order to compute optimal lake measurement strategies for the next 30 years, standard deviations must be available now. Therefore, to be conservative q_d for each tributary is replaced by the maximum daily discharge on record for that tributary. This discharge for the jth tributary is denoted by $q_{max}j$ and is computed from the maximum discharge at the closest station to the mouth of the tributary by correcting for ungauged drainage.

The error variances for the total tributary loads for year n are then computed by:

$$[\sigma_1^2 \, D(n) + \overline{\sigma}_1^2 \, (365 - D(n))] \sum_{j=1}^{27} (q_{max}^j)^2 \quad \text{and}$$

$$[\sigma_2^2 \, D(n) + \overline{\sigma}_2^2 \, (365 - D(n))] \sum_{j=1}^{27} (q_{max}^j)^2 \qquad (12)$$

where it is assumed that both chloride and total phosphorus are analyzed in the water samples collected on days D. The most appropriate number of sampling days for year n, D(n), is interpreted as a variable to be determined in association with measurement strategy optimization.

Atmospheric Loads

Precipitation samples collected by Junge and Werby (1958) suggest that the annual chloride atmospheric load for Lake Michigan is between 6.4×10^{12} and 8.7×10^{12} mg. In the mass balance model, the atmospheric load of chloride is denoted by the random variable $a_1(n)$, which is uniformly distributed between these extremes and has a mean of 7.6×10^{12} mg and a standard deviation 0.7×10^{12} mg. Murphy and Doskey (1975) collected precipitation samples from the shores of the lake. Their data are used to compute a range of 12.0×10^{14} to 18.5×10^{14} μg for the annual total phosphorus atmospheric load. This load is interpreted as a random variable, $a_2(n)$, uniformly distributed between these extremes and having a mean of 15.3×10^{14} μg and a standard deviation of 1.9×10^{14} μg. If data become available which more clearly define the phosphorus loading, then the optimal strategies reported in this paper should be recomputed.

Direct Loads

Some industries and municipalities discharge pollutants directly into Lake Michigan or discharge to a tributary below the sampling point for that tributary. The U.S. Environmental Protection Agency (1972) has computed a direct annual phosphorus load for 1971 of 17.0×10^{14} μg, which is less than the sum of 1970 tributary and direct loads of 97.0×10^{14} μg (Chapra, 1977). Corresponding data for the direct chloride load are not available, although the sum of tributary and direct loads for 1960 was computed by O'Connor and Mueller (1970). The direct loads are denoted by $\overline{d}_1(n)$ and $\overline{d}_2(n)$ and are considered to be known constants because their standard deviations are probably small compared to those of the tributary and atmospheric loads.

Summary of the Model

The approximations for the sources and sinks of chloride and total phosphorus as defined above are used to derive the stochastic difference equations for the lakewide masses:

$$m_1(n + 1) = m_1(n) - [x_1(n) + \epsilon_{s_1}(n)] \, q_s(n) - [x_1(n) + \epsilon_{c_1}(n)] \, \overline{q}_c$$

$$+ [t_1(n) + a_1(n) + \overline{d}_1(n)] \tag{13}$$

$$m_2(n + 1) = m_2(n) - [x_2(n) + \epsilon_{s_2}(n)] \, q_s(n) - [x_2(n) + \epsilon_{c_2}(n)] \, \overline{q}_c$$

$$- S(n) \, m_2(n)/\overline{h}(n) + [t_2(n) + a_2(n) + \overline{d}_2(n)] \tag{14}$$

Quinn (1975) has computed Lake Michigan water levels for the first of May for each year over the period 1900-1972. The levels have a standard deviation of 1.1 ft (0.34 m) and can be converted to volumes with an accuracy of ±0.4% (Schutze, 1977). Volumes ranged from 4.89 x 10^{15} liter to 4.97 x 10^{15} liter. Therefore, the volume has been interpreted as a constant, $\overline{V}$ = 4.93 x 10^{15} liter. Similarly, the average depth is $\overline{h}$ = 85 m. The equations for lakewide average concentrations are, therefore:

$$x_1(n + 1) = [1 - \overline{q}_s/\overline{V} - \overline{q}_c/\overline{V}] \, x_1(n)$$

$$- \frac{1}{\overline{V}} [\overline{q}_s \, \epsilon_{s_1}(n) + \overline{q}_c \, \epsilon_{c_1}(n) + x_1(n) \, \widetilde{q}_s(n)]$$

$$+ \frac{1}{\overline{V}} [t_1(n) + a_1(n) + \overline{d}_1(n)] \tag{15}$$

$$x_2(n + 1) = [1 - \overline{q}_s/\overline{V} - \overline{q}_c/\overline{V} - S(n)/\overline{h}] \, x_2(n)$$

$$- \frac{1}{\overline{V}} [\overline{q}_s \, \epsilon_{s_2}(n) + \overline{q}_c \, \epsilon_{c_2}(n) + x_2(n) \, \widetilde{q}_s(n)]$$

$$+ \frac{1}{\overline{V}} [t_2(n) + a_2(n) + \overline{d}_2(n)] \tag{16}$$

where $\widetilde{q}_s(n) = q_s(n) - \overline{q}_s$ and $\overline{q}_s$ is the mean of $q_s(n)$. The terms $\epsilon_{s_1}(n) \, \widetilde{q}_s(n)$ and $\epsilon_{s_2}(n) \, \widetilde{q}_s(n)$ are small and have been neglected.

In addition to Equations 15 and 16, or just 16, an equation is written for the phosphorus apparent sinking rate $S(n)/\overline{h}$, which is denoted by the state variable $x_3(n)$. If the rate is constant, but uncertain, then:

$$x_3(n + 1) = x_3(n) \tag{17}$$

This trivial equation can be used to estimate the accuracy of the sinking rate along with the accuracy of chloride and total phosphorus concentrations and thereby to improve the performance of the Kalman filter.

At the initial year, n = 1976, the three state variables are uncertain. From GLRD data in Table I, mean values of 7.55 mg-Cl l^{-1} and 7.17 μg-P l^{-1} are computed for x_1 (1976) and x_2 (1976), respectively. The accuracy of these estimates depends on the variance of the measurements obtained from the lake and on the number of samples employed in the computation of the means. The formula presented in the next section (Equation 18) for computing the error variance of a measurement is used to compute the error variance of the initial concentration estimates. The standard deviations of the errors are 0.05 mg-Cl l^{-1} and 0.29 μg-P l^{-1} An estimate for S(1976) is 19.3 m yr^{-1}. The standard deviation of the error in this initial estimate is the standard deviation assuming the estimate is uniformly distributed, that is 10.7 m yr^{-1}. Note that this does not equal the standard deviation of the data in Table II because these data were only used to establish a range for S.

KALMAN FILTER STATE ESTIMATION

Equations 15, 16 and 17 constitute a nonlinear stochastic model with uncertain states (x_1, x_2 and x_3), uncertain parameters (ϵ_{s_1}, ϵ_{s_2}, ϵ_{c_1}, ϵ_{c_2}, $\tilde{q}_s$, a_1, a_2, t_1 and t_2), and known constants (q_s, q_c, V, d_1 and d_2). These parameters and constants are summarized in Table III.

The Kalman filter (Sage and Melsa, 1971) is a method which can be used to calculate an unbiased minimum variance estimate of the state variables (chloride and total phosphorus). The general basis and approach of the Kalman filter is described in Appendix A. The method requires statistical characterization of both the model equations and the measurements, but does not depend on the actual measurements themselves. It is this property of the filter that allows the development of an *a priori* optimal measurement strategy. If estimates of the state variables are also desired, then measurements are required. In order to implement the linearized Kalman filter, the nonlinear state equations are linearized about nominal states which are computed by setting each uncertain parameter and initial state to its mean value and then solving the state equations (Equations 15, 16 and 17). The nominal states have been computed assuming that future loads are the same as present loads; however, the Kalman filter approach is also applicable if future loads can be represented as known functions of time.

Table III. A Summary of Parameters, Constants and Initial States

Uncertain Parameter	Description	Mean	Standard Deviation	Unit
ϵ_{s_1}	Deviation of SM concentration from lakewide average	0	0.30	mg l^{-1}
ϵ_{s_2}		0	1.3	μg l^{-1}
ϵ_{c_1}	Deviation of CSSC concentration from lakewide average	0	0.30	mg l^{-1}
ϵ_{c_2}		0	1.3	μg l^{-1}
$\tilde{q}_s$	Deviation of SM effective flow from mean $\bar{q}_s$	0	186 x 10^{11}	1
$t_1 + \bar{d}_1$	Sum of tributary load and direct load[a]	934. 10^{12}	22-67 x 10^{12}	mg
$t_2 + \bar{d}_2$		96.7 10^{14}	2.2-11.6 x 10^{14}	μg
a_1	Atmospheric loads[b]	7.6 10^{12}	0.7 x 10^{12}	mg
a_2		15.3 10^{14}	1.9 x 10^{14}	μg

Constant	Description	Value	Unit
$\bar{q}_s$	Mean effective flow through SM	547 x 10^{11}	1
$\bar{q}_c$	Flow through CSSC	28.7 x 10^{11}	1
$\bar{v}$	Volume	4.93 x 10^{15}	1

Table III (Continued)

Initial State	Description	Mean	Standard Deviation	Unit
x_1 (1976)	Chloride lakewide average concentration	7.55	0.05	mg l^{-1}
x_2 (1976)	Phosphorus lakewide average concentration	7.17	0.29	μg l^{-1}
x_3 (1976)	Phosphorus apparent settling velocity/$\overline{h}$	0.227	0.126	yr^{-1}

[a]The means are for 1960 (chloride) and 1970 (phosphorus).   The range of standard deviations corresponds to 365 $\gg D \geqslant$ 12.
[b]The means are for 1955 (chloride) and 1974 (phosphorus).

The accuracy of the model depends on the accuracy of the model parameters in a complex manner; a subject explored elsewhere (Canale *et al.*, 1978). The accuracy of a measurement of $x_1(n)$ or $x_2(n)$ depends on the number of lake water samples used in the computation of the lakewide average concentration. If the average of $L(n)$ station concentrations serves as the estimate, then it is possible to determine a formula which can be used to relate the measurement error and $L(n)$. If δ^2 is the variance due to the heterogeneous nature of the lake and γ^2 is the variance due to laboratory analysis error, then the measurement error variance is:

$$(\delta^2 + \gamma^2)/L(n) \qquad (18)$$

Equation 18 requires random sampling of the lake. In order to avoid bias due to nonrandom sampling, the following guidelines are suggested: (1) a representative concentration at each station should be computed from discrete samples at several depths; (2) only stations in the main lake should be considered; (3) stations should be visited in both the northern and southern basins; and (4) stations should be uniformly spaced along each transect.

The Kalman filter combines the model and measurements to compute an estimate of the lakewide average concentration which is theoretically more accurate than estimates based on the model or measurements alone. The estimation error variance, which is a measure of the accuracy of the Kalman estimate, can be computed and it depends on the model parameter variances and measurement error variance, but not on the actual measurements. Once the number of days, $D(n)$, on which tributary samples are collected and the number of stations, $L(n)$, at which lake samples are collected have been specified (for a number of years beyond 1976), the estimation error variances can be computed. Furthermore, $D(n)$ and $L(n)$ may be chosen to meet constraints imposed on the estimation error variances.

OPTIMIZATION OF MEASUREMENT STRATEGIES

The GLWQB (1976) has recommended a schedule for intensive sampling of the offshore waters of Lake Michigan that specifies two successive years of sampling followed by seven years during which the other Great Lakes are sampled. Lake Michigan tributaries would be sampled 26 times per year. Detection of trends in chloride and total phosphorus concentrations would be accomplished by comparing measurements of lakewide average concentrations for those years when samples are collected (1976, 1977, 1985, 1986. . . .). The estimated concentration for any one year is based solely on data from that year. No procedure has been proposed for utilizing mass balance information. If 60 stations are sampled during any single May cruise as recommended by the Great Lakes Water Quality Board, then from Equation 18, the estimation errors of the chloride and total phosphorus lakewide average concentrations have standard deviations of 0.04 mg-Cl l^{-1} and 0.2 μg-P l^{-1}, respectively.

In order to evaluate the effectiveness of this program, it is assumed that the estimates must meet tolerances of ±0.5 mg-Cl l^{-1} and ±0.5 μg-P l^{-1} with 0.90 probability. Rather than assuming that estimation errors are normal, the Chebyschev inequality (Davenport, 1970) is used to convert each tolerance to a constraint on the standard deviation of the estimation errors. To guarantee, with a probability of p, that the estimation errors will not exceed Δ, the standard deviation must not exceed $\Delta\sqrt{1 - p}$. Therefore, (for p = 0.90 and Δ = 0.5) the constraints on the standard deviations are 0.158 mg-Cl l^{-1} and 0.158 μg-P l^{-1}. The GLWQB program does not include enough stations to meet the phosphorus constraint (0.2 > 0.158 μg-P l^{-1}) but includes more than enough stations to meet the chloride constraint (0.04 < 0.158 mg-Cl l^{-1}). An increase in the number of stations from 60 to 108 would allow both constraints to be met.

If the mass balance model given by Equations 15 and 16 is used, the estimates can be improved by processing the data with a Kalman filter. The standard deviations of the Kalman filter estimation errors have been computed for 30 years based on the GLWQB lake sampling program and 26 days of tributary sampling every year (see Table IV). The chloride constraint is satisfied every year, even though the lake is not sampled every year. However, the model is not accurate enough to improve the phosphorus estimates substantially. Thus, the phosphorus constraint is still violated every year. However, an algorithm proposed by DePalma (1977) can be used to compute the minimum-cost sampling strategy that does satisfy the constraints at specified years in the future.

In particular, consider the case in which the chloride and phosphorus estimation constraints must be satisfied every nine years (1977,1986, 1995 and 2004). The number of lake stations L(n) is to be chosen between 0 and a maximum of 168 which corresponds to two ships, each making 14 transects, with 6 stations per transect. If water samples are collected at a station and analyzed for all basic chemicals, including chloride and total phosphorus, then the total cost incurred per transect is $3,320 (Richardson, 1977). The number of days of tributary sampling D(n) is to be chosen between the routine 12 and 365. If one water sample is collected from each tributary and analyzed for all basic chemistry, then the total cost incurred in sampling the system is $3,300 per day. The minimum-cost combination of lake and tributary samples, which

Table IV. Standard Deviations for the GLWQB Strategy

n	Chloride $x_1(n)$	Phosphorus $x_2(n)$	Settling Velocity $x_3(n)$
1976	0.053	0.29	0.126
1977	0.033	0.20	0.049
1978	0.046	0.53	0.049
1979	0.056	0.81	0.049
1980	0.064	1.03	0.049
1981	0.071	1.22	0.049
1982	0.078	1.37	0.049
1983	0.084	1.49	0.049
1984	0.090	1.58	0.049
1985	0.040	0.20	0.012
1986	0.031	0.17	0.011
1987	0.047	0.30	0.011
1988	0.058	0.38	0.011
1989	0.067	0.42	0.011
1990	0.076	0.46	0.011
1991	0.083	0.48	0.011
1992	0.090	0.50	0.011
1993	0.096	0.51	0.011
1994	0.037	0.19	0.008
1995	0.032	0.17	0.007
1996	0.049	0.29	0.007
1997	0.062	0.34	0.007
1998	0.072	0.38	0.007
1999	0.081	0.40	0.007
2000	0.089	0.42	0.007
2001	0.096	0.43	0.007
2002	0.103	0.44	0.007
2003	0.037	0.19	0.006
2004	0.032	0.17	0.006
2005	0.051	0.28	0.006
2006	0.065	0.33	0.006

satisfies the imposed constraints, has been calculated with a computer program described elsewhere (Canale *et al*, 1978).

Only 12 days of tributary sampling are required by the optimal strategy which represents a saving of $46,200 each year. Lake samples are required only for those years when constraints are imposed. The number of lake stations required for each of the four years when a constraint is imposed is shown in Table V as Strategy I. If the model was not used, 108 stations would be required each year resulting in a four-year cost of $239,000. The model permits a reduction in cost to $227,000. The overall savings for lake and tributary sampling is $1,398,000. The standard

Table V. Required Number of Sampling Stations and Costs for Conservative
Formulation of Tributary Load Error

Strategy	Year				Cost ($)
	1977	1986	1995	2004	
I	106	108	100	97	227,000
II	106	108	100	97	227,000
III	106	108	100	96	227,000
IV	92	107	100	96	218,000
V	84	95	94	93	201,000
VI	83	91	91	91	196,000
VII	76	78	72	69	162,000
VIII	67	68	63	61	142,000
IX	88	90	82	79	187,000
X	70	72	65	61	148,000
XI	52	54	46	43	107,000
XII	34	36	29	26	68,000
XIII	25	27	21	18	49,000

deviations corresponding to Strategy I (and all strategies discussed below)
have the following form: the chloride constraint is satisfied for every
year between 1977 and 2004, but the phosphorus constraint is only satisfied for the four years when it is imposed.

Only $12,000 is saved by utilizing the mass balance model because
the formulation of the standard deviation of the tributary load as a function of D(n) is very conservative (being based on maximum observed
discharges and Maumee River concentration data). The minimum-cost
strategy outlined above has been recomputed under somewhat more realistic assumptions which are: (1) the daily discharges equal the average
daily discharge on record for a tributary, (2) σ_1 and σ_2 are 1/50 of the
values previously assumed, and (3) $\check{\sigma}_1$ and $\check{\sigma}_2$ assume the values originally
assigned to σ_1 and σ_2. The number of lake stations now required by the
optimal strategy (Strategy I) is presented in Table VI and is substantially
less than for the conservative assumptions. The costs for the more liberal
($153,000) and conservative ($227,000) cases bracket the cost which would
be computed if actual values for the discharges and standard deviations
were available today.

The sensitivity of the optimal cost to the accuracy of the effective
SM flow, the atmospheric loads and the phosphorus apparent settling
velocity has been studied in order to ascertain the dollar value of intensive
research on these parameters. Optimal strategies have been computed for
both the conservative and liberal formulations of the tributary load errors

Table VI. Required Number of Sampling Stations and Cost for more Realistic
Formulation of Tributary Load Error

Strategy	Year				Cost ($)
	1977	1986	1995	2004	
I	106	107	45	19	153,000
II	106	107	41	14	147,000
III	106	107	33	10	141,000
IV	84	106	45	19	139,000
V	56	64	36	16	94,000
VI	55	0	0	0	30,000
VII	76	77	33	14	110,000
VIII	67	68	29	13	96,000
IX	88	89	36	14	125,000
X	70	72	27	10	98,000
XI	51	53	18	6	70,000
XII	33	36	11	3	45,000
XIII	25	27	8	2	33,000

and are summarized in Table V and VI, respectively. The effective SM
flow might be studied by an intensive current metering and sampling
program which identifies the nature and magnitude of the return flow.
Strategy II is based on a precisely known value for $\tilde{q}_s$. The corresponding
cost is therefore a lower limit on the cost achievable following such research. The savings (the difference between the costs for Strategies I
and II) is less than $6,000 (Table VI) and could be weighed against the
cost of such research.

If the atmospheric loads do not change significantly over the period
1977-2004, then frequent intensive atmospheric load sampling would not
be required. Strategy III in Tables V and VI is based on precise atmospheric loads, a_1 and a_2, resulting from intensive sampling and represents
a savings of less than $12,000.

The phosphorus apparent settling velocity has been interpreted as a
state variable, x_3, in the mass balance model. The Kalman filter calculates
this velocity based on measurements of total phosphorus concentration.
As indicated in Table IV, the standard deviation of the Kalman estimation errors for $x_3(n)$ is constant between measurement years but decreases
after a measurement is made. The accuracy of the estimate is greatly
improved after a few years of sampling. However, if the mechanisms of
phosphorus removal can be correlated with the apparent settling velocity,
then the accuracy of the apparent settling velocity might be improved
immediately. Presently, S is known to within ± 10.7 m yr^{-1}. Optimal

strategies have been computed for ranges of ±5.0 (IV), ±1.0 (V) and ±0.5 m yr^{-1} (VI), resulting in significant reductions in sampling cost. The savings could be as high as $123,000 for Strategy VI, but a reduction in the range beyond ±5.0 m yr^{-1} may be experimentally difficult to attain.

The sensitivity of the optimal cost to the standard deviation of phosphorus laboratory analysis errors is also of interest. This standard deviation affects the accuracy of a measurement of the lakewide average phosphorus concentration and is presently 1.0 μg-P l^{-1}. Optimal strategies have been computed for standard deviations equal to 0.5 (VII) and 0.1 μg-P l^{-1} (VIII). The reduction in cost can be as great as $65,000 for Strategy VII and $85,000 for Strategy VIII, as indicated by Tables V and VI.

Strategies I-VIII have been computed under the requirement that the chloride and phosphorus lakewide average concentrations be estimated to within ±0.5 mg-Cl l^{-1} and ±0.5 μg-P l^{-1} with a 0.90 probability. Optimal Strategies IX-XIII have been computed under relaxed probabilities of 0.88, 0.85, 0.80, 0.70 and 0.60 for the same statistics used in computing Strategy I. The Chebyschev inequality is used to determine the constraints which must be imposed on the standard deviation of the estimation errors. If the estimation errors were normal and 0.90 probability were demanded, then the constraint would be approximately the same as that determined from the Chebyschev inequality for 0.60 probability. From Tables V and VI, it is seen that the cost for 0.60 probability is only about 45% the cost for 0.80 probability and a sharp cost increase is encountered between 0.80 and 0.90. Specification of a performance criterion for the Kalman filter estimates will likely have to be based on scientific judgment.

CONCLUSIONS

A systematic approach for designing minimum-cost sampling strategies for trend detection in large lakes has been developed. The approach uses statistical characterizations of mass balance models and measurement techniques for chloride and total phosphorus. The accuracy of the parameters in the mass balance models for lakewide average concentrations in Lake Michigan has been examined. The model has been incorporated into a Kalman filter and acceptable variance estimates for the lakewide average concentrations have been determined with fewer lake samples than would otherwise be required. Minimum-cost surveillance plans have been computed which meet accuracy criteria of the Kalman filter estimates. The analyses indicate that tributary samples are not a cost-effective source of information for estimating long-term lakewide average concentrations, and that lake samples should be collected only for those years when the accuracy criteria must be satisfied. Furthermore, the analyses indicate that the optimal cost for long-range trend detection is more sensitive to the accuracy of the phosphorus apparent settling velocity and of the laboratory analyses than to the accuracy of the effective Straits of Mackinac flow or the atmospheric loads. Finally, the results show that substantial savings in monitoring costs over those proposed by the Great Lakes Water Quality Board can be achieved if funds are allocated for better identification of the apparent settling velocity for phosphorus and better laboratory techniques or if the accuracy requirements of the detection plan are relaxed.

APPENDIX A: KALMAN FILTER EQUATIONS

Kalman filtering techniques can be applied to linear stochastic difference equations of the form:

$$x_{k+1} = \phi_k x_k + w_k \quad (k = 0, 1, \ldots)$$

where x_{k+1} and x_k represent the values of the state variable vector at times $k + 1$ and k, ϕ_k is a state transition matrix evaluated at time k and w_k is a vector which represents random zero mean modeling (or process) errors with known covariance Q_k. The initial state of the system is characterized statistically with a mean value, x_o, and covariance P_o, at time zero. At any time k, measurements can be taken having zero mean errors with a known covariance matrix, R_k. The value of state measurements at any time k are represented mathematically as y_k. For clarity, it is assumed that y_k represents a measurement of all the state variables.

The goal of the Kalman filter is to determine the unbiased, minimum-variance estimate of the state at each time, $k = 0, 1, \ldots$ based on measurements and the behavior of the model. The Kalman filter is a recursive filter because a new estimate of the state is obtained each time a measurement is processed, and it is convenient to view the resultant calculation in two stages. Before a measurement is made at time k, the state is characterized with a mean value of x_k^- and covariance P_k^-. These values represent *a priori* estimates of the state and covariance. Following a measurement, the mean of the state is represented as x_k^+ with a covariance P_k^+. These values give *a posteriori* estimates of the state and covariance.

Given x_k^+ and P_k^+, the first stage of the Kalman filter propagates the state estimate and covariance to the next time event, $k + 1$. No measurements are made between k and $k + 1$, therefore the error in the *a priori* estimate of the state at k + 1 (x_{k+1}^-) can depend only upon the uncertainty in the state at the beginning of the interval (P_k+), the dynamics of the propagation (ϕ_k), and the uncertainty in the dynamic model (Q_k). Well-known results from state estimation theory give equations for the *a priori* value of the mean and covariance for the state at time k + 1:

$$x_{k+1}^- = \phi_k\, x_k^+$$

$$P_{k+1}^- = \phi_k\, P_k^+\, \phi_k^T + Q_k$$

These equations show exactly how the states, and the uncertainty in the states, propagate between measurements. Because Q_k is a positive semi-definite matrix, the uncertainty in the state can only increase if the model is not perfect. The dynamics of the system as defined by ϕ_k may increase or decrease the uncertainty in the state. For example, if x_k is a scalar, the variance of x_{k+1} for $x_{k+1} = ax_k$ (with $\phi_k = a$) will decrease if $|a| < 1$ and increase if $|a| > 1$, assuming $P_k^+ \neq 0$.

The second stage of the Kalman filter involves updating or improving the values of x_{k+1}^- and P_{k+1}^- to x_{k+1}^+ and P_{k+1}^+ based upon measurements involving the state at k + 1, represented as y_{k+1}. The second-stage equations are:

$$x_{k+1}^+ \quad = \quad x_{k+1}^- + K_{k+1}\,(y_{k+1} - x_{k+1}^-) \tag{19}$$

$$P_{k+1}^+ \quad = \quad (I - k_{k+1})P_{k+1}^- \tag{20}$$

$$K_{k+1} \quad = \quad P_{k+1}^-\,(P_{k+1}^- + R_{k+1})^{-1}$$

K_{k+1} is called the Kalman Gain and roughly corresponds to the ratio of the faith in the current estimate (P_{k+1}^-) to the faith in the measurement at k + 1 (R_{k+1}). If R_{k+1} is relatively large, indicating a poor measurement, then K_{k+1} will be relatively small, and Equation 19 calculates values of x_{k+1}^+ that will involve only small changes from x_{k+1}^-. If K_{k+1} is relatively large, indicating a good measurement, then x_{k+1}^+ may involve a substantial change from x_{k+1}^-. Similar arguments are true for updating the covariance matrix (Eqaution 20). If K_{k+1} is relatively small, then P_{k+1}^+ will be close to P_{k+1}^-, and if K_{k+1} is relatively large, then P_{k+1}^- is reduced by the factor $(I - K_{k+1})$ to P_{k+1}^+. Measurements cannot increase the uncertainty in the state, and thus the variance must be reduced or remain the same. The overall technique of Kalman filter estimation is summarized in Figure 1.

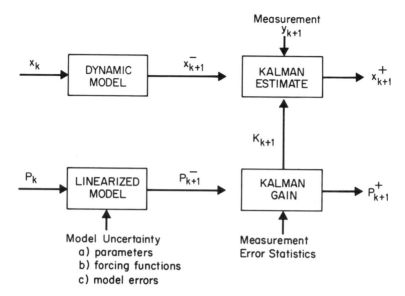

Figure 1. Simplified representation of Kalman filtering algorithm.

REFERENCES

Canale, R. P., L. M. DePalma and W. F. Powers. "Sampling Strategies for Water Quality in the Great Lakes," Report to U.S. Environmental Protection Agency, Duluth, MN (in preparation).

Chapra, S. C. "Comment on An Empirical Method of Estimating the Retention of Phosphorus in Lakes" by W. B. Kirchner and P. J. Dillon. *Water Resources Res.* 11:1033-1034 (1975).

Chapra, S. C. "Total Phosphorus Model for the Great Lakes," *J. Environ. Eng. Div.,* 103(EE2):147-161 (1977).

Davenport, W. B., Jr. *Probability and Random Processes* (New York: McGraw-Hill Book Company, 1970).

Davis, C. Great Lakes Research Division, University of Michigan, Ann Arbor, MI. Personal communication (1977).

DePalma, L. M. "A Class of Measurement Strategy Optimization Problems—with an Application to Lake Michigan Surveillance," Ph.D. dissertation, University of Michigan, Ann Arbor, MI (1977).

Dillon, P. J., and W. B. Kircher. "Reply to Chapra (1975)," *Water Resources Res.* 11:1035-1036 (1975).

Great Lakes Water Quality Board. "International Surveillance Program for the Great Lakes," in *1975 Annual Report, Great Lakes Water Quality.* (Windsor, Ontario: International Joint Commission, 1976).

Harrison, M. J., R. E. Pacha and R. Y. Morita. "Solubilization of Inorganic Phosphates by Bacteria Isolated from Upper Klamath Lake Sediment," *Limnol. Oceanog.* 17:50-57 (1972).

Hutchingson, G. E. *A Treatise on Limnology,* Vol. 1 (New York: John Wiley & Sons, Inc., 1957).

Junge, C. E., and R. T. Werby. "The Concentration of Chloride, Sodium, Potassium, Calcium and Sulfate in Rain Water over the United States," *J. Meterorol.* 15:417-425 (1958).

Kirchner, W. B., and P. J. Dillon. "An Empirical Method of Estimating the Retention of Phosphorus in Lakes," *Water Resources Res.* 11:182-183 (1975).

Larsen, D. P., and H. J. Mercier. "Phosphorus Retention Capacity of Lakes," *J. Fish. Res. Bd. Can.* 33:1742-1750 (1976).

Lorenzen, M. W., D. J. Smith and L. V. Kimmel. "A Long-term Phosphorus Model for Lakes: Application to Lake Washington," in *Modeling Biochemical Processes in Aquatic Ecosystems,* R P. Canale, Ed., (Ann Arbor, MI: Ann Arbor Science Publishers, Inc., 1976).

Michalski, M. F. P., M. G. Johnson and D. M. Veal. *Muskoka Lakes Water Quality Evaluation, Report No: 3: Eutrophication of the Muskoka Lakes* (Toronto: Ontario Ministry of the Environment, 1973).

Murphy, T. J., and P. V. Doskey. *Inputs of Phosphorus from Precipitation to Lake Michigan.* U.S. Environmental Protection Agency, Corvallis, Oregon, EPA-600/3-75-005 (1975).

O'Connor, D. J., and J. A. Mueller. "A Water Quality Model of Chloride in Great Lakes," *J. San. Eng. Div.* (ASCE) 9:955-975 (1970).

Quinn, F. H. *Lake Michigan Beginning-of-Month Water Levels and Monthly Rates of Change of Storage.* NOAA No. ERL326-CLERL2 (Ann Arbor, MI: National Oceanic Technical and Atmospheric Administration, 1975).

Quinn, F. H. "Annual and Seasonal Flow Variations through the Straits of Mackinac," *Water Resources Res.* 13:137-144 (1977).

Richardson, W. U.S. Environmental Protection Agency, Grosse Ile, MI. Personal communication (1977).

Sage, A. P., and J. L. Melsa. *Estimation Theory with Applications to Communications and Control.* (New York: McGraw-Hill Book Company, 1971).

Saylor, J. H., and P. W. Sloss. "Water Volume Transport and Oscillatory Current Flow through the Straits of Mackinac," *J. Phys. Oceanog.* 6:229-237 (1976).

Schutze, L. U.S. Army Corps of Engineers, Detroit, MI. Personal communication (1977).

Sonzogni, W. C. "Effect of Nutrient Input Reduction on the Eutrophication of the Madison Lakes," Ph.D. dissertation, University of Wisconsin, Madison, WI (1974).

Stewart, K. M., and S. J. Markella. "Seasonal Variations in Concentrations of Nitrate and Total Phosphorus and Calculated Nutrient Loading for Six Lakes in Western New York," *Hydrobiologia* 44:61-89 (1974).

U.S. Army Corps of Engineers. *Lake Erie Wastewater Management Study: Preliminary Feasibility Report.* (Buffalo, NY: U.S. Army Corps of Engineers, 1975).

U.S. Environmental Protection Agency. "Report of the Phosphorus Technical Committee," in *Proceedings of Conference on Pollution of Lake Michigan and its Tributary Basin, Illinois, Indiana, Michigan and Wisconsin* (Chicago, IL: U.S. Environmental Protection Agency, 1972).

SECTION FOUR

NEW DIRECTIONS IN ECOSYSTEM ANALYSIS

Ecosystem modeling is a relatively new area of research. Most of the work in this area has so far been directed toward developing models for specific ecosystems and toward applying these models in a straightforward manner to making predictions for management purposes. It seems to us that the field is now at the point where such models can be developed and applied in new situations with relative ease. Thus, the time is ripe for aquatic ecosystem modeling to look toward new directions for further development. The papers in this section present examples of some ways this development may occur.

Chapter 10, by C. W. Chen and D. J. Smith, explores the feasibility of coupling a relatively sophisticated ecosystem model with a complex hydro-dynamic model in order to simulate the physical, chemical and biological properties of a body of water in the three dimensions of space as well as the fourth of time. This expands the usual situation for ecosystem models where time and only one or occasionally two spatial dimensions are considered. Three-dimensional models are needed if we are to simulate the spatial heterogeneity of ecological processes in aquatic environments. A. Robertson and D. Scavia in Chapter 11 take a different tack. They explore the utility of ecosystem models as tools for increasing understanding of the ecosystems themselves. Surprisingly, such use of ecosystem models seems to have received little previous attention. The final chapter, by R. V. Thomann, considers the very important problem of developing models for simulating the distribution and dynamics of toxic substances within an ecosystem. This paper presents a completely new and different way of attacking this problem based on organism length.

Together these three papers point the way to some promising new directions for ecosystem modeling research and provide examples of the potential for expanding the application of such models beyond their present bounds.

CHAPTER 10

PRELIMINARY INSIGHTS INTO A THREE-DIMENSIONAL ECOLOGICAL-HYDRODYNAMIC MODEL

Carl W. Chen and Donald J. Smith

Tetra Tech, Inc.
Lafayette, California 94549

INTRODUCTION

As a part of the data analysis program for the International Field Year for the Great Lakes (IFYGL), a comprehensive water quality-ecological model for Lake Ontario has been developed. The model, which simulates mass transport, heat transfer, biological transformations and chemical reactions taking place in the lake, provides an integrated interpretation of physical, chemical and biological data observed in the field.

To date, the model has been conceptualized, developed and tested for simulation of a one-year period with an hourly hydrodynamic time step and a daily water quality time step. Although rigorous calibration has not been attempted, the correspondence between the model results and IFYGL data has generally been good.

This paper presents the model, the preliminary simulation results and information gained from the modeling effort.

PREVIOUS MODELS

Lorenzen *et al.* (1974) reviewed various models and evaluation methodologies that may be applicable to environmental management of the Great Lakes. A number of models have been developed to date. One class of model is represented by the phytoplankton model developed by

Di Toro *et al.* (1975) for Western Lake Erie and by similar models applied to Saginaw Bay by Richardson and Bierman (1976) and to Grand Traverse Bay by Canale *et al.* (1974). These models divide shallow embayments into horizontal segments (cells) that are considered vertically well-mixed. Flows among segments are assumed and adjusted until the conformation of observed and calculated distribution of a conservative substance such as chloride is made. Nutrient budget and phytoplankton population dynamics are then calculated with a simulation time step of about a week.

Thomann *et al.* (1975) applied the same basic methodology to Lake Ontario. The lake was segmented into two layers without horizontal subdivision. Thomann also used five layers, each segmented into several horizontal cells. It becomes extremely difficult to assume a three-dimensional model of this size, however. Estimation of these flows is complex since continuity of water mass and conservation of thermal energy must be met simultaneously.

Scavia *et al.* (1976) also developed an ecological model for Lake Ontario. The physical system was represented by three layers. The basic modeling methodology is similar to that used in others models; however, great emphasis is placed on biological comprehensiveness, *e.g.,* four types of phytoplankton, six types of zooplankton and benthic invertebrates are included in the model.

If one is concerned with horizontal heterogeneity, a hydrodynamic model is needed to calculate the water circulation and transport characteristics of the lake. Based on the results from this model, a water quality-ecological model can route the pollutants throughout the lake and compute their biological consequences.

A large number of hydrodynamic models have been developed for application to the Great Lakes (Lorenzen *et al.,* 1974). There are two time-dependent models pertinent to this work, *i.e.,* Simons' model (1973, 1974, 1975a, 1976b) and Bennett's model (1977). Both models divide the lake into four to eight horizontal layers. For each layer, an orthogonal grid system with 5- x 5-km cells is imposed. Simons' model allows a free surface; Bennett's model assumes a rigid lid. Both models employ the same equations of motion, including terms for wind stress, Coriolis force, eddy viscosity and bottom friction. The free surface model includes an equation allowing the water surface to rise or fall to maintain continuity. The rigid lid model neglects the short-term dynamic response of the water surface and employs a stream function representation to ensure continuity.

Simons (1975b, 1976b) has described a method of interfacing physical and biological-chemical models of large lakes. He has found that the effects of water transport are of approximately the same order of magnitude as the other physical processes, implying that a segmented water quality model must incorporate a water circulation model.

COMPREHENSIVE MODEL

Model Structure

The Lake Ontario model is comprised of three basic modules: hydrodynamic, interface and water quality. The relationship among the modules is shown in Figure 1.

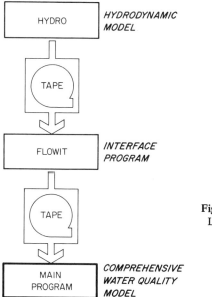

Figure 1. Overall structure of the Lake Ontario model.

The hydrodynamic module calculates lake currents throughout the annual cycle and the results are stored for subsequent use. The interface module transforms the hydrodynamic results into a form useful in the water quality-ecological module. Based on the hydrodynamic information, the water quality module calculates the mass transport of pollutants among physical segments (cells). The module also simulates the physical, chemical and biological processes occurring within each physical segment of the lake.

Grid System

Two compatible grid systems are used. For the hydrodynamic simulation, the model employs a fine grid with 250 horizontal segments at the lake surface. On the vertical axis, there are 8 layers with thicknesses, respectively, from surface to bottom of 5, 8, 12, 17, 23, 35, 50 and 70 m. Depending on the local bathymetry, the bottom layer may be thinner at certain locations. For the water quality-ecological simulation, the fine hydrodynamic grid is aggregated into a coarser grid system with the flow information for the coarse grid being generated from the hydrodynamic model results. The water quality grid has 42 horizontal segments and 7 layers (see Figures 2 and 3). For the horizontal segments, the aggregation is made in such a way that more spatial detail is provided near the lake shore. The top six layers of the water quality grid are identical to those used in the hydrodynamic grid. The bottom two layers are aggregated to form the seventh layer.

Hydrodynamic Module

The hydrodynamic module calculates the current velocities and temperature regime throughout the lake with an hourly time step. The input data (*i.e.*, the meteorological records) are updated as often as every hour. The basic equations, described by Bennett (1977) and summarized by Chen *et al.* (1975), include: (1) momentum equations, (2) stream function formulations for the vertically averaged flows, (3) continuity equations for flow balance in all directions, (4) heat budget equations, and (5) an equation of state. Auxiliary equations are those defining hydrostatic pressure, horizontal eddy viscosity, bottom friction and surface wind stress.

Bennett (1977) developed a finite difference method to solve the abovementioned equations. The program was originally developed for a uniform grid system. Since then it has been modified to include nonuniform cell sizes. Our original model applied a uniform heat flux to the entire lake surface layer. This procedure tended to predict unrealistic temperatures, *i.e.*, it made the north shore too cold and the south shore too hot (Chen *et al.*, 1975). The model was modified to include a dynamic calculation of the net rate of heat transfer at the lake surface for each model segment. This transfer was determined as a sum of individual heat transfer rates.

$$H = q_{sn} + q_{at} - q_w - q_e - q_c$$

where H = net rate of heat transfer at the model segment of the lake surface ($Kcal\ m^{-2}\ s^{-1}$)

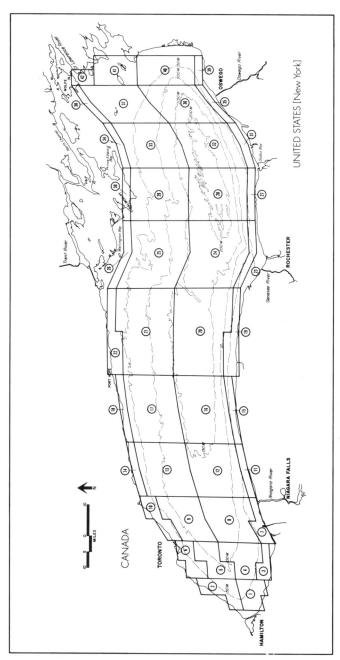

Figure 2. Surface view of coarser grid used for the water quality-ecological simulation.

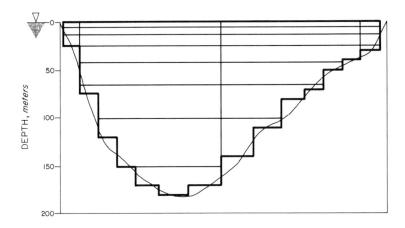

Figure 3. Vertical cross section of coarser grid used for the water quality-ecological simulation.

q_{sn} = heat gain by shortwave solar radiation
q_{at} = heat gain by atmospheric longwave radiation
q_w = heat loss by longwave back radiation
q_e = heat loss by evaporation
q_c = convective heat exchange

Shortwave solar radiation is calculated as a function of longitude, latitude, percent cloud cover and time of year. Atmospheric longwave radiation is calculated as a function of cloud cover and dry bulb temperature. The last three terms in the equation are expressed in terms of surface water temperature, dry bulb air temperature, wind speed and relative humidity. Local meteorological conditions and water temperatures are used in the heat flux computations resulting in different net heat fluxes at various locations on the lake surface. The subroutine for the heat exchange computations has previously been used in applications of heat exchange analysis for streams (Chen and Wells, 1976), reservoirs (Chen and Orlob, 1975; Orlob and Selna, 1970) and estuaries (Chen and Orlob, 1975).

Interface Module

The interface module serves as an intermediary between the hydrodynamic module and the water quality module. Input data to the module include information describing how the hydrodynamic grid system is to be aggregated for the water quality computations. Current velocities are multiplied by appropriate cross-sectional areas to calculate the flow into and out of each water quality segment. Temperatures are calculated as volume-weighted averages of the temperatures in the appropriate hydrodynamic elements. A final check on the continuity of flow is made before results are stored for input to the water quality module.

Water Quality-Ecological Module

The water quality-ecological module calculates concentrations of water quality and ecological parameters for all segments of the lake. The parameters included in the model are shown in Table I and Figure 4.

Table I. Water Quality-Ecological Parameters

1. Phytoplankton (4 groups)	Forage (alewife and smelt)
Diatoms	Eggs and larvae
Greens	Young
Flagellates	Adults
Blue-greens (nitrogen fixing)	5. Benthic Animals
2. Attached Algae	6. Detritus
Cladophora	7. Organic Sediment
3. Zooplankton (3 groups)	8. Temperature (provided by the hydro-
Herbivorous–grazer	dynamic solution)
Herbivorous–filter feeder	9. Biochemical Oxygen Demand
Carnivorous	10. Dissolved Oxygen
4. Fish	11. Coliform Bacteria
Cold water (salmonoids)	12. Plant Nutrients
Eggs and larvae	Ammonia
Young	Nitrite
Adults	Nitrate
Warm water (bass, perch)	Phosphorus
Eggs and larvae	Silicon
Young	13. Carbonate System
Adults	Carbon dioxide
Scavenger (suckers, carp)	Bicarbonate
Eggs and larvae	Alkalinity
Young	pH
Adults	14. Chloride

Differential equations, based on simple kinetic representations and the law of conservation of mass, have been derived to describe the rate of change in parameter values as a function of causative factors such as advection, dispersion, inflow (waste discharges), outflow (water export), and sinks and sources (Chen *et al.*, 1975). The first two factors are associated with hydrodynamic transport. Inflow and outflow account for mass fluxes associated with rivers, waste discharges and deposition of pollutants from the atmosphere. Sinks and sources describe the reactions taking place within each hydraulic cell.

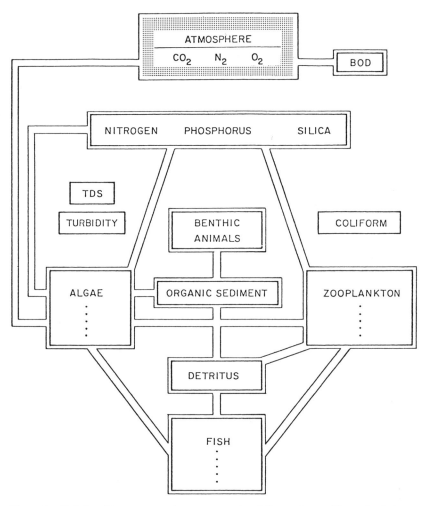

Figure 4. Relationship among major components of the water quality-ecological module.

The computer program requires 60,000 words of core for sequential execution of the hydrodynamic and water quality-ecological modules. Currently, the annual simulation of hydrodynamics with hourly time steps takes approximately 20 min of CPU time on a CDC 7600. For the water quality module, an annual simulation with daily time steps requires 10 min of CPU time.

ANNUAL SIMULATIONS

Data Preparation

The model requires four basic types of data (Table II): lake geometry, meteorology, rate coefficients and loads.

Table II. Input Needed for Model

1. Lake Geometry (time invariant)	3. Rate Coefficients (time invariant)
Longitude	Decay Rates
Latitude	Settling Rates
Shape	Specific Growth Rates
Bathymetry	Half Saturation Constants
2. Meteorological Data (time variant)	Temperature Coefficients
Wind Velocity	Etc.
Cloud Cover	4. Waste Loads (time variant)
Dry Bulb Temperature	Tributary Inflows
Humidity	Direct Waste Discharges
Atmospheric Pressure	Atmospheric Deposits
	River Outflow (St. Lawrence River)

Lake geometry was taken directly from a bathymetric chart of Lake Ontario published by the Canadian Hydrographic Service (Chart No. 881). Hourly wind data taken during IFYGL were furnished by J. R. Bennett of the Great Lakes Environmental Research Laboratory (NOAA), Ann Arbor, MI, who received the original data set from F. C. Elder of the Canada Centre for Inland Waters, Burlington, Ontario. Other meteorological data were obtained from technical memoranda published by the Canadian Atmospheric Environmental Service (Phillips and McCulloch, 1974). Rate coefficients were estimated from typical literature values. They are similar to the values used previously by DiToro *et al.* (1975), Canale *et al.* (1974), Scavia *et al.* (1976) and Chen *et al.* (1975). Waste load data were taken primarily from the work of Casey *et al.* (1976).

Results from Hydrodynamic Model

Water currents are known to be sensitive to wind conditions. With winds that change in speed and direction on an hourly basis in the model, it would be very difficult to select model output at a specific time and compare it to observed data which only are available averaged over a longer period of time. For the purposes of comparison in our context, the hydrodynamic simulation was performed with steady wind through Julian day 190 ("onset" of stratification). Figures 5 and 6 present the

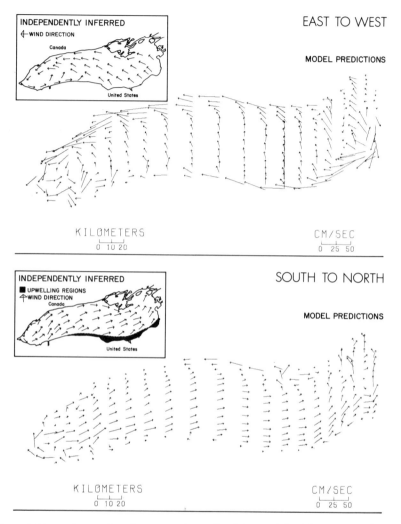

Figure 5. Comparisons of surface currents predicted by model with currents inferred from independent data for wind blowing from east to west and wind blowing from south to north.

model results for a steady wind of 3.5 m sec^{-1} coming from four differ-ent directions. The currents shown in the small boxes were inferred from field data (International Joint Commission, 1969).

Note that the circulation patterns have generally been reproduced by the model. Winds from the west and northwest (Figure 6) move water in a southeasterly direction until it reaches the south shore of the lake.

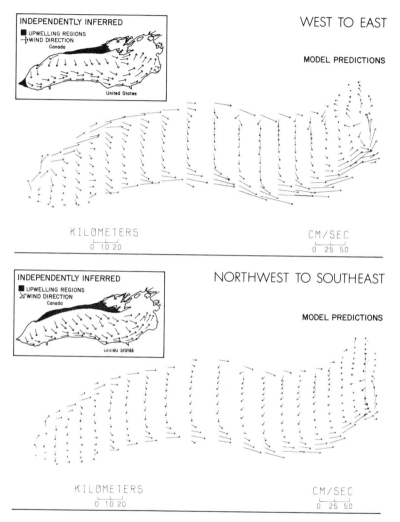

Figure 6. Comparisons of surface currents predicted by model with currents inferred from independent data for wind blowing from west to east and wind blowing from northwest to southeast.

There, a very strong coastal jet is generated forcing water to flow eastward along the shore. Along the north shore, the model predicts a strong current flowing eastward in the shallow water. Upwelling occurs in the nearshore area. These latter two predictions are not consistent with the inferred currents which flow from east to west against the wind along the northwest above. Under the influence of a wind coming from the east, it is interesting to note that the model generates a surface current flowing eastward against the wind, west of the Niagara River entrance. This calculated circulation pattern is consistent with that observed. It is to be noted that the steady wind was used only to show the reasonableness of the hydrodynamic solutions. Actual wind data collected during IFYGL were applied in all subsequent annual simulations.

The simulated temperature regime of the lake has been analyzed in several ways. First, the surface water isotherms were plotted and compared with data obtained through infrared (IR) measurements. Second, two-dimensional temperature plots along selected longitudinal transects were made to illustrate known phenomena, such as upwelling, downwelling, the thermal bar and the thermocline. Finally, vertical temperature profiles were constructed for stations where temperature data were available.

Several Airborne Radiation Thermometer (ART) flights were made during the IFYGL year (Irbe and Mills, 1976). Even though ART measures only the temperature of the surface film, the results are presumed comparable to the average temperature of the top five meters computed by the model. The comparisons are made in Figures 7 and 8. The calculated surface temperatures are seen to follow the observed distribution and seasonal succession of the surface temperature of the lake, except for June.

The longitudinal transects examined were from Olcott to Oshawa and from Rochester to Presqu'ile. Calculated temperature distributions for these two transects are presented in Figure 9. Several observations can be made with respect to the transect plots:

1. The model illustrated development of the thermal bar during April-May.
2. The thermal bar eroded and disappeared by the end of June. Lakewide thermal stratification was established at that time.
3. In June (day 181), the north shore water was slightly warmer than the south shore water. At that time there was no upwelling. The surface water appeared to gain heat as it traveled east along the south shore. The warm water returned to the west on the surface along the north shore in a counterclockwise circulation pattern.
4. In July (day 212) and August (day 243), the south shore water was warmer than the north shore water in the western portion of the lake. In the eastern part, the north shore was slightly warmer. During this period upwelling along the western portion of the north shore appears to have reduced the surface temperature.
5. In September, the lake entered the cooling cycle. The surface temperature gradually decreased but the lake remained thermally stratified.

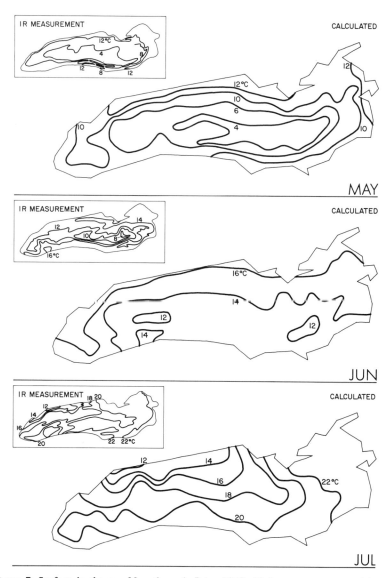

Figure 7. Surface isotherms, May through July, 1972–IR imagery versus model calculated values.

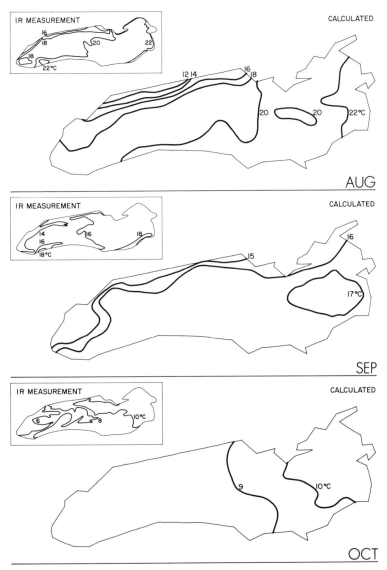

Figure 8. Surface isotherms, August through October, 1972—IR imagery versus model calculated values.

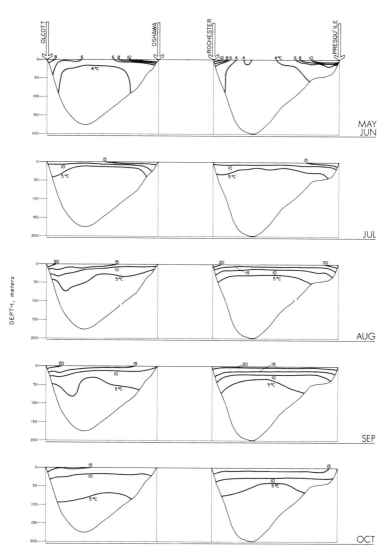

Figure 9. Model calculated longitudinal temperature transects, May through October, 1972 (Olcott to Oshawa and Rochester to Presqu'ile).

Calculated and observed temperature profiles for specific stations and dates are plotted in Figures 10 and 11. The observed data were obtained from STORET, the data management system of U.S. Environmental Protection Agency. The model results follow the observed data reasonably well.

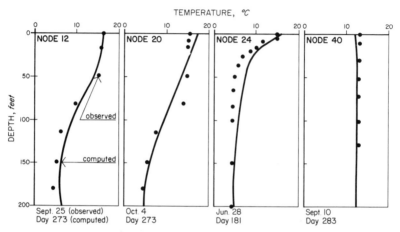

Figure 10. Model calculated and observed temperature profiled are nodes 12, 20, 24 and 40.

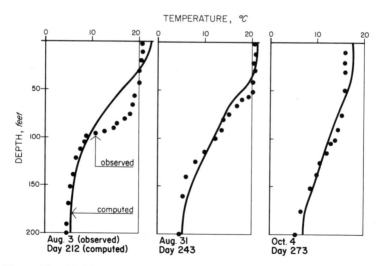

Figure 11. Model calculated and observed temperature profiles at node 36.

Results from Water Quality-Ecological Model

Figure 12 presents calculated and observed (from STORET) nutrient profiles for one location located 14 miles offshore from Gilbralter Point in Western Lake Ontario (79° 16' 48" W, 43° 25' 12"). The corresponding model nodal point number is 8. Similar results are shown in Figure 13 for node 20 (78° 00' 36" W, 43° 35' 24" N), located 16 miles offshore from Norton. It should be noted that, in general, the model correctly predicted the nutrient profiles observed in the field; however, it overestimated the concentration of NO_3-N and PO_4-P occurring during the summer.

Figure 14 compares calculated and observed dissolved oxygen profiles at nodes 20, 13 (78° 48' 00" W, 43° 39' 00" N), and 32 (76° 58' 48" W, 43° 29' 24"). The dissolved oxygen data were furnished by Brian J. Eadie of the Great Lakes Environmental Research Laboratory (NOAA), Ann Arbor, MI. They were collected by the use of descending probes and some question exists regarding the accuracy of the readings, as the probe was lowered rapidly.

The observed and calculated dissolved oxygen (DO) profiles indicate that the lake water was not low in dissolved oxygen. As the lake becomes thermally stratified, the surface DO decreases with rising temperature. The high DO content of the cooler hypolimnion waters appears to be "locked inside" the lake system. According to the model, the typical summer DO profiles show a maximum at 10 m and a minimum at 20 m. The DO maximum corresponds to the maximum algal density discussed in the next paragraph. From 10 to 20 m, the respiration rate of the algae appears to have greatly exceeded the photosynthesis rate. As a result, both dissolved oxygen and algal density decrease. Below that, community respiration becomes so small that it does not cause any further DO depression.

Since algal biomass contains approximately 1% chlorophyll *a*, a comparison of calculated algal biomass and observed chlorophyll *a* can be made at a 100:1 ratio. Figures 15 to 17 provide such comparisons. The conformation is reasonable. It is interesting to note that the model correctly predicts the high algal density observed at a depth of approximately 10 m. A detailed rationale for this will be discussed later.

The model includes four groups of phytoplankton, three groups of zooplankton and four groups of fish, the latter with three life stages each. Each group of organisms has its own temperature preference and growth-limiting food (or nutrient). During the simulation, one would expect to see the ecological succession of organisms on an annual cycle. Figures 18 and 19 show this annual succession. Diatoms are seen to

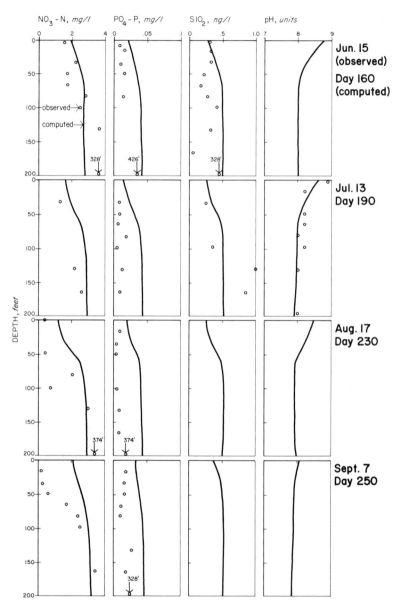

Figure 12. Model calculated and observed profiles of nitrate, phosphate, silicate concentration and pH at node 8.

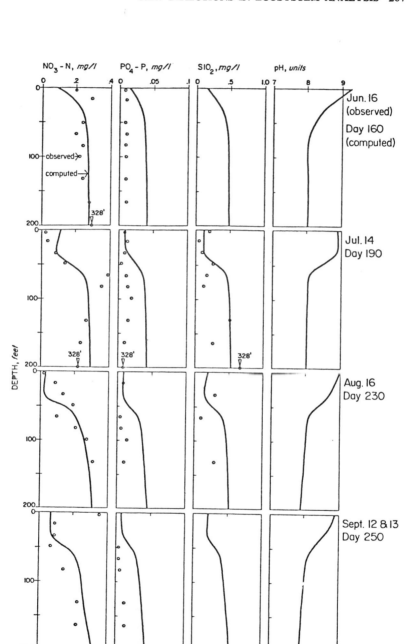

Figure 13. Model calculated and observed profiles of nitrate, phosphate, silicate concentration and pH at node 20.

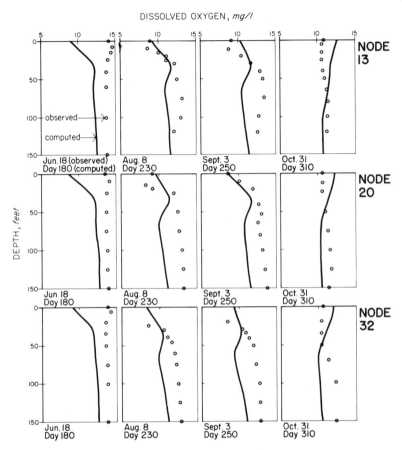

Figure 14. Model calculated and observed dissolved oxygen concentration profiles for nodes 13, 20 and 32.

peak early in the spring. Nondiatoms, including flagellates and green algae, peak later in the summer. Herbivorous zooplankton, being algae grazers, also grow rapidly early in the spring. Carnivorous zooplankton, in turn, graze on the herbivorous zooplankton and increase in number during the summer. The seasonal succession curve for salmonoid fishes is seen to be fairly flat. The alewife young increase more drastically in the summer after spring spawning. The succession curves for both fishes are shown here only to illustrate the capability of the model. At this point, the values of all the coefficients necessary for modeling fish have been assumed. Effort has not yet been expended in a literature search to verify the reasonableness of the coefficients.

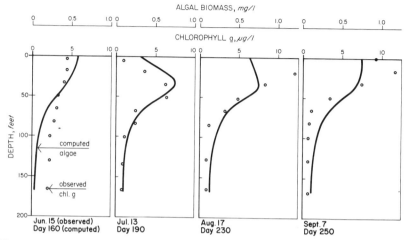

Figure 15. Model calculated and observed chlorophyll *a* concentration profiles at node 8.

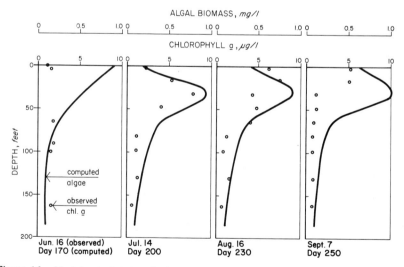

Figure 16. Model calculated and observed chlorophyll *a* concentration profiles at node 20.

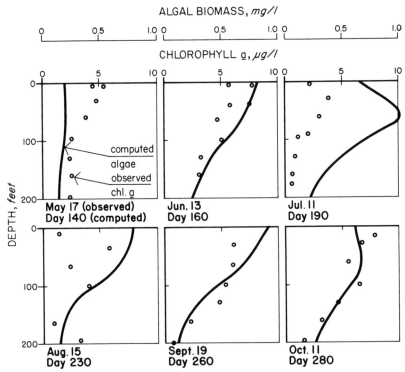

Figure 17. Model calculated and observed chlorophyll *a* concentration profiles at node 38.

SPECIAL ANALYSES

Vertical Profiles of Algae

Vertical profiles of algae in Lake Ontario water have maxima at a depth of about 10 m. Plausible explanations for this phenomenon are: (1) light intensity near the water surface is so high that it inhibits algal growth; (2) algae settle as they grow and the dynamic equilibrium of growth and settling may result in the maximum concentration occurring at a distance below the surface; and (3) certain nutrients, assimilated by algae near the surface, can only be resupplied from below and therefore low nutrient concentration may limit algal growth at the surface.

To find out if light was a probable cause, two formulations were used for the light-limitation term for algae:

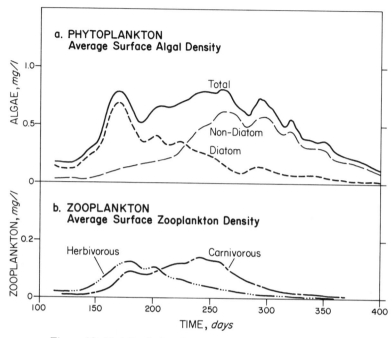

Figure 18. Model calculated annual succession of plankton.

1. the Michaelis-Menten type formulation

$$F(I) = \frac{I}{I + K_I}$$

2. the Steele formulation

$$F(I) = \frac{I}{I_s} \exp\left[-\frac{I}{I_s} + 1 \right]$$

where F(I) = light-limiting factor
  I = light intensity
  K_I = light half-saturation intensity
  I_s = optimum light intensity.

  The results from using these two formulations are compared in Figures 20 and 21. It is interesting to note that Steele's formulation, which allows for inhibition at high light intensities, does not produce as strong a maximum at depth as does the Michaelis-Menten type formulation. In fact, detailed analysis indicates that Steele's formulation does not cause light inhibition on the surface. Rather, its use results in a smaller growth

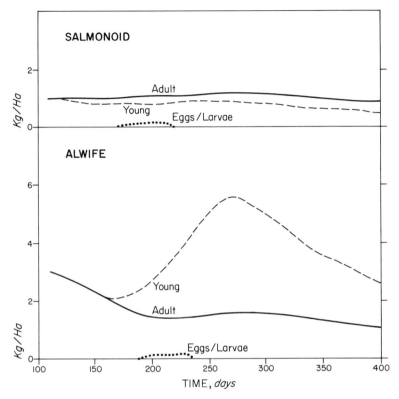

Figure 19. Model calculated annual succession of fishes.

rate for suboptimal light intensities at depth. Light inhibition occurs usually only in the top few centimeters of the water column. Because the model growth rate in the top layer is averaged over the top five meters, the light inhibition effect is "washed out." Using 300 langleys per day (or 0.035 kcal m^{-2} sec^{-1}) for the optimal light intensity as suggested by DiToro *et al.* (1975), Steele's formulation yields a lower specific growth rate and therefore a lower standing crop than the Michaelis-Menten formulation; however, an adjustment of coefficient values for Steele's formulation may produce the vertical profiles of algae as desired.

The conclusions to be made from the analysis are:

1. To produce a maximum algal density at depth rather than at the surface, one does not require a light inhibition formulation.
2. Proper selection of coefficients for reasonable formulations can produce desired responses in the model output. The fact that the observed data are reproduced by the model output does not guarantee that the formulations in the model are correct.

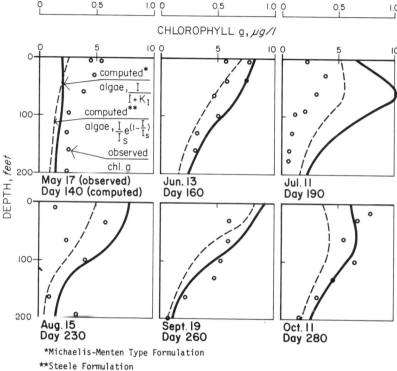

Figure 20. Computed algal densities using the Steele and Michaelis-Menten formulations vs observed values for node 20.

The vertical profiles of algae in Lake Ontario probably result from a combination of all the plausible mechanisms, *i.e.,* light inhibition, settling algae and nutrient limitations on the surface. However, from this analysis, it appears that nutrient limitations on the surface and settling of algae may be more significant than light inhibition.

Dissolved Oxygen Resources

One of the most important environmental factors in the Great Lakes is the dissolved oxygen content, particularly that of the hypolimnion. Eutrophic lakes may be more objectionable because of the development of anoxic conditions in the hypolimnion than because of the establishment of high algal densities at the lake surface. To display the state of dissolved oxygen resources in Lake Ontario, the seasonal variations of surface and

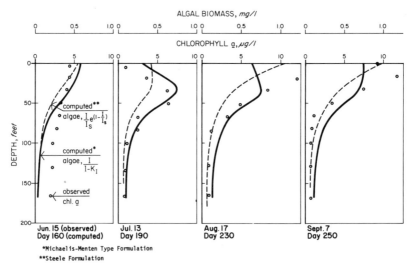

Figure 21. Computed algal densities using the Steele and Michaelis-Menten formulations vs observed values for node 8.

hypolimnion dissolved oxygen have been plotted in Figures 22 and 23. The associated water temperatures are also shown in the figures. Based on these figures, it is apparent that the surface dissolved oxygen is inversely correlated with surface water temperature. As the surface water temperature reaches a maximum of 20°C in June, July and August, the surface dissolved oxygen concentration reaches a minimum of approximately 9 mg l^{-1}. This is, of course, expected since there is an inverse relationship between temperature and the solubility of dissolved oxygen.

In the hypolimnion, the temperature maximum and the dissolved oxygen minimum may not occur at the same time. This is due to the lack of free exchange of oxygen between hypolimnion waters and the atmosphere. The rate of temperature rise in the hypolimnion is a function of the thermal diffusivity. The rate of dissolved oxygen decreases, however, appears to be more strongly related to the community respiration rate. Based on Figure 23, the dissolved oxygen in hypolimnion appears to reach a minimum near the end of August (Julian day 260). For a time period of 160 days, the hypolimnion dissolved oxygen decreases from an initial value of approximately 14 mg l^{-1} at a rate of approximately 0.055 mg l^{-1} day^{-1} or approximately 1.6 mg l^{-1} month^{-1}. This is approximately half the oxygen depletion rate observed in Lake Erie (Dobson and Gilbertson, 1971).

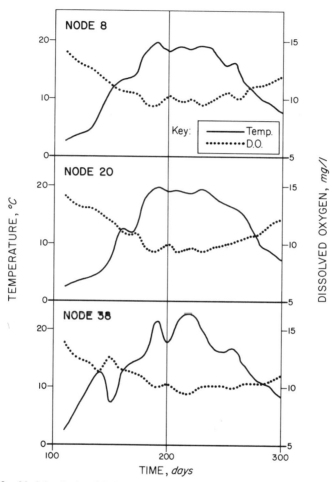

Figure 22. Model calculated DO and temperature values for surface water at nodes 8, 20 and 38.

DISCUSSION

Based on the experience of developing the comprehensive model described herein, several observations can be made:

1. The model calculates three-dimensional aspects of lake water quality on a daily basis. The results can be converted to water quality profiles that are directly comparable to the field data which are normally observed at a discrete location and time. There is no need for averaging

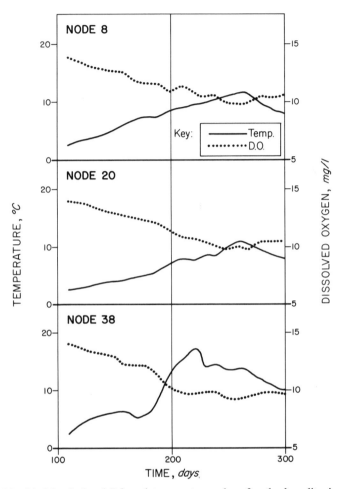

Figure 23. Model calculated DO and temperature values for the hypolimnion (-50 m) at nodes 8, 20 and 38.

water quality observations as would be necessary in order to make the field data comparable to results from a one-dimensional model.

2. While developing an appropriate time-averaging period for the hydrodynamic model output, it was found that the water quality response of the lake depended a great deal on the prevailing weather conditions and therefore on the hydrodynamics of the lake. The predictive capability of any model that does not take the variability of meteorology into account cannot be too precise.

3. It is intriguing that the same set of rate coefficients have been found applicable to many prototype systems, including Lake Ontario. This simplifies parameter estimation for a model requiring a large number of rate coefficients.

4. The model requires very simple input data, *i.e.,* bathymetric data, time varying meteorological records and waste loads, all of which are readily available. Intrinsic lake data such as currents and temperature regimes are not required as model input.

5. The model simulates a number of parameters, *e.g.,* thermal regimes, coliform contamination, DO resources, phytoplankton concentrations and fish populations. It is designed to serve multiple purposes, such as hypothesis testing, evaluation of waste disposal alternatives, and assessment of the impact of power plants. It provides a framework from which one can gain insight into complex lake dynamics, the understanding of which is prerequisite to the formulation of management alternatives.

ACKNOWLEDGMENT

The work was performed under the auspices of the Great Lakes Environmental Research Laboratory of the National Oceanic and Atmospheric Administration. Dr. Andrew Robertson and Mr. Donald Scavia served as project officers. Dr. John R. Bennett furnished the hydrodynamic model and assisted in its adaptation to the overall model program.

REFERENCES

Bennett, J. R. "A Three-Dimensional Model of Lake Ontario's Circulation—I. Comparison With Observation," *J. Phys. Oceanog.* 7:591-601 (1977).

Canale, R. P., D. F. Hineman and S. Nachiappan. "A Biological Producduction Model for Grand Traverse Bay," Technical Report No. 37, (Ann Arbor, MI: Sea Grant Program, University of Michigan, February 1974).

Casey, D. J., P. A. Clark and J. Sandwick. "Comprehensive IFYGL Materials Balance Study for Lake Ontario," Vol. I, U.S. Environmental Protection Agency, Rochester, NY (May 1976).

Chen, C. W., and G. T. Orlob. "Ecologic Simulation for Aquatic Environments," in *System Analysis and Simulation in Ecology, Vol. III,* B. C. Patten, Ed. (New York: Academic Press, Inc., 1975), pp. 475-588.

Chen, C. W. M. W. Lorenzen and D. J. Smith. "A Comprehensive Water Quality Ecological Model for Lake Ontario," report to Great Lakes Environmental Research Laboratory, National Oceanic and Atmospheric Administration, Ann Arbor, MI (1975).

Chen, C. W., and J. T. Wells, Jr. "Boise River Ecological Modeling," in *Modeling Biochemical Processes in Aquatic Ecosystems,* R. P. Canale, Ed. (Ann Arbor, MI: Ann Arbor Science, 1976), pp. 171-204.

DiToro, D. M., D. J. O'Connor and R. V. Thomann. "Phytoplankton-Zooplankton-Nutrient Interaction Model for Western Lake Erie," in *Systems Analysis and Simulation in Ecology, Vol. III,* B. C. Patten, Ed., (New York: Academic Press, Inc., 1975), pp. 424-473.

Dobson, H.H., and M. Gilbertson. "Oxygen Depletion in the Hypolimnion of the Central Basin of Lake Erie, 1929 to 1970," *Proc. 14th Conf. Great Lakes Res.* (International Association of Great Lakes Research, 1971), pp. 743-748.

International Joint Commission. *Pollution of Lake Erie, Lake Ontario and the International Section of the St. Lawrence River, Volumes I, II, and III,* (Windsor, Ontario: 1969).

Irbe, J. G., and R. J. Mills. "Aerial Surveys of Lake Ontario Water Temperature and Description of Regional Weather Conditions During IFYGL—January, 1972 to March, 1973," Report No. CLI 1-76, (Environment Canada, 1976).

Lorenzen, M. W., C. W. Chen, E. K. Noda and L. S. Hwang. "Review of Evaluation Methodologies for Lake Erie Wastewater Management Study," Report to the U.S. Army Corps of Engineers District, Buffalo, NY (1974).

O'Connor, D. J., and Mueller, J. A. "Water Quality Model of Chlorides in the Great Lakes," *J. San. Eng. Div., ASCE,* 96:955-975 (1970).

Orlob, G. T., and L. G. Selna. "Temperature Variations in Deep Reservoirs," *J. Hydraul. Div., ASCE,* 96:391-410 (1970).

Phillips, D. W., and J. A. W. McCulloch. "Monthly Means and Deviations from Normal Temperature and Precipitation in the Lake Ontario Basin During IFYGL," Technical Memoranda, (Burlington, Ontario: Canada Centre for Inland Waters, 1974).

Richardson, W. L., and V. J. Bierman, Jr. "A Mathematical Model of Pollutant Cause and Effect in Saginaw Bay, Lake Huron," in *Water Quality Criteria Research of the U.S. EPA,* EPA-600/3-76-079 (Corvallis, OR: U.S. Environmental Protection Agency, 1976).

Scavia, D., B. J. Eadie and A. Robertson. "An Ecological Model for Lake Ontario, Model Formulation, Calibration, and Preliminary Evaluation," Great Lakes Environmental Research Laboratory, NOAA, Tech. Report ERL-371-GLERL 12, (Boulder, CO: Environmental Research Laboratory, NOAA, 1976).

Simons, T. J. "Development of Three-Dimensional Numerical Models of the Great Lakes," Canada Centre for Inland Waters Scientific Series No. 12, (Burlington, Ontario: Canada Centre for Inland Waters, 1973).

Simons, T. J. "Verification of Numerical Models of Lake Ontario: Part I. Circulation in Spring and Early Summer," *J. Phys. Oceanog.* 4:507-523 (1974).

Simons. T. J. "Verification of Numerical Models of Lake Ontario. II. Stratified Circulations and Temperature Changes," *J. Phys. Oceanog.* 5:98-110 (1975a).

Simons, T. J. "Interfacing Physical and Biochemical Models of Large Lakes," reprint of a paper presented at the 6th Triennial IFAC World Congress, August 24-30, Boston, MA (1975b).

Simons, T. J. "Verification of Numerical Models of Lake Ontario. III Long-Term Heat Transports," *J. Phys. Oceanog.* 6:372-378 (1978a).

Simons, T. J. "Analysis and Simulation of Spatial Variations of Physical and Biochemical Processes in Lake Ontario," *J. Great Lakes Res.* 2(2): 215-233 (1976b).

Thomann, R. V., D. M. DiToro, R. D. Winfield and D. J. O'Connor. "Mathematical Modeling of Phytoplankton in Lake Ontario 1. Model Development and Verification," EPA-660/3-75-005 (Corvallis, OR: U.S. Environmental Protection Agency, 1975), 177 p.

CHAPTER 11

THE EXAMINATION OF ECOSYSTEM PROPERTIES OF LAKE ONTARIO THROUGH THE USE OF AN ECOLOGICAL MODEL*

Andrew Robertson and Donald Scavia

U.S. Department of Commerce
National Oceanic and Atmospheric Administration
Environmental Research Laboratories
Great Lakes Environmental Research Laboratory
Ann Arbor, Michigan 48104

INTRODUCTION

Most of the development of mathematical models of Great Lakes' ecosystems has been motivated by an attempt to provide a tool for resource managers (*e.g.,* Thomann *et al.,* 1975, 1976; Canale *et al.,* 1974, 1976). Such efforts are obviously of great potential value for that purpose. There is, however, another major reason for developing such models. They present a way of synthesizing existing knowledge of these ecosystems and furnish insights into ecosystem structure and function. In other words, the models should aid in understanding the basic ecological processes in the Great Lakes. There has so far been little study of the usefulness of these models for this purpose.

We have developed an ecosystem model for Lake Ontario (Scavia *et al.,* 1976) that simulates the general features of the Ontario ecosystem quite well. Using an updated version of this model, we have investigated the utility of employing the modeling approach to study certain basic ecological properties of the system; properties that are, in many cases, almost impossible to study in an ecosystem the size of Ontario with direct field

*GLERL Contribution No. 126.

measurements. This paper presents examples of properties that can be investigated and attempts to evaluate the validity and usefulness of this approach for gaining insights into large ecosystems.

METHODS

The model used in this study is a modification of the one previously developed for Lake Ontario by Scavia et al. (1976). It includes constructs in two vertical layers for the seasonal cycles of: (1) five groups of phytoplankton, (2) five groups of herbivorous zooplankton, (3) one group of carnivorous zooplankton, (4) detritus, (5) available phosphorus, (6) dissolved organic nitrogen, (7) ammonia, (8) nitrite plus nitrate, (9) soluble reactive silicon, (10) detrital silicon, (11) dissolved oxygen and (12) the carbonate system. Forcing functions for the model are solar radiation, water temperature, winds, allochthonous loads and hydraulic flushing rates. Simulated sinking rates determine the transport of particulate organic matter into the sediment compartment where the dynamics of particulate organic carbon, phosphorus, nitrogen and silicon; dissolved organic nitrogen; dissolved inorganic carbon; ammonia; nitrite-nitrate; soluble reactive phosphorus and silicon; dissolved oxygen; and benthic macroinvertebrates are simulated. Transport of dissolved materials across the sediment-water interface is modeled as a diffusive process. The model equations and a comparison of model output with data collected during 1972-1973 have been presented by Scavia (1978a).

The model includes compartments representing the most important plant nutrients, most of the primary producers and the major primary consumers in the offshore region of Lake Ontario. Decomposers are not modeled explicitly, but rather their function is included as first-order decay terms. Secondary consumers are parameterized to include all carnivorous zooplankton. All higher trophic levels are included implicitly as first-order loss terms; therefore feedback to these higher trophic levels is not simulated.

For the purposes of this work, the model equations are solved numerically with a fixed-step (0.5 day) second-order Runge-Kutta algorithm. The state variable solutions, as well as the terms in the differential equations, are stored at each integration step at the time when the overall derivatives are calculated. The rates and state variables are then summed over the appropriate compartments and time periods. The aggregated state variables represent calculated concentrations of the following categories in each layer: (1) dissolved inorganic carbon, (2) producers, (3) herbivores, (4) carnivores and (5) detritus. The aggregated rates represent the carbon flows between categories or trophic levels. For example, all the individual terms used in the model to calculate ingestion of each phytoplankton type by each herbivore type are summed to form the link between producers and herbivores. Transport between the epilimnion and hypolimnion includes sinking, diffusion and changes in concentration due to changes in thermocline depth. The terms epilimnion and hypolimnion are used loosely in this work. During periods of thermal stratification, the depth of the calculated thermocline determines the thickness of these layers. During nonstratified periods the epilimnion is defined as the top 10 m.

RESULTS

Carbon Cycling Diagrams

An example of the results obtained when the model output is used to calculate the concentrations of carbon in the different categories and its flows among these categories is presented in Figure 1. The values in the boxes are the average carbon concentrations (μg l^{-1}), while the values in the pipes are the average daily flowrates (μg l^{-1} day^{-1}); both concentrations and flows are averages for an entire year in the epilimnion.

With the exception of the inorganic carbon compartment, the sizes of the boxes are scaled to the relative amounts of carbon in the different categories, while the widths of the connecting pipes are scaled to the relative amounts of carbon flow. At present, the influences of fish are included in the model only as constant percent losses from the herbivores and carnivores, and so a compartment of indeterminate size is included for them in the figure.

From this figure, one can gain an overall view of the cycling of carbon through the upper waters of the Lake Ontario ecosystem. For example, note that the model indicates that herbivores acquire more carbon from the detritus pool than from producers and also lose more to this pool than to carnivores. This suggests that exchange of materials between herbivores and detritus is a significant, although somewhat inefficient, element in this ecosystem. Also it can be noted that substantial amounts of particulate matter, including both producers and detritus, are lost to the hypolimnion on an annual basis, thus furnishing a major source of food for the organisms of that region.

A similar annual average carbon cycling diagram has been constructed for the hypolimnion (Figure 2). A comparison with the epilimnion diagram shows the decreased significance of the producers and the much reduced flows in this deeper colder zone.

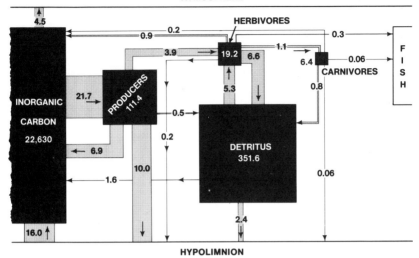

Figure 1. A cycling diagram of the yearly average concentrations of carbon (μg l^{-1}) and yearly average flows of carbon (μg l^{-1} day^{-1} in the epilimnion of Lake Ontario.

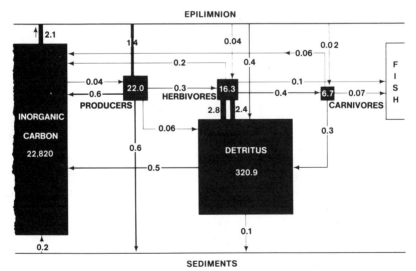

Figure 2. A cycling diagram of the yearly average concentrations of carbon (μg l^{-1}) and yearly average flows of carbon (μg l^{-1} day^{-1}) in the hypolimnion of Lake Ontario.

Such diagrams have also been constructed for each season of the year. Starting at January 1, the results from running the model for 364 days were combined in four 91-day groups. Although these grouping do not completely correspond to the calendar seasons, we consider the agreement to be close enough to designate them by the conventional season names. Examples of these seasonal diagrams showing the carbon cycling in the epilimnion for summer and for winter are presented in Figures 3 and 4, respectively. A comparison of these two figures clearly illustrates the expected large seasonal changes that occur in this ecosystem.

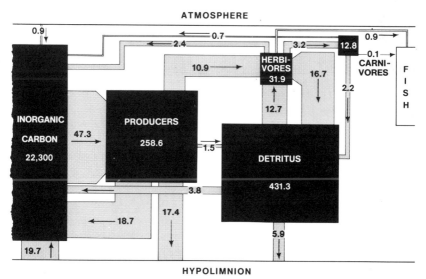

Figure 3. A cycling diagram of the average concentrations of carbon ($\mu g\ l^{-1}$) and average flows of carbon ($\mu g\ l^{-1}\ day^{-1}$) during summer in the epilimnion of Lake Ontario.

These cycling diagrams can be used to gain an overall perspective of the Lake Ontario ecosystem. Even a cursory examination will show whether the system operates as a detrital- or producer-based food web. Major trophic paths and relations are indicated, as well as the relative abundance of each trophic level. All these properties, and others, can be examined and compared on any meaningful time frame. These diagrams contain so much information, in fact, that it becomes somewhat difficult to focus on specific aspects. Thus, in the following subsections certain properties have been singled out for consideration as a further illustration of the types of properties that can be examined using the model results.

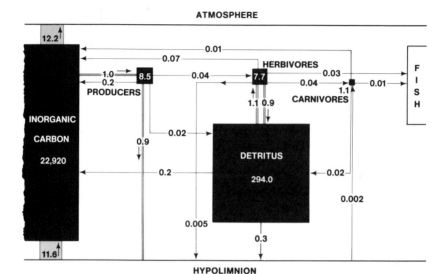

Figure 4. A cycling diagram of the average concentrations of carbon (μg l^{-1}) and average flows of carbon (μg l^{-1} day^{-1}) during winter in the epilimnion of Lake Ontario.

Relative Concentrations

The carbon concentrations of the different categories in the epilimnion are compared in Table I on both a seasonal and an annual basis. According to these results, detritus is the major reservoir of particulate carbon and always makes up more than 50% of this material. Its importance diminishes during summer when producers account for more than one-third of the total. The herbivores plus carnivores never account for even 10% of the total seston. All the categories reach their highest concentration in summer and all but detritus reach their lowest value in winter. Detritus reaches its low point in spring. In general, this category tends to lag behind the others in its response to the seasons, as might be expected from its place in the food chain.

Hypolimnion/Epilimnion Concentration Ratios

The relative concentrations of carbon in the different categories have also been calculated for the hypolimnion. Although these are not presented here, they have been divided by the comparable values for the epilimnion to obtain concentration ratios in the different categories for these two layers (Table II).

Table I. Distribution of Particulate Carbon (μg l^{-1}) in the Epilimnion
(percentage of total seston in parentheses)

	Seasonal Averages				Average for Year
	Winter	Spring	Summer	Fall	
Producers	8.5	114.3	258.6	64.0	111.4
	(2.7)	(26.3)	(35.2)	(13.5)	(22.8)
Herbivores	7.7	23.9	31.9	13.4	19.2
	(2.5)	(5.5)	(4.3)	(2.8)	(3.9)
Carnivores	1.1	5.6	12.8	6.2	6.4
	(0.4)	(1.2)	(1.7)	(1.3)	(1.3)
Detritus	294.0	291.0	431.3	390.1	351.6
	(94.4)	(66.9)	(58.7)	(82.4)	(72.0)
Total Seston	311.3	434.8	734.6	473.7	488.6
	(100.0)	(99.9)	(99.9)	(100.0)	(100.0)

Table II. Hypolimnion: Epilimnion Concentration Ratios

	Seasonal Averages				Average for Year
	Winter	Spring	Summer	Fall	
Producers	0.67	0.15	0.17	0.33	0.20
Herbivores	1.23	1.02	0.62	0.87	0.85
Carnivores	1.00	1.00	1.17	0.81	1.05
Detritus	1.01	0.98	0.77	0.95	0.91
Total Seston	1.01	0.76	0.56	0.86	0.75

Except during the thermally stratified summer season, the ratios for herbivores, carnivores and detritus are close to one, indicating little effect of depth on concentration. Producers, on the other hand, are always more concentrated in the upper lighted layers even during the periods of strong vertical mixing in spring and fall. Thus, according to the model, the vertical distribution of the producers is controlled more by the fact that photosynthesis is limited to the upper layers than by vertical water movement.

Carbon Budgets for Separate Categories

Another aspect that can be investigated with the model is the relative significance within each category of the various gains and losses of carbon. As an example of this, Table III presents the carbon budget for the epilimnetic herbivores, both for each season separately and for the entire year.

Both producers and detritus provide important sources of carbon for the herbivores. However, detritus is always the more important source and furnishes almost the entire source during winter. It should be borne in mind that we are dealing here with a comparison of the relative significance of carbon sources. When herbivores consume producer biomass, they usually assimilate a higher percent of the material than when they consume detritus. Thus, the relative significance of these two categories as energy sources may be substantially different from that as carbon sources.

Detritus is also the major sink for material leaving the herbivores. Indications are that loss to this category is much more important than loss to the carnivores. This is somewhat at odds with the commonly held idea that the major pathway for material is to move up the food chain

Table III. Gains and Losses of Carbon (μg l^{-1} day $^{-1}$) for the Herbivores in the Epilimnion (percentage of gain or loss in parentheses)

| | Seasonal Averages | | | | Average for Year |
	Winter	Spring	Summer	Fall	
Gains from:					
Producers	0.04	3.0	10.9	1.6	3.9
	(3.5)	(41.7)	(46.2)	(32.0)	(42.4)
Detritus	1.1	4.2	12.7	3.4	5.3
	(96.5)	(58.3)	(53.8)	(68.0)	(57.6)
Losses to:					
Carnivores	0.04	0.7	3.2	0.6	1.1
	(3.8)	(9.9)	(13.8)	(11.5)	(12.1)
Fish	0.03	0.06	0.9	0.4	0.3
	(2.9)	(0.8)	(3.9)	(7.7)	(3.3)
Detritus	0.9	5.2	16.7	3.6	6.6
	(85.7)	(73.7)	(72.0)	(69.2)	(72.5)
Inorganic Carbon	0.07	0.6	2.4	0.3	0.9
	(6.7)	(8.5)	(10.3)	(5.8)	(9.9)
Hypolimnion	0.01	0.5	0	0.3	0.2
	(1.0)	(7.1)	(0)	(5.8)	(2.2)

with the losses along the way due primarily to respiration (*i.e.,* inorganic carbon in Table III). The model results suggest that production and cycling of detritus may be a very important role for herbivores and that, in this way, they pass material to the bacteria as much as, or more than, to the carnivores.

Turnover Times

Turnover times for the various categories, that is, the amount of time it would take to replace the material in each category (Odum, 1971), can be calculated by dividing the average concentration in each category by the average total flow into that category. Both seasonal and annual turnover times have been calculated for the epilimnion (Table IV).

Table IV. Epilimnion Turnover Times (days)

	Seasonal Averages				Average for Year
	Winter	Spring	Summer	Fall	
Producers	8.5	3.7	5.5	8.4	5.1
Herbivors	6.8	3.3	1.4	2.7	2.1
Carnivores	27.5	8.0	4.0	10.3	5.8
Detritus	312.8	48.5	21.1	90.7	44.5
Total Seston	311.3	14.1	15.5	62.3	22.5

As expected, the turnover times are generally longest in winter and shortest in summer. This is because, in the model as well as the real world, the rates of metabolic processes, within the temperature range tolerated by an organism, generally increase as temperature increases. Interestingly, the turnover times for the producers do not completely conform to the pattern. The shortest turnover time for this category is predicted to be in spring, not summer. In the environment, this may well be related to the fact that the nutrients these organisms need for growth are relatively more available during spring. Thus, even though temperatures are higher in summer, growth is limited by lack of nutrient availability.

The herbivores display the shortest turnover times. This is undoubtedly related to their use of detritus as a major food source. As discussed earlier, this material is less efficiently assimilated than living biomass, and so the herbivores must have a higher flow-through rate to meet their nutritional requirements.

Trophic Level Efficiencies of Carbon Intake

The efficiencies of carbon transfer at various places in the food web can be estimated by calculating the percent of the material entering one trophic level or category that is passed on to another level. For example, the flow of carbon from producers to herbivores divided by the total carbon uptake of producers (photosynthesis) is, when converted to percent, the ecological trophic level efficiency of the producers from the viewpoint of the herbivores (Odum, 1971). Such efficiencies have been calculated on both a seasonal and annual basis for feeding in the epilimnion of herbivores on producers and of carnivores on herbivores (Table V).

Both types of efficiency are highest during summer and lowest during winter, suggesting a correlation with temperature. The efficiency of transfer from herbivores to carnivores is generally lower than that from producers to herbivores. However, it should be remembered that only carnivorous zooplankton are modeled explicitly. This relation might change if planktivorous fish were included in the model.

Table V. Trophic Level Carbon Intake Efficiencies (%) in the Epilimnion

	Seasonal Averages				Average for Year
	Winter	Spring	Summer	Fall	
Producers to Herbivores	4.0	9.7	23.0	21.1	18.0
Herbivores to Carnivores	3.5	9.7	12.6	12.0	12.0

DISCUSSION

The foregoing illustrates the type of basic ecological properties that can be explored with ecosystem models. Obviously, the treatment is not exhaustive. There are many other properties that could be considered. Further, the inclusion in the model of additional or more detailed categories could provide added insight. Having the model based on the exchange of other nutrients besides carbon or even of energy would also be instructive. In this regard, a model simulating the movement of energy would have special value as this is the property usually considered in experimental studies on ecosystem structure (*e.g.,* Odum, 1971).

The question remains, however—are the results of such modeling efforts valid for investigating ecosystem properties? Most of the properties discussed above have received very little study as of yet in Lake Ontario or, for that matter, in the Great Lakes. Thus, it is not possible to verify rigorously the results by comparison to actual measurements. However, it has been possible to compare the model results to actual data for a number of the state variables and certain of the processes. There is quite good agreement (Scavia *et. al.*, 1976; Scavia, 1978a,b) and so we feel that the extension to investigate unmeasured processes, as illustrated in this paper, is at least partially warranted. Additionally, the model results for the system properties discussed above generally show patterns that are in agreement with, or at least do not conflict with, the expectations of ecological theory. Thus, it seems reasonable to consider the results for these properties as at least working hypotheses on which further studies can be based.

This, in fact, points up what may well be the major value of models in increasing ecological understanding. As pointed out previously, there have been few studies carried out on the basic ecosystem properties of Lake Ontario. Actually these properties have received little study for large ecosystems in general. This is undoubtedly due to the great amounts of time and money that must be expended to carry out such work. The models provide a relatively inexpensive way to synthesize existing information and to develop hypotheses concerning these properties. Thus, they should stimulate development of field and experimental studies on these properties and on the basic structure of large ecosystems in general.

SUMMARY

This paper has illustrated the value of ecosystem models for increasing our fundamental understanding of such systems. By using an existing ecosystem model of Lake Ontario to calculate the values for a few of the many basic properties that can be studied, it has been shown that such a model provides a useful way of investigating such properties. It is concluded that ecosystem models have substantial potential as tools for basic ecology and their use for such a purpose well deserves further investigation.

REFERENCES

Canale, R. P., L. M. DePalma and A. H. Vogel. "A Plankton-Based Food Web Model for Lake Michigan," in *Modeling Biochemical Processes in Aquatic Ecosystems,* R. P. Canale, Ed. (Ann Arbor, MI: Ann Arbor Science Publishers, Inc., 1976), pp. 33-74.

Canale, R. P., D. F. Hineman and S. Nachiappan. "A Biological Production Model for Grand Traverse Bay," Technical Report No. 37, (Ann Arbor, MI: University of Michigan Sea Grant Program, 1974), 116 p.

Odum, E. P. *Fundamentals of Ecology*, 3rd ed. (Philadelphia: W. B. Saunders Company, 1971), 574 p.

Scavia, D. "An Ecological Model of Lake Ontario," manuscript, U.S. Dept. Commerce, NOAA, Great Lakes Environmental Research Laboratory, Ann Arbor, MI (1978a).

Scavia, D. "Examination of Phosphorus Cycling and Control of Phytoplankton Productivity in Lake Ontario during IFYGL," manuscript, U.S. Dept. Commerce, NOAA, Great Lakes Environmental Research Laboratory, Ann Arbor, MI (1978b).

Scavia, D., B. J. Eadie and A. Robertson. "An Ecological Model for Lake Ontario: Model Formulation, Calibration, and Preliminary Evaluation," NOAA Technical Report ERL 371 (Boulder, CO: U.S. Department of Commerce, NOAA, Environmental Research Laboratories, 1976), 63 p.

Thomann, R. V., D. M. DiToro, R. P. Winfield and D. J. O'Connor. "Mathematical Modeling of Phytoplankton in Lake Ontario. 1. Model Development and Verification," EPA-660/3-75-005 (Duluth, MN: Environmental Research Laboratory, U.S. Environmental Protection Agency, 1975), 218 p.

Thomann, R. V., R. P. Winfield, D. M. DiToro and D. J. O'Connor. "Mathematical Modeling of Phytoplankton in Lake Ontario. Part 2. Simulations Using Lake 1 Model," EPA-600/3-76-065 (Duluth, MN: Environmental Research Laboratory, U.S. Environmental Protection Agency, 1976), 88 p.

CHAPTER 12

AN ANALYSIS OF PCB IN LAKE ONTARIO USING A SIZE-DEPENDENT FOOD CHAIN MODEL

Robert V. Thomann

Department of Environmental Engineering
and Science
Manhattan College
Bronx, New York 10471

INTRODUCTION

The presence of hazardous substances in the aquatic food chain is a problem of rapidly developing dimensions and magnitude. Passage of the Toxic Substances Act of 1976, unprecedented monetary fines and continual development of data on lethal and sublethal effects attest to the expansion of control on the production and discharge of such substances. On the other hand, new product production and the ever-present potential for insect and pest infestations with attendant effects on man and animals result in continuing demand for product development. Lake Ontario is a specific example of this confrontation. High levels of polychlorinated biphenyls (PCB) have been measured in the upper trophic levels (*e.g.,* coho salmon); this has resulted in closing the lake to fishing for several species. As a result of such situations, considerable effort has been devoted in recent years to the development of analysis schemes that would provide an understanding of the mechanisms of food-chain transport and accumulation. Such schemes ideally would result in meaningful *a priori* judgments of the effects of a given substance on the environment.

In this paper, a generalized model is presented that is focused specifically on the aquatic ecosystem and the transport and accumulation of hazardous substances in the aquatic food chain. The PCB problem in

Lake Ontario is used as the problem setting. Specific toxicological effects are not included in this work; the emphasis is on the fate of the substance in the aquatic ecosystem. The primary motivation for the development of the model is to place in a generalized framework, the large amount of work from laboratory and field experiments on hazardous substances in the food chain. The model is presented as an example of a framework in an early developmental stage in contrast to the more complex models of phytoplankton biomass dynamics. Consequently, the thrust of the model is on simplified analytical structures requiring only simple computations. Reports of studies on the transport of hazardous substances generally begin by assigning to the ecological system a series of compartments positioned in space and time. The concept of a compartment arises from a grouping of ecological properties, species or types (e.g., "phytoplankton" or "fish"). The continuum of the environment is replaced by finite, discrete, interacting trophic levels. The details of each compartment need not be specified and internal mechanisms are not necessarily examined. Attention is usually directed towards a portion of the ecosystem, the degree of specificity of compartments depending heavily on the aims of the investigator. The ecological concepts of compartment analyses have been reviewed by Dale (1970) and Patten (1971). Examples of compartment models in food chain studies of hazardous substances include Gillet (1974), Hill *et al.* (1976), Lassiter *et al.* (1976) and Hoefner and Gillet (1976).

Compartment analyses provide a great deal of flexibility and can incorporate reasonably complicated food webs. However, the basic difficulty with viewing the food chain problem in a compartment sense is that any general underlying theoretical framework may be masked by the choice of the compartments. The discretization of the aquatic food chain obscures the "continuous" nature of the ecosystem. Furthermore, if there are m ecological variables (ecological compartments) positioned at n spatial locations, there are a total of m x n compartments and m x n equations (differential or algebraic) to be solved. Thousands of equations can easily result. In this work, therefore, a formalization is proposed for a generalized unified framework for the transport of substances in the food chain. Discretization of ecological space into compartments would be a special case of the general theory.

THE GENERALIZED MODEL

The principal questions that prompted the development of a generalized hazardous substance model are: "If compartments represent discrete, homogenous entities in ecological space, what is the continuous analog of this approximation? Since compartment analyses often represent differential equations around previously chosen trophic levels, what is the suitable mathematical framework that would represent ecological space as a continuum rather than discretized levels?" Figure 1 illustrates the first question by analogy to the often-made discrete approximation to a continuous water body for analysis of dissolved substances in the water. Continuous space is replaced by a series of discrete regions in space and mass balance equations are written around each region in space for the substance of interest. Similarly in ecological compartment analysis, mass balance equations are written around a series of individual "regions" thereby describing the transport and cycling of the substance in the food chain. For a toxicant, the relevant measure is the milligrams toxicant per unit biomass at that level.

Compartment Approach

Therefore, for compartment analyses, one lets:

$$\nu_{ij} = (\frac{\text{mass toxicant}}{\text{mass trophic level i}})$$

at location j and:

$$M_{ij} = (\frac{\text{mass of trophic level i}}{\text{volume of water}})$$

also at location j. Then $\nu_{ij}M_{ij} = s_{ij}$ is the mass of toxicant of level i relative to the volume of water at location j. In one-dimensional food chains (excluding food webs), for a volume of water V_j, the rate of change of the mass of the toxicant is given by:

$$V_j \frac{d(\nu_{ij}M_{ij})}{dt} = (K_{i-1,i}M_{i-1}\nu_{i-1})_j \, V_j - (K_{ii}M_i\nu_i) \, _jV_j \tag{1}$$
$$+ \text{advection} + \text{dispersion} + \text{sources} - \text{sinks}$$

where $K_{i-1,i}$ and K_{ii} are transfer rates $[T^{-1}]$ into and out of level i.

If it is assumed temporarily that the mass/volume of each trophic level is constant and that there is a single completely mixed volume, then Equation 1 becomes:

$$M_i' \frac{d\nu_i}{dt} = S_{i-1,i} \, \nu_{i-1} - S_{ii} \, \nu_i + W_i \tag{2}$$

where $M_i' = M_i V$ = trophic level mass
$S_{i-1,i} = K_{i-1,i}M_{i-1}$
$S_{ii} = K_{ii}M_i$
W_i = external sources and sinks

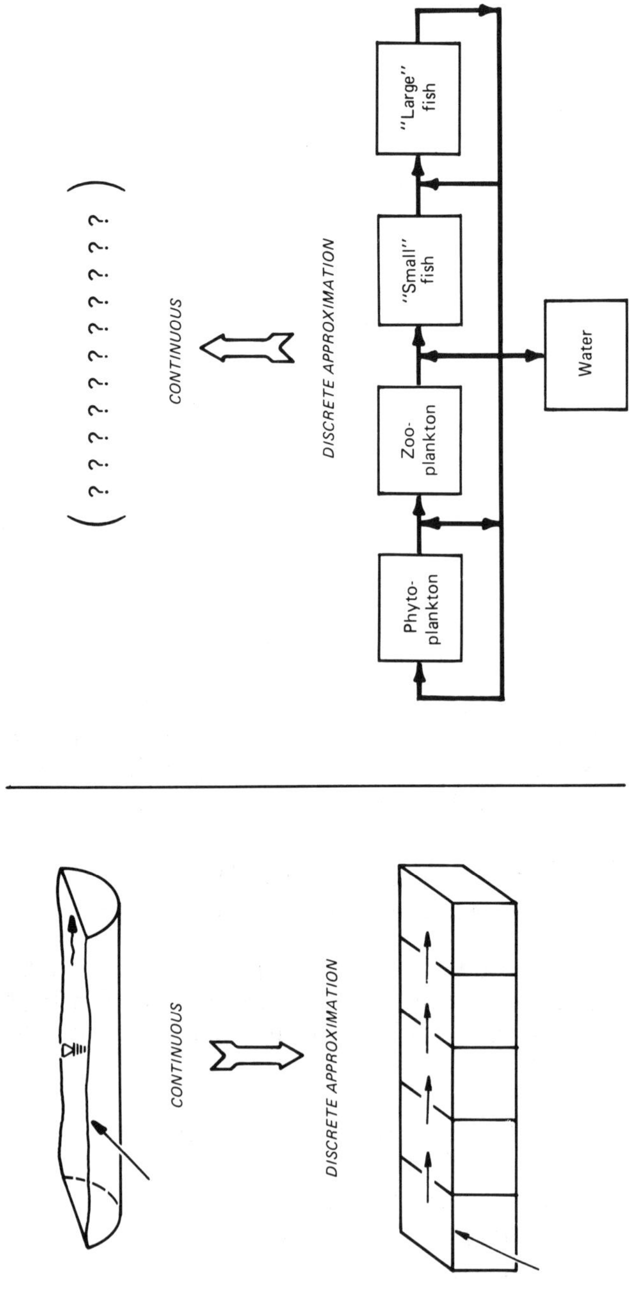

(a) Transport of substance in water.

(b) Transport of substance in food chain.

Figure 1. The continuous environment and discrete approximation.

(Advective and dispersive terms can, of course, be included). Now, it is noted that Equation 2 is identical to the mass balance equation that results for a finite difference approximation for water quality variables in physical space. In Equation 2, however, physical space is replaced by trophic space or ecological space. The question posed earlier is thus made specific by Equation 2, *i.e.,* what is the underlying continuous equation of which Equation 2 is an approximation?

The subscript i in Equation 2 essentially represents some ecological variable in continuous trophic space. Therefore, in reality, Equation 2 represents a mass balance around some region in ecological space represented by an ecological coordinate. Such a coordinate would be distinct from the usual physical coordinates of length, width and depth of a water body. A generalized model would proceed by writing a mass balance equation that incorporates both physical and ecological space. Such a derivation is part of a class of models generally referred to as "Population Balance Models" because of their use in describing age distributions. Detailed reviews of such models are given by Himmelblau and Bischoff (1968) and Rotenberg (1972).

Continuous Model

Following these authors, consider a function $\Psi (\Delta x, \Delta y, \Delta z, \Delta \xi)$ where $\Delta x, \Delta y, \Delta z$ represent, in the usual sense, the physical coordinates and $\Delta \xi$ represents some other property of the system. For example, $\Delta \xi$ may be the age of a population or the size of suspended particles. Then, $\Psi (\Delta x \Delta y \Delta z \Delta \xi)$ represents the fraction of some quantity (in this work, the PCB concentration) that is in the physical region volume element $V = \Delta x \Delta y \Delta z$ and also is in the property range of $\Delta \xi$. If a general mass balance of the toxicant throughout both physical and "property" space is carried out (see Himmelblau and Bischoff), one obtains:

$$\frac{\partial \Psi}{\partial t} + \frac{\partial}{\partial x} (v_x \Psi) + \frac{\partial}{\partial y} (v_y \Psi) + \frac{\partial}{\partial z} (v_z \Psi) + \frac{\partial}{\partial \xi} (v_\xi \Psi) = \text{sources and sinks} \quad (3)$$

where
$$v_x = dx/dt$$
$$v_y = dy/dt$$
$$v_z = dz/dt$$
$$v_\xi = d\xi/dt, \text{ the time rate of change of the property i}$$

Equation 3 can be compared to the usual three-dimensional mass balance equation written only in physical space.

In the case of food chain modeling, a relevant metric must be chosen that can represent position in trophic space. One such measure is the size or length of the organism; through suitable allometric relationships this can be related to the mass of the organism (see Eberhardt, 1969). The choice of organism length as a suitable metric in trophic space is advantageous from a field sampling point of view. That is, it is considerably less difficult to fractionate samples by size than by weight. Nets and filters can be easily used to divide the organisms in a food chain by size. An example of the continuous nature of the Lake Michigan ecosystem is shown in Figure 2. The overlapping size preferences for food of the various stages of the alewife illustrate the role that organism size plays in food chain transfer, as well as the continuous nature of the food chain. Furthermore, if there is no change in species composition, the length metric may be related to the lipid content of organisms; and this relation may be of value in the modeling of the dynamics of lipophylic compounds.

Of course, the use of an overall length metric does not permit a distinction to be made between organism types. Therefore, fish larvae and large crustaceans look the same to this metric—"a centimeter is a centimeter." It should be recognized also that introducing size dependence as an independent variable greatly oversimplifies a complex ecosystem. Such dependence implies that larger organisms prey on smaller organisms; this implication is obviously not always true. The framework given below, in principle, permits different size class interactions; however, this is not pursued in this work. For certain problem contexts, organism size may be very roughly correlated to trophic position, but it is recognized that, in many cases, trophic status and organism size are not related. The approach here is to introduce an ecologically reasonable simplification into existing model structures to provide additional analytical insight.

Conceptually, incorporation of other ecological properties is readily accomplished in Equation 3 but obviously at the sacrifice of analytical simplicity. The idea then is to describe the transport of a hazardous substance such as PCB by how much of the total mass of the substance is located in various size ranges of organisms and in various locations in physical space. For organism size and one physical dimension, say longitudinal distance, Figure 3 shows the nature of Ψ. As indicated, Ψ is a density function of the toxicant for positions x and property ξ. If the property ξ is now considered to be organism size (L) then Ψ has units similar to μg-PCB l^{-1} μm^{-1} where the organism size is given in micrometers (μm). Note that since Ψ is a continuous density function, the mass of toxicant at a specific organism size is zero; only over some range of trophic length is there a nonzero amount of toxicant. This is a consequence of the assumption that biological space is described by a continuum of organism sizes. The special case of single organism aquaria experiments is easily derived from this general view.

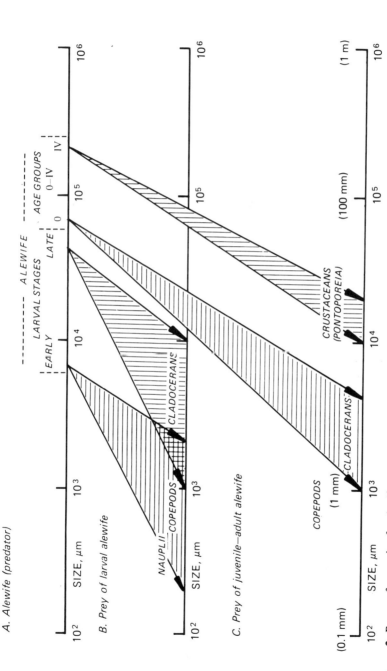

Figure 2. Range of prey size for alewife larval stage and juvenile-adult age groups 0-IV, Lake Michigan (adapted from Morsell and Norden, 1968; Norden, 1968).

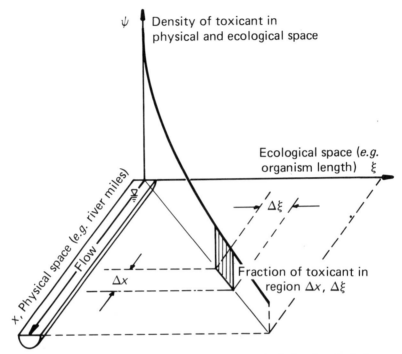

Figure 3. Distribution of toxicant in one-dimensional physical and ecological space.

For the property ξ, now considered as organism size, and using the standard dispersion and advective terms in physical space, Equation 3 becomes:

$$\frac{\partial \Psi}{\partial t} + \frac{\partial}{\partial x} (v_x \Psi) + \frac{\partial}{\partial y} (v_y \Psi) + \frac{\partial}{\partial z} (v_z \Psi) - \frac{\partial}{\partial x} (E_x \frac{\partial \Psi}{\partial x}) - \frac{\partial}{\partial y} (E_y \frac{\partial \Psi}{\partial y})$$

$$- \frac{\partial}{\partial z} (E_z \frac{\partial \Psi}{\partial z}) + \frac{\partial}{\partial L} (v_L \Psi) = S (x,y,z,L,t) \tag{4}$$

where v_x, v_y and v_z are net advective velocities in x, y and z respectively; E_x, E_y and E_z are dispersion coefficients in xyz space; S represents the sources and sinks of Ψ; and $v_L = dL/dt$, the "transfer velocity" with which the toxicant is transported up the food chain which is represented by the metric L.

A separate equation is necessary for the water phase of the toxicant (c) and is given by:

$$\frac{\partial c}{\partial t} + \frac{\partial}{\partial x}(v_x c) + \frac{\partial}{\partial y}(v_y c) + \frac{\partial}{\partial z}(v_z c) - \frac{\partial}{\partial x}(E_x \frac{\partial c}{\partial x}) - \frac{\partial}{\partial y}(E_y \frac{\partial c}{\partial y})$$

$$- \frac{\partial}{\partial z}(E_z \frac{\partial c}{\partial z}) = W(x,y,z,t) \pm {}_0\!\int^\infty S'(x,y,z,L,t)\, dL \tag{5}$$

where W is the direct input of the toxicant to the water body from external sources (independent of Ψ) and $\pm S'$ is the source or sink of c due to uptake and excretion mechanisms in biological space. Note that Equation 5 does not include the L dimension as an independent variable and also that, if sediment interactions are to be included, a separate sediment equation must be written to provide input to the water phase equation of Equation 5. Figure 4 shows that the mass of the toxicant is given by the area between L_1 and L_2 or the cumulative mass distribution along L is given by s_c as:

$$s_c = {}_0\!\int^L \Psi(L')\, dL' \quad [ML^{-3}] \tag{6}$$

The total toxicant mass in the entire trophic space (s_T) is then given by:

$$s_T = {}_0\!\int^\infty \Psi(L')\, dL' \tag{7}$$

to which must be added the mass in the water and sediment for the total mass of toxicant in the entire system. The relationship between Ψ and the body burden of PCB in μg-PCB g^{-1} (wet wt) is given through use of a biomass density function $[m(L)]$ (g (biomass) l^{-1} μm^{-1}). Thus:

$$\frac{\int_L^{L+\Delta L} \Psi\, dL}{\int_L^{L+\Delta L} m\, dL} \cong \frac{\Psi}{m} = \nu(L) \tag{8}$$

where ν is the concentration of PCB per unit biomass. Figure 5 shows the relationship between Ψ, m and ν.

For Lake Ontario, a completely mixed water volume is assumed in order to develop the concepts and explore the meaning and determination of the model parameters and behavior. Figure 6 schematically shows the various interactions in the model for a single completely mixed volume of water. As indicated, the water concentration in the volume is interactive with a continuous distribution along the L dimension, receives excreted toxicant from the entire food chain, and loses toxicant due to uptake over the entire biomass.

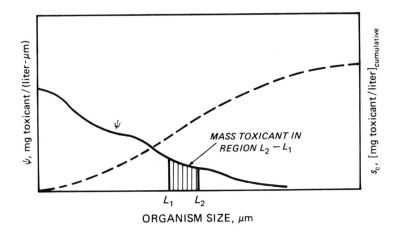

Figure 4. Density function Ψ and distribution function c.

THE COMPLETELY MIXED WATER
VOLUME CASE

If, in Equations 4 and 5, the water body is assumed to be well mixed with no gradients in concentration, then the governing equation for Ψ, the PCB density function, is:

$$V \left(\frac{\partial \Psi}{\partial t} + \frac{\partial v_L \Psi}{\partial L} \right) = Q\Psi_{in} - Q\Psi + W'(L) - S'(L) \tag{9}$$

where Q = mass transport through volume V
    Ψ_{in} = the input density function
    $W'(L)$ = sources of Ψ
    $S'(L)$ = sinks of Ψ

Equation 9 can be recognized as an advective equation, or maximum gradient equation, identical to that used in analysis of river systems but here derived for continuous trophic space in a completely mixed volume. Since the equation has no dispersion terms, there is no "mixing" in the food chain. The left-side terms represent the rates of change of the density of PCB with time and organism size. The first term on the right side of Equation 9 is the mass transport of the PCB density entering the

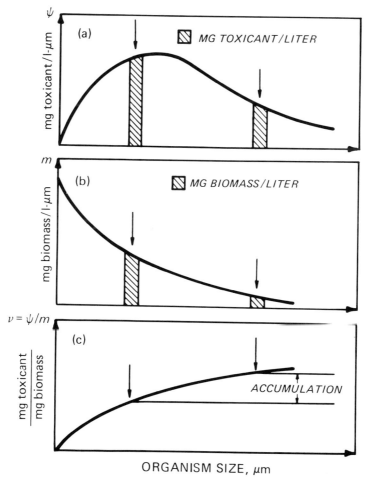

Figure 5. (a) The toxicant density function, (b) the biomass density function, and (c) the bioaccumulation function.

volume. This would include the mass of PCB distributed in the phytoplankton, zooplankton and fish entering the completely mixed volume V.

The mass transport Q is a quantity that reflects the transport of the entire food chain and, in general, may be a function of L. For the passive regions of the food chain (*e.g.,* nonmotile phytoplankton), Q represents the water flow. For active regions (*e.g.,* migratory fish), Q represents the transport of toxicant out of the volume due to the velocity of those particular trophic regions. Further, it can be assumed that:

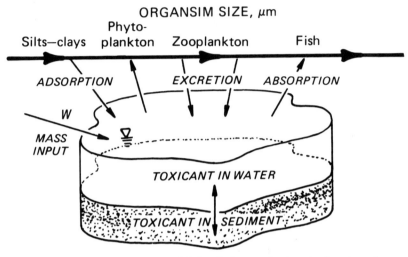

Figure 6. Schematic of general model for single completely mixed water volume.

1. the uptake of PCB is proportional to the water concentration (c), and the biomass along L,
2. excretion of PCB occurs from the entire food chain according to first-order kinetics on Ψ, and
3. there is no input of PCB mass associated with the food chain, *i.e.*, $\Psi_{in} = 0$.

The assumption of uptake proportional to the water concentration is particularly important since it implies linearity with input loads. The literature on laboratory experiments for PCB in the upper levels of the food chain appears to justify this assumption as shown in Figure 7. The body burden of PCB for the two organisms shown in Figure 7 is approximately linear with respect to the water concentration of PCB over a reasonably wide range. However, in the micron and submicron range of particles (*e.g.*, clays), the physical adsorption phenomenon is not necessarily linear and often varies logarithmically over wide ranges.

First-order excretion rates for PCB also appear to be justified on the basis of laboratory experiments (see, for example, Hansen *et al.*, 1974).

Under these assumptions, Equation 9 becomes:

$$V \left(\frac{\partial \Psi}{\partial t} + \frac{\partial v_L \Psi}{\partial L} \right) = -Q(L)\Psi - K(L)V\Psi + k_u(L)\ m(L)Vc \qquad (10)$$

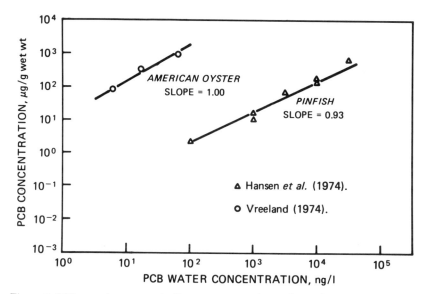

Figure 7. PCB organism concentration as an approximate linear function of PCB water concentration.

and the water equation is:

$$V \frac{dc}{dt} = W - Qc - \lambda Vc + V \left({}_{L_1} \int^{\infty} K(L) \Psi dL \right) - \left[{}_{L_1} \int^{\infty} k_u(L)m(L) \, dL \right] Vc \qquad (11)$$

where K is the excretion rate $[T^{-1}]$, k_u is the uptake rate $[(T \, M \, (\text{biomass}) \, L^{-3})^{-1}]$, m(L) is the biomass density function $[M \, (\text{biomass}) \, L^{-3} \, L^{-1}]$, and λ is the decay rate $[T^{-1}]$ for c due to photooxidation, vaporization or other mechanisms of breakdown. All parameters and inputs may be complicated functions of L. For example, the PCB excretion rate may vary with organism size as discussed bleow.

Steady State

If steady state is assumed in Equations 10 and 11, then:

$$\frac{dv_L \Psi}{dL} + K' \Psi = k_u(L)m(L)c \qquad (12)$$

and

$$c = \frac{1}{G} \left[\frac{W}{V} + {}_{L_1} \int^{\infty} K \Psi dL \right] \qquad (13)$$

where
$$K' = Q/V + K = 1/t_o + K \tag{14}$$
$$\text{(for } t_o = V/Q, \text{ the detention time)}$$

and
$$G = Q/V + \lambda + {}_{L_1}\!\int^{\infty} k_u m \; dL. \tag{15}$$

A separate equation should also be written for the biomass density function [m(L)] which represents the distribution of biomass with organism size. In this work, however, it is assumed that [m(L)] is given and not affected by the concentration of PCB in the food chain. Thus, consider the range of organism sizes to be divided into a series of continuous regions where each region is defined by the model parameters and the biomass density function for each region i is given by:

$$m_i(L) = (m_o)_i \; exp \; (-(b/v_L)_i L) \tag{16}$$

where $b[T^{-1}]$ is an effective "biomass respiration" along the food chain. (The case of variable coefficients with organism size is discussed in Thomann, 1978.) For fixed water PCB concentration, the solution to Equation 12 for a given region is:

$$\Psi = \frac{k_u cm_o}{K'-b} \; (exp(-bL/v_L) - (exp \; (-K'L/v_L)) + \Psi_o exp(-K'L/v_L) \tag{17}$$

and since m is given by Equation 16:

$$\frac{\Psi}{m} = \nu = \frac{k_u c}{K'-b} \; (1-exp(-((K'-b)/v_L)L)) + \nu_1 exp(-((K'-b)/v_L)L) \tag{18}$$

Equations 18 and 13 form the basic steady-state model. Equation 13, which includes the external mass input of the toxicant (W) permits calculation of the concentration of the "maximum permissible" discharge so as not to exceed toxicant levels at any point in the food chain. Furthermore, these equations incorporate physical as well as chemical and biological interactions as, for example, between water detention time, selective uptake and variable excretion. Yet, the entire food chain is easily computed via Equation 18 in contrast to the special case of homogeneous compartments which requires simultaneous solution of algebraic equations.

PCB Data Analysis

Laboratory Data

Uptake and clearance rates of PCB directly from water can be estimated from laboratory experiments on bioconcentration by single species in aquaria. For a single organism, Equations 10 and 11 reduce to:

$$V \frac{ds}{dt} = -Qs - KVs + k_u MVc \tag{19}$$

$$V \frac{dc}{dt} = W - Qc - \lambda Vc - k_u MVc + KVs \tag{20}$$

where s is the mass concentration of PCB in the organism per liter of water (μg - PCB l^{-1}). For aquaria experiments, the water concentration is often held constant. The test organism is held in the tank (Q = 0 in Equation 19). Then, Equation 19 becomes:

$$\frac{ds}{dt} = -Ks + k_u Mc \tag{21}$$

for s = 0 at t = 0, the solution for the PCB mass in the test organism becomes:

$$s = \frac{k_u Mc}{K} (1 - e^{-Kt}) \tag{22}$$

and since s = νM:

$$\nu = \frac{k_u c}{K} (1 - e^{-Kt}). \tag{23}$$

Equation 23 indicates that the rate of buildup of PCB in the laboratory organisms provides an estimate of the excretion rate (assuming this buildup is reversible). Further, at equilibrium:

$$\nu = \frac{k_u c}{K}$$

or

$$\nu/c = N = k_u/K \tag{24}$$

where N is the bioconcentration factor $[(\mu g\text{-PCB})(g \text{ wet wt})^{-1} (\mu g\text{-PCB}/l)^{-1}]$ for the absorption of PCB directly from the water. Thus, N is a biological partition coefficient for PCB. For depuration experiments:

$$\nu = \nu_o e^{-Kt} \tag{25}$$

where ν_o is the body burden of PCB when the organism is placed in a tank of zero PCB concentration. Estimates of the excretion rate from Equations 23 or 25 may not always agree due to factors such as accumulation and retention of substances in the skeletal structures or other less easily cleared components of a laboratory animal. With estimates of K and N at equilibrium, then k_u can be calculated from Equation 24.

It is known that the metabolism of organisms is related to the mass of the organism (Eberhardt and Nakatani, 1968; Eberhardt, 1969). Norstrom *et al.* (1976) in their work on a model of bioaccumulation recognized and incorporated, for example, a log-log relationship of methylmercury and PCB clearance rates for fish over the weight range of 1-300 mg. Additional extensions of these and other data are discussed below, where it is indicated that a log-log relationship of clearance rate with organism length is justified. Further, since the velocity of transfer up the food chain is related to the grazing or predation rate, it is reasonable to assume that v_L also depends on the organism length. The uptake rate also would be expected to be related to the size of the organism, principally through the mass of lipids, which in turn would result in differential solubilization of a substance.

Data have been compiled on the variation of uptake and clearance rates of PCB. The relationship with length of the organisms is shown in Figures 8 and 9 and Tables I and II. Figure 8 indicates a range of bioconcentration factors with organism size and, except for one point, it does not appear to show any functional relationship with size. If that one point is accepted, however, then a rough approximation for the bioconcentration factor is:

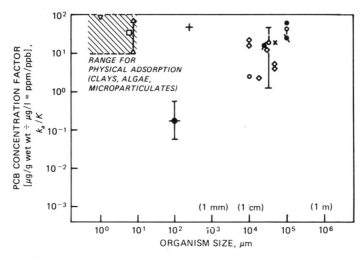

Figure 8. Variation in PCB uptake with organism size for uptake from water only.

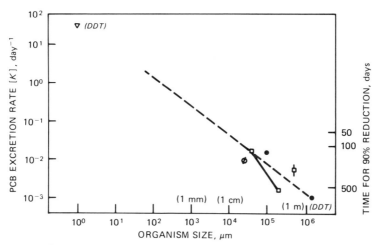

Figure 9. Variation of PCB excretion with organism size.

$$N_{PCB} = (\frac{k_u}{K})_{PCB} \simeq 0.01L^{0.7} \qquad (26)$$

for L in μm and k_u/K in $(g/l)^{-1}$ and N_{PCB} in $[(\mu g\text{-PCB})(g \text{ wet wt})^{-1} (\mu g/l)^{-1}]$. The bioconcentration factor could also be taken as a constant over the entire size range at a value of approximately 50 $[(\mu g\text{-PCB})g^{-1} (\mu g/l)^{-1}]$. This implies that k_u and K have the same slope with respect to L. Also, as an approximation, from Figure 9:

$$(K)_{PCB} \simeq 80 \ L^{-0.8} \qquad (27)$$

for K in $(day)^{-1}$ and L in μm. Thus:

$$(k_u)_{PCB} \simeq 0.8 \ L^{-0.1} \qquad (28)$$

for the case using Equation 26, and:

$$(k_u)_{PCB} \simeq 4000 \ L^{-0.8} \qquad (29)$$

for a constant N of 50. These are only approximate relationships since the available data are somewhat variable. The model framework does not require any specific functional form; these forms essentially provide a first estimate of the rate parameters k_u and K. It is interesting to note though that the data in Figures 8 and 9 include a variety of organisms such as molluscs, larger crustaceans, small and large fish, and even some terrestial organisms. The data on physical adsorption of PCB are included to illustrate the extension of the theory to include concentration on microparticulates, including phytoplankton. Indeed, for substances such as PCB with

Table I. Explanation of Symbols for Figure 8 (PCB Concentration Factors)[a]

Symbol	Organism or Organism Group	PCB Analyzed	Reference
●	Spot *(Leiostromus xanthurus)*	Aroclor 1254	Moriarty (1975)
○	Mayfly Nymph *(Ephemera danica)*	Clophen A 50	Sodergren and Svensson (1973)
⬍	Oyster *(Crassostrea virginica)*	Isomer Mixture	Vreeland (1974)
⬗	Pinfish *(Lagodon rhombiodes)*	Aroclor 1016	Hansen *et al.* (1974)
◖	Spot *(Leiostromus xanthurus)*	Aroclor 1254	Duke and Dumas (1974)
♀	Minnow *(Cyprinodon variegatus)*	Aroclor 1254	Hansen *et al.* (1973)
◇	Aquatic Invertebrates; midge, scud, mosquito larvae, stonefly, dobsonfly, crayfish	Aroclor 1254	Sanders and Chandler (1972)
+	*Daphnia magna*	Aroclor 1254	Sanders and Chandler (1972)
×	Pink Shrimp *(Penaeus duorarum)*	Aroclor 1254	Nimmo *et al.* (1971)
⊕	Protozoa *(Tetrahymena pyriformis)*	Aroclor 1254	Cooley *et al.* (1972)
□	Algae *(chlorella)*	Clophen A 50	Sodergren (1971)
▲	Marine Phytoplankton— Natural Assemblage	Isomer Mixture	Biggs *et al.,* (1977)
▼	Particulates—Dewamish Estuary, Hudson River, Puget Sound		Hafferty *et al.* (1977) Unpublished Dexter (1976)

[a]All tests on whole organisms or corrected to whole organisms.

Table II. Explanation of Symbols for Figure 9 (PCB Excretion Factors)[a]

Symbol	Organism or Organism Group	Excretion Measurement	PCB Analyzed	Reference
●	Spot *(Leiostromus xanthurus)*	Whole organism	Aroclor 1254	Moriarty (1975)
●	Man	Adipose tissue	DDT	Moriarty (1975)
⬦	Rhesus Monkey			Allen *et al.* (1974)
∅	Pinfish *(Lagodon rhombiodes)*	Whole organism	Aroclor 1016	Hansen *et al.* (1974)
□[b]	Yellow Perch *(Perca flavescens)*	Whole organism	Aroclor 1254	Norstrom *et al.* (1976)
▽	Bacteria *(Aerobacter aerogenes, Bacillus subtilis)*	Whole organism	DDT	Johnson and Kennedy (1973)

[a]Note that for some organisms DDT excretion factors have been included.
[b]Estimated from field data.

relatively low water solubility, adsorption onto particles in the range of 1-100 μm is quite common. As reviewed by Wilson and Forester (1973), such particles also include particulate organic detritus and phytoplankton, although one might suspect different adsorption properties for organic particles in contrast to inorganic particles such as clays. Adsorption experiments such as reported by Haque et al. (1974) indicate decreasing concentration factors with increasing particule size and show that the adsorption phenomenon is fundamentally a phenomenon of physical chemistry as opposed to the interactions occurring in the process of bioaccumulation. Dexter (1976) examined the partitioning of PCB between the two phases using laboratory experiments, field data from the Puget Sound area and a theoretical partitioning model. For phytoplankton, values of 14-75 $[(\mu g/g)(\mu g/l)^{-1}]$ are reported, with the higher value associated with a higher degree of chlorination. Similar results were obtained from the laboratory experiments on sediment. Overall, the range of field and laboratory results was reported to be from 13-75 $[(\mu g/g)(\mu g/l)^{-1}]$, which was compared by Dexter to his predicted range of 11-160 $[(\mu/g)(\mu g/l)^{-1}]$.

Field Data Analysis—PCB in Lake Ontario

The laboratory data discussed above provide estimtes of k_u and K as functions of trophic position. In the basic steady-state model of Equation 18, the parameters v_L and b must also be known. At present, it appears that experimental data are not available to permit independent estimation of the velocity transfer coefficient (v_L). To make such an estimation would require feeding experiments over a range of trophic lengths. With data available on the transfer coefficient and the toxicant concentration for various levels in the food chain, an estimate can be made of the components of Equation 18, although such an estimate will not be completely independent of the data themselves. The analysis of the modeling framework at this stage, therefore, does not constitute a verification of the model, but a type of first calibration. The primary benefit of such an analysis is that it places the available field and laboratory data in an analytical framework to gain further insight into the nature of the accumulation of PCB in the food chain of Lake Ontario.

As part of the IFYGL Program, Haile et al. (1975) have obtained data on the PCB concentrations in Lake Ontario fish, water, sediment, net plankton, Cladophora and benthos. Their results, together with data obtained by the Ontario Ministry of the Environment (1976), are summarized in Table III. The data show a general bioaccumulation of PCB from the smaller organisms to the larger coho salmon. The data on the salmon are generally from inshore stations while the data on net plankton and the alewife group include both inshore and open-lake stations. As noted

Table III. Some PCB Data for Lake Ontario PCB Concentrations
(Arochlor 1254 equivalent)

Component	Average Concentration	Standard Deviation	Assumed Range of Organism Size (μm)
	(μg g^{-1} wet wt.)		
Net Plankton[a] (64 μm mesh)	0.72	0.35	100-10^4
Alewife, Smelt[a], Slimy Sculpin	3.24	2.14	10^4-$2.5 \cdot 10^5$
Coho Salmon[b]	6.47	2.85	$2.5 \cdot 10^5$-10^6
Water[a,c]	55.4	21.2	

[a]Haile *et al.* (1975).
[b]Ont. Min. Env. (1976).
[c]Concentration in ng l^{-1}, dissolved + particulate, $<$64 μm.

by Haile *et al.* (1975), the water data include both dissolved and micro-particulate fractions.

Suspended solids in the open waters of Lake Ontario are relatively low, however, and average about 1 mg l^{-1}. For a known solids level, one can estimate the relative fraction of dissolved and particulate PCB for a given partition coefficient. If a linear adsorption isotherm is assumed, then for

$$\nu = N_p c, \quad c_p = \nu M, \quad \text{and} \quad c_p + c = c_T$$

then

$$c_p = (N_p M)c \qquad (29)$$

and

$$c = c_T/(1 + N_p M) \qquad (30)$$

where N_p = equilibrium partition coefficient for PCB adosrbing and desorbing between particulate and dissolved phases, [(μg-PCB) (g^{-1} solids)(μg-PCB/l)$^{-1}$], M = mass of solids (g l^{-1}), c_p = concentration of PCB in particulate phase (μg · PCB l^{-1}), c = concentration of PCB in dissolved phases. For N_p = 100 [(μg/g) (μg/l)$^{-1}$], M = 0.001 g l^{-1}, and c_T = 55 ng l^{-1}, the dissolved concentration is 50 ng l^{-1}, well within the variability of the data. Further, approximately 50% of the remaining 5 ng l^{-1} on the solids is associated with the phytoplankton and would also be available to the food chain. Finally, the work of Zitko (1974) showed that PCB on suspended solids can be bioaccumulated by juvenile Atlantic salmon (although the nature of the experiment obscures the relative uptake from the solids and the water). As a result, it is assumed that the total PCB concentration in the water is available to the food chain. The length ranges in Table III that were assigned to each component are somewhat arbitrary, although realistic.

Data on biomass density throughout the entire food chain are somewhat limited especially in the upper trophic levels. This is due partly to the difficulty of obtaining reliable estimates for fish biomass. It should be noted, however, that the absolute magnitude of the biomass is not required by Equation 18; only the slope (b/v). Nevertheless, Christie *et al.* (1976) have estimated fish biomass in Lake Ontario to be 767,000 metric ton (alewife, rainbow smelt and sculpin) during 1972-73. When equally distributed throughout the lake, this is equivalent to about 0.5 (mg wet wt)l^{-1}, a relatively significant amount when compared to about 2 (mg wet wt)l^{-1} for the phytoplankton. At the lower end of the organism size scale, data are available from particle counts given by Stoermer *et al.* (1975) and from acoustical measurements made by McNaught *et al.* (1975). Stoermer's data cover the range from 5 to 150 μm and McNaught's data span the region from 800 to about 5,000 μm. The lower range therefore provides data on particle density in the region <64 μm, the mesh size used by Haile *et al.* (1975). The data for 800 to 5,000 μm provide biomass information on the net plankton.

The estimated mass density function is obtained from:

$$m_{i+1} = \frac{c_{i,\,i+1}}{L_{i+1} - L_i} \qquad i = 1 \ldots n \qquad (31)$$

where $c_{i,\,i+1}$ is the concentration (*e.g.,* number particles/100 ml) in the size range L_i to L_{i+1} (μm), n is the total number of size ranges, and m_{i+1} is the mass density for the region in concentration units/μm; the mass density is plotted at the end of the region below. Figure 10 shows semi-log plots of the mass density estimates as functions of particle size. As shown, the results of this limited analysis of the particle and acoustical data indicate regions where one might assume an exponential decrease in biomass density as given by Equation 16. Note also that the biomass density drops by several orders of magnitude over the trophic length range of the data. The slope of the density function, $b/v_L\,[\mu m^{-1}]$, is a ratio of model coefficients required by Equation 18. The quotient represents the ratio of respiration to the rate at which biomass is transferred up to higher trophic levels via predation. The model requires the slope of the density function over the entire length scale. Thus, the data of Figure 10 were used to extrapolate the slope of the density function out to the upper limit of 10^6 μm. This is admittedly a hazardous extrapolation and illustrates the need to obtain biomass data throughout the food chain. The relationship for the mass slope using Figure 10 is therefore:

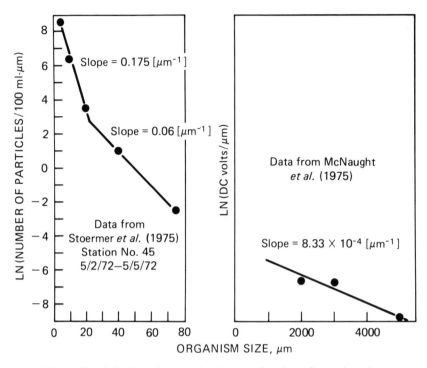

Figure 10. Lake Ontario mass density as a function of organism size.

$$b/v_L = 6.3 \, L^{-1.05} \qquad (32)$$

for b/v_L in μm^{-1} and L in μm. Equation 32 was used to estimate b/v_L for the food chain Equation 18.

Organism size for Lake Ontario was divided into three continuous regions: 10^2-10^4 μm, 10^4-$2.5 \cdot 10^5$ μm and $2.5 \cdot 10^5$-10^6 μm. These regions are not compartments in the classical sense but are rather continuous domains with constant coefficients. Estimates of k_u and K for PCB were obtained from Figures 8 and 9 and used in Equation 18 to obtain a good correlation to the observed data. This provided estimates of the coefficient groupings $k_u/(K' \cdot b)$ and $(K' \cdot b)/v_L$. The first group, $k_u/(K' \cdot b)$, was used to obtain an estimate of $K' \cdot b$ which together with the second coefficient, $(K' \cdot b)/v_L$, was used to obtain v_L. Then, given b/v_L from the mass density data, b was estimated. This value was then used with the original estimate of $K' \cdot b$ to close the estimation procedure on K'. The "free" parameter was therefore v_L, but an external check on the magnitude of v_L was provided by the b/v_L ratio from the biomass data.

The results of applying Equation 18 to the PCB data of Table III are shown in Figure 11 and the associated coefficients are given in Table IV. The ratio b/v from Table IV results in a biomass estimate at $2.5 \cdot 10^5$ μm of 0.03 mg l^{-1} or about one order of magnitude below the estimated 0.5 mg-wet wt l^{-1} of fish biomass in that region. This implies that the b/v ratio is too low for the first region. Only a slight adjustment to b/v of $5.6 \cdot 10^{-4}$ $(\mu m)^{-1}$ (from $8.3 \cdot 10^{-4}$ $(\mu m)^{-1}$) results in the higher biomass level at $2.5 \cdot 10^{-5}$ μm. The two principal mechanisms of the steady-state model are shown in Figure 11. The uptake directly from the water (at $v = 0$) increases rapidly with organism size principally because of the decreased excretion rate. The uptake of PCB due to trophic-level transfer through the food chain results from increased transfer velocity and the reduced mass density at the high levels.

For the salmon, the calculations indicate that about 30% of the observed 6.5 μg-PCB g^{-1} fish is due to transfer from lower levels in the food chain and about 70% from direct uptake. Therefore buildup of PCB in the Lake Ontario ecosystem is postulated to be due to a complex interaction between water uptake rate, excretion rate, rate of transfer through the food chain and decrease of mass density with trophic length. It should also be recalled that the toxicant concentration in the food chain is linearly related to the water concentration which, in turn (Equation 13), is linearly related to the input mass loading of PCB. In principle, then, one could estimate the required input loading such that a given level of PCB concentration would not be exceeded in any region of the food chain. However, sufficient data are not yet available on the present input to Lake Ontario to permit such a definitive calculation. Before such a calculation is made, it would be preferable to estimate the water concentration independent of the measured values given input mass loading and estimates of removal rates.

However, one can estimate the total water (dissolved and microparticulate) concentration that would be necessary in order not to exceed certain levels in the food chain, if the coefficients used herein are considered reasonable. If the U.S. "action level" of 5 μg-PCB g^{-1} wet wt is used at the $5 \cdot 10^5$ μm level, then the water concentration would have to be about 42 ng l^{-1} or a reduction of about 33% of the present 55 ng l^{-1}. If 2 μg-PCB g^{-1} (the Canadian action level) is used, then the water concentration would have to be about 17 μg l^{-1} or an equivalent input PCB reduction of about 70%.

Evidence has recently been presented to indicate that the Lake Ontario PCB water concentration data for 1972 (Haile *et al.,* 1975) may have been erroneously measured. Analyses of samples collected in 1976 as reported in testimony by Risebrough (1976) and Veith (1976) indicate

Table IV. Coefficients Used in Analysis of PCB Data with
Steady-State Size-Dependent Model

	Trophic Region		
	1	2	3
Coefficient	10^2-10^4 μm	10^4-$2.5 \cdot 10^5$ μm	$2.5 \cdot 10^5$-10^6 μm
Water Uptake—k_u $(g\, l^{-1}\, day^{-1})^{-1}$	0.5	0.38	0.23
Excretion—K $(day)^{-1}$	0.05	0.0072	0.0025
Transfer Velocity—v_L $(\mu m\, day^{-1})$	12	190	728
Biomass Respiration— b $(day)^{-1}$	0.01	0.0024	0.001
$(K'\text{-}b)/v_L$ $(\mu m)^{-1}$	$3.0 \cdot 10^{-3}$	$2.5 \cdot 10^{-5}$	$2.0 \cdot 10^{-6}$
$k_u/(K'\text{-}b)$ $(g\, l^{-1})^{-1}$	13.4	79	158

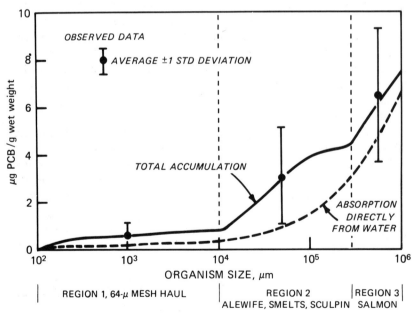

Figure 11. Analysis of PCB in Lake Ontario.

total water PCB concentrations for several stations and at the Rochester intake of about 1 ng/l. If this level is a more correct representation of the concentration of 1972, then the above analysis indicates that nearly all the accumulation at the upper trophic level is due to food-chain transfer as opposed to absorption directly from the water.

SUMMARY

In order to incorporate both bioaccumulation of toxic substances directly from the water and subsequent transfer up the food chain, a mass balance model is constructed that introduces organism size as an additional independent variable. The model represents an ecological continuum through size dependence; classical compartment analyses are therefore a special case of the continuous model.

The principal factors that influence the total toxicant concentration in various regions of the food chain include excretion and uptake rates, the rate of decrease of biomass density with organism size and the food chain transfer velocity, a parameter that reflects average predation along the food chain. The model behaves linearly with respect to external mass loading of the toxicant and hence can be used in principal to estimate the input that can be allowed without exceeding given levels in various regions of trophic space.

The analysis of some PCB data from Lake Ontario is used as an illustration of the theory. A completely mixed water volume is used. Trophic space is considered from 100 to 10^6 μm representing a wide range of organism lengths. PCB data were available for the material from hauls of a net with 64-μm mesh size and for alewife, smelt, sculpin and coho salmon. Laboratory data from the literature together with the field data were used for preliminary estimates of the model coefficients. The analysis indicated that about 30% of the observed 6.5 μg-PCB (g^{-1} fish) at the coho salmon size range is due to transfer from lower levels in the food chain and about 70% from direct water intake. The model shows rapid accumulation of PCB with organism size due principally to decreased excretion rates (which may be due to an increased lipid pool) and decreased biomass at higher trophic levels.

The introduction of organism size as an independent variable in the mass balance of a toxicant provides a generalized analysis framework. Such a theoretical structure permits the use in an integrated way of diverse laboratory experiments on uptake and excretion as well as an interpretive framework for field data of toxicant concentrations. Finally, the framework should ultimately provide a basis for estimating the effect of external mass loading on concentration in the food chain.

ACKNOWLEDGMENTS

This work was completed during a sabbatical from Manhattan College to which grateful acknowledgment is made. Additional thanks are given to Hydroscience, Inc., Westwood, New Jersey, for providing financial and staff support. Further funding was provided by the U.S. Environmental Protection Agency through the Large Lakes Research Station at Grosse Isle under a grant to Manhattan College. Drs. Wayland Swain and Nelson Thomas and Mr. Bill Richardson provided much encouragement and insight.

Special thanks are due Drs. Dominic DiToro and Donald J. O'Connor of Manhattan College and John L. Mancini and John St. John of Hydroscience, Inc., who provided a valuable forum for discussion of model behavior.

REFERENCES

Allen, J. R., L. A. Carstens and D. A. Barsotti. "Residual Effects of Short-Term, Low-Level Exposure of Non-Human Primates to Polychlorinated Biphenyls," *Toxicol. Appl. Pharmacol.* 30(3):440-451 (1974).

Biggs, D. C., C. D. Powers, R. G. Rowland, H. B. O'Connor and C. F. Wurster. "Uptake and Accumulation of ^{14}C-PCB by Natural Assemblages of Marine Phytoplankton," Marine Sciences Research Center, SUNY, Stonybrook, NY (in preparation).

Christie, W. J., J. H. Kutcuhn and W. A. Pearce. "Overview of the International Survey of Fish Stocks of Lake Ontario During IFYGL," unpublished manuscript, Ontario Ministry of Natural Resources, Picton, Ontario (1974), 21 p.

Cooley, N. R., J. M. Keltner and J. Forester. "Mirex and Aroclor 1254: Effect and Accumulation by *Tetrahymena pyriformis* Strain W." *J. Protozool.* 19(4):636-638 (1972).

Dale, M. G. "Systems Analysis and Ecology," *Ecology* 51(1):1-16 (1970).

Dexter, R. N. "An Application of Equilibrium Adsorption Theory to the Chemical Dynamics of Organic Compounds in Marine Ecosystems," Ph.D. dissertation, University of Washington, Seattle (1976).

Duke, T. W., and D. P. Dumas. *Implications of Pesticide Residues in the Coastal Environment.* Pollution and Physiology of Marine Organisms. (New York: Academic Press, Inc., 1974), pp. 137-164.

Eberhardt, L. L., and R. E. Nakatani. "A Postulated Effect of Growth on Retention Time of Metabolites," *J. Fish. Res. Bd. Can.* 35(13): 591-596 (1968).

Eberhardt, L. L. "Similarity, Allometry, and Food Chains," *J. Theor. Biol.* 24:43-55 (1969).

Gillettt, J. W., J. Hill, A. W. Jarvinen and W. D. Schnoor. *A Conceptual Model for the Movement of Pesticides through the Environment,* EPA-660/3-74-024 (Corvallis, OR: U.S. Environmental Protection Agency, 1974).

Hafferty, A. J., S. P. Pavlou and W. Hom. "Release of Polychlorinated Biphenyls (PCB) in a Salt Wedge Estuary as Induced by Dredging of Contaminated Sediments," in *The Science of Total Environment,* Vol. 8 (Amsterdam: Elsevier, 1977).

Haile, C. L., G. D. Veith, G. F. Lee and W. C. Boyle. *Chlorinated Hydrocarbons in the Lake Ontario Ecosystem (IFYGL),* EPA-660/3-75-002 (Corvallis, OR: U.S. Environmental Protection Agency, 1975).

Hansen, D. J., P. R. Parish and J. Forester. "Aroclor 1016: Toxicity to and Uptake by Estuarine Animals," *Environ. Res.* 7:363-373 (1974).

Hansen, D. J., S. C. Schimmel and J. Forester. "Aroclor 1254 in Eggs of Sheepshead Minnows: Effect on Fertilization Success and Survival of Embryos and Fry," in *Proc. 22nd Annual Conference Southeast Association of Game and Fish Commission* (1973), pp. 420-426.

Haque, R., D. W. Schmedding and V. H. Freed. "Aqueous Solubility, Adsorption and Vapor Behavior of Polychlorinated Biphenyl Aroclor 1254," *Environ. Sci. Techol.* 8(2):139-142 (1974).

Hill, J., H. P. Kallig, D. F. Paris, N. L. Wolfe and R. G. Zepp. *Dynamic Behavior of Vinyl Chloride in Aquatic Ecosystems,* EPA-600/3-76-001 (Athens, GA: U.S. Environmental Protection Agency, 1976), 64 p.

Himmelblau, D. M., and K. B. Bischoff. *Process Analysis and Simulation, Deterministic Systems* (New York: John Wiley & Sons, Inc., 1968), 348 p.

Hoefner, J. W., and J. W. Gillett. "Aspects of Mathematical Models and Microcosm Research," in *Proc. Conference in Environmental Modeling and Simulations,* EPA 600/9-76-016 (Washington, DC: U.S. Environmental Protection Agency, 1976), pp. 624-628.

Johnson, B. T., and J. O. Kennedy. "Biomagnification of p, p'-DDT and Methoxychlor by Bacteria," *Appl. Microbiol.* 26(1):66-71 (1973).

Lassiter, R. R., J. L. Malanchuk and G. L. Baughman. "Comparison of Processes Determining the Fate of Mercury in Aquatic Systems," in *Proc. Conference on Environmental Modeling and Simulation,* EPA 600/9-76-016 (Washington, DC: U.S. Environemtnal Protection Agency, 1976), pp. 619-623.

McNaught, D. C., M. Buzzard and S. Levine. *Zooplankton Production in Lake Ontario as Influenced by Environmental Perturbations,* EPA 660/3-75-021 (Corvallis, OR: U.S. Environmental Protection Agency, 1975), 156 p.

Moriarty, F. "Organochlorine Insecticides; Persistent Organic Pollutants," in *Exposure and Residues* (New York: Academic Press, Inc., 1975), pp. 29-72.

Morsell, J. W., and C. W. Norden. "Food Habits of the Alewife *Alosa psuedoharengus* (Wilson) in Lake Michigan," in *Proc. 11th Conference of Great Lakes Research* (International Association of Great Lakes Research, 1968), pp. 96-102.

Nimmo, D. R., R. R. Blackman, A. J. Wilson, Jr. and J. Forrester. "Toxicity and Distribution of Aroclor 1254 in the Pink Shrimp *Penueus duorarum," Intl. J. Life in Oceans and Coastal Waters,* 11(3):191-197 (1971).

Norden, C. R. "Morphology and Food Habits of the Larval Alewife *Alosa pseudoharengus* (Wilson) in Lake Michigan," in *Proc. 11th Conference of Great Lakes Research* (International Association of Great Lakes Research, 1968), pp. 103-110.

Norstrom, R. J., A. E. McKinnon and A. S. W. DeFreitas. "A Bioener-getics-Based Model for Pollutant Accumulation by Fish. Simulation of PCB and Methylmercury Residue Levels in Ottawa River Yellow Perch (*Perca flavescens*)," *J. Fish. Res. Bd. Can.* 33:248-267 (1976).

Ontario Ministry of the Environment. "Ontario Research Program to Monitor PCB Levels," news release, Ontario, Canada, July 2, 1976.

Patten, B. D., Ed. *Systems Analysis and Simulation in Ecology,* Vol. 1 (New York: Academic Press, Inc., 1971) 607 p.

Risebrough, R. W. Testimony for "In the Matter of Proposed Toxic Pollu-Effluent Standards for Polychlorinated Biphenyls (PCB's)," FWPCA (307), Docket No. 4, Exhibit A, Washington, DC, 15 pp. + Attachments (1976).

Rotenberg, M. "Theory of Popultaion Transport," *J. Theor. Biol.* 37: 291-305 (1972).

Sanders, H. O., and J. H. Chandler. "Biological Magnification of a Poly-chlorinated Biphenyl (Aroclor 1254) from Water by Aquatic Inverte-brates," *Bull. Environ. Contam. Toxicol.* 7:257-263 (1972).

Sodergren, A. "Accumulation and Distribution of Chlorinated Hydrocarbons in Cultures of *Chlorella pyrenoidosa* (Chlorophyceae)," *Oikos* 22: 215 (1971).

Sodergren, A., and B. J. Svensson. "Uptake and Accumulation of DDT and PCB by *Ephemera danica* (Ephemeroptera) in Continuous-Flow Systems," *Bull. Environ. Contam. Toxicol.* 9(6):345-350 (1973).

Stoermer, E. F., M. M. Bowman, J. C. Kingston and A. L. Schaedel. *Phytoplankton Composition and Abundance in Lake Ontario During IFYGL,* EPA-660/3-75-004 (Corvallis, OR: U.S. Environmental Protec-tion Agency, 1975).

Thomann, R. V. *Size Dependent Model of Hazardous Substances in Aquatic Food Chain,* EPA-600/3-78-036 (Duluth, MN: U.S. Environ-mental Protection Agency, 1978), 39 p.

Veith, G. D. Testimony for "In the Matter of Proposed Toxic Pollutant Effluent Standards for Polychlorinated Biphenyls (PCB's)," FWPCA (307), Docket No. 4, Exhibit B, Washington, DC, 34 pp. + Attachments (1976).

Wilson A. J., and J. Forrester. "Methods and Problems in Analysis of Pesticides in the Estuarine Environment," in *Proc. Seminar on Metho-dology for Monitoring the Marine Environment* (Washington, DC: U.S. Environmental Protection Agency, 1973), pp. 108-124.

Zitko, V. "Uptake of Chlorinated Paraffins and PCB from Suspended Solids and Food by Juvenile Atlantic Salmon," *Bull. Environ. Contam. Toxicol.* 12(4):406-412.

INDEX